UNIFORM ALGEBRAS

PRENTICE-HALL SERIES IN MODERN ANALYSIS

R. CREIGHTON BUCK, editor

UNIFORM ALGEBRAS

THEODORE W. GAMELIN

Department of Mathematics
The University of California at Los Angeles

PRENTICE-HALL, INC.

Englewood Cliffs, N.J.

Library of Congress Catalog Card Number 70-84036

Current Printing (last digit):
10 9 8 7 6 5 4 3 2 1

PRINTED IN THE UNITED STATES OF AMERICA

13-937805-7

To my family:

HELEN, SHELLY, ANDY, DANNY

PREFACE

The central problem in the field of uniform algebras is to decide whether a given complex-valued function can be uniformly approximated by members of a prescribed algebra of functions. The present theory of uniform algebras is an outgrowth of an attempt to attack this problem using the tools provided by the modern theory of functional analysis. For instance, one is led to determine the maximal ideal space of the algebra, to study functionals orthogonal to the algebra, and to investigate other properties of the algebra which have a bearing on the problem.

The functional-analytic approach to uniform algebras is inextricably interwoven with the theory of analytic functions. The deepest general theorems in uniform algebras involve the successful application of the theory of analytic functions of several complex variables. Moreover, concrete algebras of analytic functions provide models for the study of uniform algebras, and one is led to relate the behavior of functions in a uniform algebra to familiar properties of analytic functions. One of the major problems is to determine under what conditions an analytic structure can be introduced into the maximal ideal space of a uniform algebra. In the other direction, the concepts and techniques introduced to deal with these problems, such as "peak points" and "parts," provide new insights into the classical theory of approximation by analytic functions. In some cases, elegant proofs of old results are obtained by abstract methods. The new concepts also lead to new problems in classical function theory, which serve to enliven and freshen that subject. In short, the relation between the functional analysis and the analytic function theory is both fascinating and complex, and it serves to enrich and deepen each of the respective disciplines. It is the purpose of this book to gather the principal techniques of uniform algebras under one roof, and to examine the interplay between these functional-analytic concepts and analytic function theory.

Uniform algebras form a branch of the theory of commutative Banach algebras. For a variety of reasons, it did not begin to develop in earnest until more than a decade after Gelfand first introduced his elegant theory. The contributions signaling the beginning were perhaps the papers of Shilov appearing about 1950. In the early 1950's, when the idea of a representing measure was still a novelty, a profound influence on the direction the subject was to take was exerted by R. Arens and I. M. Singer. The work of A. M. Gleason during the first decade also left an indelible impression on the field. For its development over the years, the subject undoubtedly owes its greatest debt to E. Bishop and J. Wermer, who were always at the forefront of activity, and who solved a number of the hard problems involved in bringing the subject to its current state. This past decade has seen the field flourish and expand rapidly, with the entrance of many productive mathematicians whose contributions have extended its boundaries in many directions.

The prerequisites for reading this book are the fundamentals of real variable theory (Lebesgue convergence theorems, Fubini theorem), the basic principles of Banach spaces (uniform boundedness principle, Hahn-Banach theorem), and some knowledge of weak topologies in Banach spaces (Alaoglu's theorem). Many arguments will hinge on the Riesz representation theorem, which allows one to identify the conjugate space of $C(X)$ with the space of measures on X. (Here, and forevermore, a "measure" is a "finite complex Baire measure," or equivalently, a "finite complex regular Borel measure.") Occasionally additional tools are brought in to handle certain problems.

Chapters I, II, and III are designed for use in an intermediate level graduate course or seminar on commutative Banach algebras. The Gelfand transform is discussed in the first half of chapter I, and is prerequisite to all that succeeds. However, the interdependence of the remaining sections of the first three chapters is rather loose, allowing the lecturer a good deal of maneuverability.

Chapter IV provides an introduction to the theory of abstract Hardy spaces. Chapter V deals with that portion of the Hardy space theory most closely resembling the function theory of the unit disc. Chapter VI concerns itself with introducing (at least the remnants of) an analytic structure in the maximal ideal space of a uniform algebra. And chapter VII treats an important class of examples, the algebras of analytic almost periodic functions. Chapters V, VI, and VII are by and large independent of one another, although they depend to varying degrees on information contained in chapter IV.

In chapter VIII, we consider the theory of rational approximation on planar sets, utilizing the constructive techniques developed especially by various Soviet mathematicians and culminating in the work of A. G. Vitushkin and M. S. Melnikov. Chapter VIII depends only on certain results in chapter II and on elementary analytic function theory. Part of the subject matter dovetails with the abstract function algebraic approach, a notable example being Vitushkin's theorem VIII. 5.1 on approximation on nowhere dense sets. However, the relationship between the abstract and the constructive approaches is not completely understood and will form an active field of research for some time.

I have attempted to follow a policy of labeling a lemma or theorem only when a name seems to have been attached to it by common usage. Consequently, the appellation "The S. Gootch Theorem" does not necessarily mean that S. Gootch proved the theorem, although it can be assumed that S. Gootch had at one time or another something to do with some version of the theorem.

An attempt has been made to cite the sources for theorems in the notes following each chapter. However, the evolution of many theorems involves too many contributors at too many stages to permit an accurate accounting. Consequently, the notes should be regarded as only fragmentary. In no event should it be assumed that an uncited result is due to the author.

Exercises are provided at the end of each chapter. These range from the trivial to the quite difficult, part of the game being to distinguish which are which. Most of them are drawn from the literature cited in the bibliography. Rather than attempt to cite accurately the sources of all the exercises or of just some of the exercises, it was decided to suppress the references completely. The reader who has found a tantalizing exercise will be able to familiarize himself with the literature while locating it.

This book evolved from a seminar which I gave at the Universidad Nacional de Buenos Aires while I was a member of the faculty of the Universidad Nacional de La Plata in Argentina. Seminar notes were being prepared, with the aid of several of the participants, to be published in the Buenos Aires series *Cursos y Seminarios de Matemática*. However, midway through the project the government elected to intervene in the university. On "la noche de los bastones largos," the Buenos Aires police took a disciplinary measure against certain professors and students, during which one of my colleagues and seminar collaborators was injured. The seminar was forced by the events to postpone its activity. In the ensuing disheartening months, each of the seminar members eventually came to the decision to leave the country, and the continuance of the project became an impossibility.

Upon assuming my new position at M.I.T., I returned to the project, this time with the idea of enlarging the scope to book form. At this point, I was fortunate in being able to collaborate with John Garnett and Larry Zalcman on various aspects of the subject, and especially in hashing out the work of Vitushkin. I would like to acknowledge my deep debt of gratitude to John Garnett, who read preliminary versions of the bulk of the manuscript and who gave valuable advice on every aspect of the book. And I would like to give special thanks to Larry Zalcman, who read portions of the original manuscript, and whose preparation of notes for the Springer Lecture Notes

Series formed an invaluable stimulus and source for the writing of chapter VIII.

I would like to acknowledge here my debt to my principal mentor K. Hoffman, whose overall guidance has been invaluable, and whose assistance with the volume at hand has been crucial. I would also like to record my thanks to the following people for various helpful remarks and corrections: A. Browder, S. Fisher, I. Glicksberg, D. Wilken, K. Yale, and many others. For all her hard work in typing the bulk of the manuscript, I would like to give special thanks to Jane Jordan. And I would like to thank Sophie Kouras, who assisted on the typing at some critical points.

THEODORE W. GAMELIN

Pacific Palisades, California

CONTENTS

CHAPTER I

COMMUTATIVE BANACH ALGEBRAS

In this chapter, we will develop the rudiments of Gelfand's theory of commutative Banach algebras.

A **Banach algebra** is a complex Banach space A which is also an (associative) algebra, in which the multiplication and the norm are linked by the inequality

$$\|fg\| \leq \|f\| \, \|g\|, \qquad f, g \in A.$$

The Banach algebra A is **commutative** if $fg = gf$ for all $f, g \in A$. The Banach algebra A has an **identity** if there is an element $1 \in A$ such that $\|1\| = 1$ and $1f = f1 = f$ for all $f \in A$. We will be interested only in commutative Banach algebras with identity, which we will refer to simply as Banach algebras.

We will assume throughout this chapter that A is a commutative Banach algebra with identity.

1. Spectrum and Resolvent

An element $f \in A$ is **invertible** if there exists $g \in A$ such that $fg = 1$. The inverse g of f is evidently unique. It is denoted by f^{-1}, or $1/f$. The family of invertible elements of A is denoted by A^{-1}.

A complex number λ is in the **resolvent** of $f \in A$ if $\lambda - f$ is invertible. Otherwise, λ is in the **spectrum** of f. The spectrum of f is denoted by $\boldsymbol{\sigma}(f)$.

1.1 Theorem. For any $f \in A$, the spectrum of f is a non-empty compact subset of the complex plane. The function $(\lambda - f)^{-1}$ depends analytically on λ, for λ belonging to the resolvent of f.

Proof. Here "analytic" means "locally expressible by a convergent power series." Actually, any two reasonable definitions of analyticity of vector-valued functions coincide.

If $|\lambda| > \|f\|$, the series $\sum_{n=0}^{\infty} f^n/\lambda^{n+1}$ converges to a function $g(\lambda)$ which is analytic at ∞. A direct computation shows that $g(\lambda)(\lambda - f) = 1$, that is,

$g(\lambda) = (\lambda - f)^{-1}$. Hence $\sigma(f)$ is contained in the closed disc of radius $\|f\|$.

Suppose that λ_0 belongs to the resolvent of f. The series $\sum_{n=0}^{\infty} (\lambda_0 - \lambda)^n(\lambda_0 - f)^{-(n+1)}$ converges to a function $h(\lambda)$, which is analytic in the disc $\{|\lambda - \lambda_0| < 1/\|(\lambda_0 - f)^{-1}\|\}$. Again a direct calculation shows that $h(\lambda) = (\lambda - f)^{-1}$. Consequently, the resolvent of f is open, and $(\lambda - f)^{-1}$ is analytic on the resolvent of f.

For each continuous linear functional L on A, $L((\lambda - f)^{-1})$ is an analytic function of λ on the resolvent of f which vanishes at ∞. If $\sigma(f)$ were empty, then $L((\lambda - f)^{-1})$ would be identically zero, by Liouville's theorem. By the Hahn-Banach theorem, $(\lambda - f)^{-1}$ would be identically zero, an impossibility. This shows that $\sigma(f)$ is not empty, proving the theorem.

In the course of the proof of 1.1, we established the following two facts.

1.2 Theorem. If λ belongs to the spectrum $\sigma(f)$ of f, then $|\lambda| \leq \|f\|$.

1.3 Theorem. If λ belongs to the resolvent of f, and $d(\lambda,\sigma(f))$ is the distance from λ to $\sigma(f)$, then

$$d(\lambda,\sigma(f)) \geq 1/\|(\lambda - f)^{-1}\|.$$

The following theorem is crucial to the theory.

1.4 Theorem (Gelfand-Mazur Theorem). A commutative Banach algebra with identity which is a field must be isometrically isomorphic to the field of complex numbers.

Proof. Every Banach algebra with identity contains a subalgebra which is isometrically isomorphic to the field of complex numbers, namely, the complex multiples of the identity. So it suffices to show that if A is a field, then every $f \in A$ is a complex multiple of the identity.

Suppose $f \in A$. By 1.1, there is a complex number λ such that $\lambda - f$ is not invertible. Since A is a field, $\lambda - f$ must be 0, and $f = \lambda$. That proves the theorem.

2. The Maximal Ideal Space

An ideal J of A is **maximal** if $J \neq A$, while J is contained in no other proper ideal of A. The set of maximal ideals of A is called the **maximal ideal space** of A, and is denoted by M_A. Later we will introduce a topology for M_A.

The following lemma is elementary, and is valid, more generally, for commutative rings with identity.

2.1 Lemma. Every proper ideal of A is contained in a maximal ideal. An ideal J is maximal if and only if A/J is a field.

2.2 Theorem. Every maximal ideal of A is closed. If J is a maximal ideal of A, then A/J is isometrically isomorphic to the field of complex numbers.

Proof. First we note that every $f \in A$ satisfying $\|1 - f\| < 1$ is invertible. In fact, if $\|1 - f\| < 1$, then 1 belongs to the resolvent of $1 - f$, by 1.2, and $f = 1 - (1 - f) \in A^{-1}$.

Now if I is any proper ideal, and $f \in I$, then $f \notin A^{-1}$. So $\|1 - f\| \geq 1$. This same inequality persists for f belonging to the closure $\bar{I}$ of I. Hence the closure of any proper ideal is a proper ideal, and all maximal ideals must be closed.

Now let J be a maximal ideal of A. Since J is closed, A/J becomes a Banach space, with the norm

$$\|f + J\| = \inf\{\|f + g\|: g \in J\}.$$

It is easy to show that

$$\|fg + J\| \leq \|f + J\| \, \|g + J\|, \qquad f,g \in A.$$

So A/J becomes a commutative Banach algebra. Since $\|1 + g\| \geq 1$ for all $g \in J$, $\|1 + J\| = 1$. Consequently $1 + J$ is an identity for A/J.

By the Gelfand-Mazur theorem, A/J is isometrically isomorphic to the field of complex numbers. This completes the proof.

Suppose that J is a maximal ideal in A. The projection $A \rightarrow A/J$ is an algebra homomorphism with kernel J. The Gelfand-Mazur theorem allows us to identify A/J with the complex field. So J is the kernel of a non-zero complex-valued homomorphism ϕ. If $f \in A$, then we can define $\phi(f)$ explicitly as that unique complex number λ such that $f + J = \lambda + J$, that is, such that $f - \lambda \in J$.

Conversely, if ϕ is a non-zero complex-valued homomorphism of A, and A_ϕ is the kernel of ϕ, then A/A_ϕ is a field, so A_ϕ is a maximal ideal in A. This proves the following theorem.

2.3 Theorem. The correspondence $\phi \rightarrow A_\phi$, A_ϕ the kernel of ϕ, is a bijective correspondence of non-zero complex-valued homomorphisms of A and maximal ideals in A.

It is customary to identify each maximal ideal in M_A with the complex-valued homomorphism that it determines. We will practice this convention without further comment.

The following lemma will allow us to define a topology for M_A. Note that the assertion regarding the continuity of ϕ also follows from 2.2, since linear functionals are continuous if and only if their kernels are closed.

2.4 Lemma. Let ϕ be a non-zero complex-valued homomorphism of A. Then ϕ is continuous, and $\|\phi\| = 1 = \phi(1)$.

Proof. Since $\phi(1)^2 = \phi(1)$, either $\phi(1) = 1$ or $\phi(1) = 0$. The latter case is excluded, or else ϕ would be identically zero. Hence $\phi(1) = 1$.

If $f \in A$, and $|\lambda| > ||f||$, then $\lambda - f$ is invertible. From the relation $\phi(\lambda - f)\phi((\lambda - f)^{-1}) = \phi(1) = 1$, we obtain $\phi(\lambda - f) \neq 0$, or $\phi(f) \neq \lambda$. It follows that $|\phi(f)| \leq ||f||$. Since this is true for all $f \in A$, ϕ must be continuous, and $||\phi|| \leq 1$. Since $\phi(1) = 1$, we obtain $||\phi|| = 1$.

Lemma 2.4 allows us to identify M_A with a subset of the unit sphere of the conjugate space A^* of A. We define the topology of M_A to be the weak-star topology that M_A inherits from A^*. In other words, a net $\{\phi_\alpha\}$ in M_A converges to ϕ if and only if $\phi_\alpha(f) \to \phi(f)$ for all $f \in A$. A basis of open neighborhoods of $\psi \in M_A$ is given by sets of the form

$$N(\psi; f_1, \ldots, f_n; \epsilon) = \{\phi \in M_A: |\phi(f_j) - \psi(f_j)| < \epsilon, 1 \leq j \leq n\},$$

where $\epsilon > 0$, n is a positive integer, and $f_1, \ldots, f_n \in A$.

2.5 Theorem. The maximal ideal space M_A of A is a compact Hausdorff space.

Proof. The weak-star limit of homomorphisms satisfying $\phi(1) = 1$ is again a non-zero homomorphism. Hence M_A is a weak-star closed subset of the unit ball of A^*. By Alaoglu's theorem, the unit ball of A^* is weak-star compact. Consequently, M_A is compact.

The **Gelfand transform** of $f \in A$ is the complex-valued function $\hat{f}$ on M_A defined by $\hat{f}(\phi) = \phi(f)$.

2.6 Theorem. The Gelfand transform is a homomorphism of A onto an algebra $\hat{A}$ of continuous functions on M_A. The algebra $\hat{A}$ separates the points of M_A, and $\hat{A}$ contains the constants. The Gelfand transform is norm-decreasing:

$$||\hat{f}||_{M_A} \leq ||f||, \quad f \in A.$$

Proof. One verifies immediately that $f \to \hat{f}$ is an algebra homomorphism. From the definition of the topology of M_A, it is clear that each $\hat{f} \in \hat{A}$ is continuous on M_A. Since $||\phi|| = 1$ for all $\phi \in M_A$, it follows that $|\hat{f}(\phi)| \leq ||f||$ for all $f \in A$. Hence $||\hat{f}||_{M_A} \leq ||f||$.

The Gelfand transform of the identity of A is the function which is identically one on M_A. Hence $\hat{A}$ contains the constants.

If $\hat{f}(\phi_1) = \hat{f}(\phi_2)$ for all $f \in A$, then $\phi_1(f) = \phi_2(f)$ for all $f \in A$, and $\phi_1 = \phi_2$. Hence $\hat{A}$ separates the points of M_A.

2.7 Theorem. If $f \in A$, then the spectrum $\sigma(f)$ of f coincides with the range $\hat{f}(M_A)$ of $\hat{f}$ on M_A.

Proof. Suppose $\lambda \in \sigma(f)$. Then $\lambda - f$ is not invertible, so $(\lambda - f)A$ is a proper ideal. By 2.1, there is a maximal ideal J containing $\lambda - f$. If J is the kernel of ϕ, then $\phi(\lambda - f) = 0$, and $\hat{f}(\phi) = \lambda$. Hence $\lambda \in \hat{f}(M_A)$.

Conversely, suppose $\lambda \in \hat{f}(M_A)$. Choose $\phi \in M_A$ such that $\hat{f}(\phi) = \lambda$. Then $\phi(\lambda - f) = 0$, so $\lambda - f$ is not invertible, and $\lambda \in \sigma(f)$. That proves the theorem.

3. Examples

3.1 Example. The algebra $C(X)$ of continuous complex-valued functions on a compact Hausdorff space X is a Banach algebra, with the usual supremum norm

$$\|f\|_X = \sup_{x \in X} |f(x)|, \qquad f \in C(X).$$

Every $x \in X$ determines the evaluation homomorphism $\phi_x \in M_{C(X)}$, defined by

$$\phi_x(f) = f(x), \qquad f \in C(X).$$

Theorem. Every $\phi \in M_{C(X)}$ is the evaluation homomorphism ϕ_x at some point $x \in X$.

Proof. Suppose $\phi \in M_{C(X)}$ is distinct from each ϕ_x, $x \in X$. Then for every $x \in X$, there is $f_x \in C(X)$ such that $f_x(x) \neq 0$ while $\phi(f_x) = 0$. Then $|f_x|^2$ is positive in a neighborhood of x, and $\phi(|f_x|)^2 = \phi(f_x)\phi(\bar{f}_x) = 0$. Choose $x_1, \ldots, x_n \in X$ such that $|f_{x_1}|^2 + \ldots + |f_{x_n}|^2 = g$ is positive on X. Then g is invertible in $C(X)$. This contradicts $\phi(g) = 0$, proving the theorem.

This theorem shows that X and $M_{C(X)}$ are homeomorphic. In particular, the space X is completely determined by the Banach algebra structure of $C(X)$.

3.2 Example. Any uniformly closed subalgebra A of $C(X)$ which contains the constants is a commutative Banach algebra, again with the supremum norm. If A separates the points of X, the correspondence $x \to \phi_x$ is an embedding of X as a closed subset of M_A. The following special case shows how maximal ideals which do not belong to X may arise.

Here, and throughout the book, Δ will denote the closed unit disc $\{|z| \leq 1\}$ in the complex plane. Its boundary $b\Delta$ is the unit circle $\{|z| = 1\}$. The subalgebra of functions in $C(b\Delta)$ which can be approximated uniformly on $b\Delta$ by polynomials in z is denoted by $P(b\Delta)$.

The kth Fourier coefficient of a function $f \in C(b\Delta)$ is given by

$$c_k = \frac{1}{2\pi} \int_0^{2\pi} f(e^{i\theta}) e^{-ik\theta}\, d\theta$$
$$= \frac{1}{2\pi i} \int_{b\Delta} f(z) z^{-k-1} dz.$$

By Fejer's theorem, f is the uniform limit of the functions

$$\sigma_n = (f_0 + \ldots + f_n)/(n + 1),$$

where

$$f_m = \sum_{k=-m}^{m} c_k e^{ik\theta} = \sum_{k=-m}^{m} c_k z^k, \qquad z = e^{i\theta}.$$

If the negative Fourier coefficients of a function $f \in C(b\Delta)$ vanish, then the f_m and the σ_n are polynomials in z. So $f \in P(b\Delta)$. Moreover, the polynomials σ_n converge uniformly on Δ, by the maximum modulus principle, to a continuous extension $\tilde{f}$ of f to Δ which is analytic on int Δ.

Conversely, if $f \in C(b\Delta)$ can be extended continuously to Δ to be analytic on int Δ, then the negative Fourier coefficients of f vanish, by Cauchy's theorem. In particular, the negative Fourier coefficients of any $f \in P(b\Delta)$ vanish.

This shows that the following three assertions are equivalent, for a function $f \in C(b\Delta)$:

(*) $f \in P(b\Delta)$,
(**) f extends continuously to Δ to be analytic on int Δ,
(***) the negative Fourier coefficients of f vanish.

Every $\lambda \in \Delta$ determines a homomorphism ϕ_λ of $P(b\Delta)$, obtained by evaluating the analytic extensions of functions in $P(b\Delta)$ at λ. The correspondence $\lambda \to \phi_\lambda$ embeds Δ as a closed subset of $M_{P(b\Delta)}$. It is customary to identify Δ with its embedded image in $M_{P(b\Delta)}$.

Suppose $\phi \in M_{P(b\Delta)}$ is arbitrary. Let $\lambda = \phi(z)$, where z is the coordinate function. Since $\|z\|_{b\Delta} = 1$, we have $|\lambda| \leq 1$, that is, $\lambda \in \Delta$. Also, $\phi(p(z)) = p(\lambda) = \phi_\lambda(p)$ for all polynomials p. Since polynomials are dense in $P(b\Delta)$, ϕ coincides with ϕ_λ.

Consequently the maximal ideal space of $P(b\Delta)$ coincides with Δ. The Gelfand transform $\hat{f}$ of $f \in P(b\Delta)$ is the continuous extension of f to Δ which is analytic on int Δ.

3.3 Example. For n a positive integer, let $C^{(n)}[0, 1]$ denote the vector space of continuous complex-valued functions on the interval $[0, 1]$ which are n-times continuously differentiable. The space $C^{(n)}[0, 1]$ is a Banach algebra, with pointwise multiplication, and the norm

$$\|f\|_{C^{(n)}} = \sum_{k=0}^{n} \sup_{0 \leq t \leq 1} |f^{(k)}(t)|.$$

The maximal ideal space of $C^{(n)}[0, 1]$ is $[0, 1]$.

3.4 Example. For $0 < \alpha \leq 1$, let $\mathrm{Lip}_\alpha[0, 1]$ be the set of continuous complex-valued functions on $[0, 1]$ which satisfy a Lipschitz condition of order α. The norm in $\mathrm{Lip}_\alpha[0, 1]$ is defined by

$$\|f\|_\alpha = \sup_{0 \leq t \leq 1} |f(t)| + \sup_{0 \leq s < t \leq 1} \frac{|f(s) - f(t)|}{|s - t|^\alpha}.$$

The space $\text{Lip}_\alpha[0, 1]$ is a Banach algebra, the multiplication being the usual pointwise multiplication of functions. The maximal ideal space of $\text{Lip}_\alpha[0, 1]$ is $[0, 1]$.

3.5 Example. Let l^1 be the set of two-tailed sequences $a = \{a_n\}_{n=-\infty}^{\infty}$, whose norm $\|a\| = \sum_{n=-\infty}^{\infty} |a_n|$ is finite. If a and b belong to l^1, the convolution $c = a * b$ of a and b is defined by

$$c_n = \sum_{k=-\infty}^{\infty} a_{n-k} b_k.$$

With convolution as multiplication, l^1 is a commutative Banach algebra.

Let $\epsilon_n \in l^1$ be the sequence whose nth entry is 1 and whose other entries are zero. Then ϵ_0 is an identity for l^1. Every $a \in l^1$ can be expressed in the form

$$a = \sum_{n=-\infty}^{\infty} a_n \epsilon_n,$$

the series converging in l^1.

Now $\epsilon_n * \epsilon_k = \epsilon_{n+k}$ for $-\infty < n,k < \infty$. In particular, each ϵ_n is invertible and has inverse $\epsilon_n^{-1} = \epsilon_{-n}$. Moreover, $\epsilon_n = (\epsilon_1)^n$, $-\infty < n < \infty$. So the expression for $a \in l^1$ becomes

$$a = \sum_{n=-\infty}^{\infty} a_n \epsilon_1^n.$$

In particular, ϵ_1 and ϵ_1^{-1} generate l^1.

For every complex number λ of unit modulus, the continuous linear functional ϕ_λ on l^1 is defined by

$$\phi_\lambda(a) = \sum_{-\infty}^{\infty} a_n \lambda^n.$$

One verifies that ϕ_λ is multiplicative on l^1, so that $\phi_\lambda \in M_{l^1}$.

Let $\phi \in M_{l^1}$ be arbitrary. Then $|\phi(\epsilon_1)| \leq \|\epsilon_1\| = 1$, and $1/|\phi(\epsilon_1)| = |\phi(\epsilon_{-1})| \leq \|\epsilon_{-1}\| = 1$. So $\phi(\epsilon_1) = \lambda$ has unit modulus. Since ϕ coincides with ϕ_λ on the generators ϵ_1 and ϵ_1^{-1} of l^1, $\phi = \phi_\lambda$.

This shows that the maximal ideal space of l^1 can be identified with the unit circle $b\Delta$. The Gelfand transform of $a \in l^1$ is the function

$$\hat{a}(e^{i\theta}) = \sum_{-\infty}^{\infty} a_n e^{in\theta}, \qquad 0 \leq \theta \leq 2\pi.$$

The algebra $\widehat{l^1}$ of Gelfand transforms is precisely the family of continuous functions on $b\Delta$ which have absolutely convergent Fourier series.

In his seminal series of papers on Banach algebras, Gelfand gave an elegant proof of Wiener's theorem on absolutely convergent Fourier series, which we will now present. The proof served to draw a good deal of attention to commutative Banach algebras.

Theorem. If f is a continuous function on $b\Delta$ with an absolutely convergent Fourier series, and f does not vanish on $b\Delta$, then $1/f$ has an absolutely convergent Fourier series.

Proof. By assumption, $f = \hat{a}$ for some $a \in l^1$, where a belongs to no maximal ideal of l^1. Hence a is invertible in l^1. So $\widehat{a^{-1}} = 1/f$ has an absolutely convergent Fourier series.

3.6 Example. The preceding example can be generalized as follows. Let G be a locally compact abelian group with Haar measure σ. The Banach space $L^1(\sigma)$, together with the convolution multiplication defined by

$$(f * g)(x) = \int_G f(x - y)g(y)d\sigma(y),$$

is a commutative Banach algebra, denoted by $L^1(G)$. The algebra $L^1(G)$ does not have an identity, unless G is discrete.

A **character** of G is a continuous homomorphism of G into the circle group. The set $\hat{G}$ of characters of G is a group, whose operation is pointwise multiplication. With the topology of uniform convergence on compacta, $\hat{G}$ becomes a locally compact abelian group, called the **character group** or **dual group** of G. The celebrated Pontryagin duality theorem states that the dual group of $\hat{G}$ is G.

Every character $\chi \in \hat{G}$ determines a continuous homomorphism of $L^1(G)$ via the formula

$$\hat{f}(\chi) = \int_G f(x)\overline{\chi(x)}d\sigma(x).$$

Every continuous non-zero homomorphism of $L^1(G)$ arises in this manner from a character of G. The maximal ideal space of $L^1(G)$ is homeomorphic to $\hat{G}$.

Suppose G is the group of real numbers R. Every character of R is of the form $s \to e^{ist}$, for some real number t. Hence $\hat{R} = R$. The Gelfand transform becomes essentially the usual Fourier transform:

$$\hat{f}(t) = \int_{-\infty}^{\infty} f(s)e^{-ist}ds.$$

3.7 Example. Let G be a locally compact abelian group, and let $M(G)$ be the Banach space of finite Baire measures on G, with the total variation norm. The convolution $\mu * \nu$ of two measures $\mu, \nu \in M(G)$ is defined on Baire sets E by

$$(\mu * \nu)(E) = \int_G \mu(E - x)d\nu(x).$$

With convolution as multiplication, $M(G)$ becomes a commutative Banach algebra. The algebra $M(G)$ has an identity, the point mass at the identity of the group. The correspondence $f \to f d\sigma$ embeds $L^1(G)$ isometrically as a closed ideal in $M(G)$. The maximal ideal space of $M(G)$ is horrible, unless G is discrete.

3.8 Example. Let H be a Hilbert space, and let $B(H)$ be the algebra of continuous linear operators on H. Then $B(H)$ is a Banach algebra, with the norm

$$||T|| = \sup\{||Tx||: x \in H, ||x|| \leq 1\}.$$

The algebra $B(H)$ has an identity, the identity operator.

Except in trivial cases, $B(H)$ is not commutative. However, of special interest are subalgebras of $B(H)$ which are commutative. This includes, for example, any subalgebra of $B(H)$ generated by a single operator.

4. The Shilov Boundary

A subset E of M_A is a **boundary** for A if every function $\hat{f} \in \hat{A}$ assumes its maximum modulus on E.

4.1 Lemma. Let $f_1, \ldots, f_n \in A$, and let V be the basic open subset of M_A defined by

$$V = \{\psi: |\hat{f}_j(\psi)| < 1, 1 \leq j \leq n\}.$$

Either V meets every boundary for A, or else $E \backslash V$ is a closed boundary for A whenever E is a closed boundary for A.

Proof. Suppose there is a closed boundary E for A such that $E \backslash V$ is not a boundary for A. We must show that V meets every boundary for A.

Choose $f \in A$ such that $||\hat{f}||_{M_A} = 1$, while $||\hat{f}||_{E \backslash V} < 1$. By replacing f by a suitable power f^m, we can assume that $|\hat{f}\hat{f}_j| < 1$ on $E \backslash V$. Since also $|\hat{f}\hat{f}_j| < 1$ on V, we have $|\hat{f}\hat{f}_j| < 1$ on E, and hence on M_A. It follows that the set on which $\hat{f}$ attains its maximum modulus 1 is contained in V. So V meets every boundary for A. This proves the lemma.

The following theorem is due to Shilov.

4.2 Theorem. The intersection of all closed boundaries for A is a boundary for A.

Proof. Let ∂_A denote the intersection of all closed boundaries for A. Evidently $\phi \in \partial_A$ if and only if for every neighborhood U of ϕ, there is $f \in A$ such that the set on which $\hat{f}$ attains its maximum modulus is contained in U.

Suppose $f \in A$ satisfies $|\hat{f}| < 1$ on ∂_A. We must show that $|\hat{f}| < 1$ on M_A.

If $\phi \in M_A$ satisfies $|\hat{f}(\phi)| \geq 1$, then there is a basic open neighborhood of ϕ in M_A which is disjoint from some closed boundary for A. Let $V_1, \ldots, V_k$ be such basic neighborhoods which cover the compact set $\{|\hat{f}| \geq 1\}$. By 4.1, $M_A \backslash (\bigcup_{j=1}^{k} V_j)$ is a boundary for A. Since $|\hat{f}| < 1$ on this boundary, we obtain $|\hat{f}| < 1$ on M_A. This proves the theorem.

The intersection of all closed boundaries for A is called the **Shilov boundary** of A, and is denoted by ∂_A. It is the smallest closed boundary of A.

By the maximum modulus principle, the circle $b\Delta$ is a boundary for $P(\Delta)$. If $\lambda \in b\Delta$, the function $1 + \bar{\lambda} z$ assumes its maximum modulus only at $z = \lambda$. Hence $b\Delta$ must be contained in any boundary for $P(\Delta)$. So the Shilov boundary of $P(\Delta)$ is the topological boundary of Δ.

In connection with this example, the following tantalizing question remains open: If M_A is homeomorphic to the disc Δ, does ∂_A contain the topological boundary $b\Delta$ of Δ? Does ∂_A even meet the topological boundary of Δ?

4.3 Theorem. If $f \in A$, then $\hat{f}(\partial_A)$ contains the topological boundary of $\hat{f}(M_A)$.

Proof. Suppose there exists $\phi \in M_A$ such that $\hat{f}(\phi) \in b\hat{f}(M_A)$, while $\hat{f}(\phi)$ is at a positive distance δ from $\hat{f}(\partial_A)$. Choose a complex number λ such that $\lambda \notin \hat{f}(M_A)$, while $|\lambda - \hat{f}(\phi)| < \delta/2$. By 2.7, $\lambda - f$ is invertible.

Let $g = (\lambda - f)^{-1}$. Then $\hat{g} = 1/(\lambda - \hat{f})$. Since $|\hat{g}(\psi)| \leq 2/\delta$ for $\psi \in \partial_A$, we have $\|\hat{g}\|_{M_A} \leq 2/\delta$. However, $|\hat{g}(\phi)| > 2/\delta$. This contradiction establishes the theorem.

5. Two Basic Theorems

In this section, we wish to prove two basic results. The first concerns the application of certain analytic functions to Banach algebra elements. The second is a formula for the spectral radius.

5.1 Theorem Let $f \in A$. Let h be a complex-valued function which is defined and analytic in a neighborhood of $\hat{f}(M_A) = \sigma(f)$. Then there exists $g \in A$ such that $\hat{g} = h \circ \hat{f}$.

Proof. The Cauchy integral formula states that

$$h(z_0) = \frac{1}{2\pi i} \int_\Gamma \frac{h(z)}{z - z_0} dz, \qquad z_0 \in \sigma(f),$$

for an appropriate contour Γ surrounding $\sigma(f)$. We define

$$g = \frac{1}{2\pi i} \int_\Gamma h(z)(z - f)^{-1} dz.$$

The integral exists as an ordinary Riemann integral.

Approximating the integral by finite Riemann sums, we see that

$$\phi(g) = \frac{1}{2\pi i}\int h(z)(z - \phi(f))^{-1}dz = h(\phi(f)), \qquad \phi \in M_A.$$

Consequently $\hat{g} = h \circ \hat{f}$.

The **spectral radius** of $f \in A$ is defined to be

$$\sup\{|\lambda| : \lambda \in \sigma(f)\}.$$

By 2.7, the spectral radius of f coincides with $\|\hat{f}\|_{M_A}$.

5.2 Theorem. The spectral radius of $f \in A$ is given by

$$\|\hat{f}\|_{M_A} = \lim_{n\to\infty} \|f^n\|^{1/n}.$$

Proof. For any positive integer n and any $\phi \in M_A$,

$$|\hat{f}(\phi)| = |\hat{f}^n(\phi)|^{1/n} \leq \|f^n\|^{1/n}.$$

Consequently,

$$\|\hat{f}\|_{M_A} \leq \liminf_{n\to\infty} \|f^n\|^{1/n}.$$

Now, let L be a continuous linear functional on A. Define $h(\lambda) = L((\lambda - f)^{-1})$, for $\lambda \notin \sigma(f)$. Then h is analytic off $\sigma(f)$, and

$$h(\lambda) = \sum_{n=0}^{\infty} \frac{L(f^n)}{\lambda^{n+1}}$$

for large λ. Since h is analytic for $|\lambda| > \|\hat{f}\|_{M_A}$, the series representation must converge for all λ satisfying $|\lambda| > \|\hat{f}\|_{M_A}$. Hence,

$$\sup_n |L(f^n)|/|\lambda|^{n+1} < \infty$$

whenever $|\lambda| > \|\hat{f}\|_{M_A}$.

Now fix λ satisfying $|\lambda| > \|\hat{f}\|_{M_A}$. The supremum above must be finite for all continuous linear functionals L on A. By the principle of uniform boundedness,

$$\sup_n \|f^n\|/|\lambda|^{n+1} = M < \infty.$$

Hence

$$\limsup_{n\to\infty} \|f^n\|^{1/n} \leq \limsup_{n\to\infty} M^{1/n}|\lambda|^{1+1/n} = |\lambda|.$$

Since this is true whenever $|\lambda| > \|\hat{f}\|_{M_A}$, we obtain

$$\limsup_{n\to\infty} \|f^n\|^{1/n} \leq \|f\|_{M_A}.$$

That completes the proof.

5.3 Corollary. The Gelfand transform $f \to \hat{f}$ is an isometry if and only if $\|f^2\| = \|f\|^2$ for all $f \in A$.

Proof. If $f \to \hat{f}$ is an isometry, then $\|f^2\| = \|\hat{f}^2\|_{M_A} = \|\hat{f}\|^2_{M_A} = \|f\|^2$. Conversely, if $\|f^2\| = \|f\|^2$ for all $f \in A$, then $\|f^{2^n}\| = \|f\|^{2^n}$ for all $n \geq 1$. So $\|f\| = \|f^{2^n}\|^{1/2^n} \to \|\hat{f}\|_{M_A}$.

6. Hulls and Kernels

Let J be an ideal in A. The **hull** of J is the set of maximal ideals in M_A which contain J. In other words, the hull of J consists of all $\phi \in M_A$ such that $\phi(f) = 0$ for all $f \in J$. Evidently the hull of J is a closed subset of M_A.

6.1 Theorem. Let J be a closed ideal in A, and let B be the linear span of J and the identity. Then B is a closed subalgebra of A, and M_B is obtained from M_A by identifying the hull of J to a point.

Proof. We can assume that J is a proper ideal. Since the sum of a closed subspace and a finite dimensional subspace of a Banach space is closed, B is a closed subalgebra of A. Also, J is a maximal ideal in B.

For every $\phi \in M_A$, let $\tau(\phi)$ be the restriction of ϕ to B. Then τ is a continuous map of M_A into M_B. Evidently $\tau(\phi)$ is the maximal ideal J of B if and only if ϕ belongs to the hull of J, so τ is one-to-one off the hull of J.

Suppose $\psi \in M_B$ is distinct from the maximal ideal J. Then there is $f \in J$ such that $\psi(f) = 1$. For $g \in A$, define $\phi(g) = \psi(gf)$. Then ϕ is linear, and $\phi(1) = 1$. If $g_1, g_2 \in A$, then $\phi(g_1g_2) = \psi(g_1g_2f) = \psi(g_1g_2f^2) = \psi(g_1f)\psi(g_2f) = \phi(g_1)\phi(g_2)$. Hence $\phi \in M_A$. Evidently $\tau(\phi) = \psi$. This shows that τ maps M_A onto M_B.

So τ determines a continuous one-to-one map of the space M_A, with the hull of J identified to a point, onto M_B. This map must then be a homeomorphism. This proves the theorem.

6.2 Theorem. Let J be a closed ideal in A. Then the maximal ideal space of the quotient algebra A/J is the hull of J.

Proof. The adjoint σ of the quotient map $A \to A/J$ is defined by

$$\sigma(\psi)(f) = \psi(f + J), \qquad f \in A, \psi \in M_{A/J}.$$

One easily checks that σ is the desired homeomorphism.

Let E be a subset of M_A. The **kernel** of E, denoted by **ker(E)**, is the intersection of the maximal ideals corresponding to the points of E. In other words, ker(E) is the set of $f \in A$ such that $\hat{f}$ vanishes on E. Evidently ker(E) is a closed ideal in A.

A subset E of M_A is a **hull** if E is the hull of some ideal in A. If E is a hull, then E is the hull of ker(E).

Finite unions of hulls are hulls, and arbitrary intersections of hulls are hulls. So the hulls can be taken as the closed subsets of a topology for M_A, the **hull-kernel topology.** The closure of a subset F of M_A in the hull-kernel topology is the hull of ker(F). A point $\phi \in M_A$ does not belong to the hull-kernel closure of F if and only if there is an $f \in A$ such that $\hat{f}$ vanishes on F, while $\phi(f) \neq 0$.

The hull-kernel topology is a T_1-topology (points are closed) which is smaller than the usual topology of M_A. The hull-kernel topology is a Hausdorff topology if and only if the hull-kernel topology and the usual topology of M_A coincide. This occurs if and only if every closed subset of M_A is a hull. And this occurs if and only if $\hat{A}$ is regular on M_A, that is, for every closed subset E of M_A and every $\phi \in M_A \backslash E$, there is $f \in A$ such that $\hat{f}(\phi) \neq 0$, while $\hat{f}$ vanishes on E. If a Banach algebra A has one of these equivalent properties, A is said to be a **regular Banach algebra.** Examples of regular Banach algebras are provided by the group algebras $L^1(G)$.

6.3 Theorem. If A is a regular Banach algebra, then $\hat{A}$ is normal on M_A, that is, for every pair of compact disjoint subsets E and F of M_A, there is an $f \in A$ such that $\hat{f} = 0$ on E and $\hat{f} = 1$ on F.

Proof. Let $\phi \in F$. First we will find $g \in A$ such that $\hat{g} = 0$ on E, while $\hat{g} = 1$ in a neighborhood of ϕ.

Choose $h \in \ker(E)$ such that $\hat{h}(\phi) = 1$. The set $W = \{\psi \in M_A\colon |\hat{h}(\psi)| \geq \frac{1}{2}\}$ is a compact neighborhood of ϕ which does not meet E.

Let J be the kernel of $E \cup W$. By hypothesis, the hull of J is $E \cup W$. By 6.2, the maximal ideal space of A/J is $E \cup W$.

The coset $H = h + J$ satisfies $\hat{H} = 0$ on E, while $|\hat{H}| \geq \frac{1}{2}$ on W. The function Q on the complex plane, defined to be 0 in a neighborhood of $z = 0$ and 1 in a neighborhood of $\{|z| \geq \frac{1}{2}\}$, is analytic in a neighborhood of the spectrum of H. By 5.1, there is $G = g + J \in A/J$ such that $G = Q \circ H$. That is, there is $g \in A$ such that $\hat{g} = 0$ on E and $\hat{g} = 1$ on W.

By compactness of F, we can find $g_1, \ldots, g_n \in A$ and subsets $W_1, \ldots, W_n$ of M_A such that $F \subseteq \bigcup_{j=1}^{n} W_j$, $\hat{g}_j = 0$ on E, and $\hat{g}_j = 1$ on W_j.

Now the Gelfand transform of $g_1 + g_2 - g_1 g_2$ is 0 on E and 1 on $W_1 \cup W_2$. Proceeding in this manner, we obtain $f \in A$ such that $\hat{f} = 0$ on E while $\hat{f} = 1$ on $W_1 \cup \ldots \cup W_n$. This proves the theorem.

7. Commutative B^*-Algebras

In this section, we will axiomatize the algebras $C(X)$, X a compact Hausdorff space. In order to do this, we will introduce an abstract version of the complex conjugation operator, sending a function f into its complex conjugate $\bar{f}$.

An **involution** of A is an operation $f \rightarrow f^*$ from A to A, which satisfies

(i) $f^{**} = f$,
(ii) $(f+g)^* = f^* + g^*$,
(iii) $(\lambda f)^* = \bar{\lambda} f^*$,
(iv) $(fg)^* = f^* g^*$,

where f and g belong to A, while λ is complex. A **commutative B^*-algebra** is a commutative Banach algebra A with an involution $f \to f^*$ which satisfies

$$||f^*f||^2 = ||f||^2, \qquad f \in A.$$

Our object is to prove the following theorem.

7.1 Theorem (Gelfand-Naimark Theorem). Let A be a commutative B^*-algebra. Then the Gelfand transform is an isometric isomorphism of A and $C(M_A)$ satisfying

$$\widehat{f^*} = \bar{\hat{f}}, \qquad f \in A.$$

The important part of the theorem is that the Gelfand transform carries the involution into complex conjugation. Before starting the proof, we prove some lemmas.

7.2 Lemma. If $f \to f^*$ is an involution of A, then $1^* = 1$.

Proof. $1^* = 11^* = 1^{**}1^* = (1^*1)^* = 1^{**} = 1$.

7.3 Lemma. If A is a commutative B^*-algebra, and $f \in A$, then $||f^2|| = ||f||^2$ and $||f|| = ||f^*||$.

Proof. $||f^2||^2 = ||(f^2)^*f^2|| = ||(f^*f)^*(f^*f)|| = ||f^*f||^2 = ||f||^4$, so $||f^2|| = ||f||^2$. Also, $||f||^2 = ||f^*f|| = ||f^{**}f^*|| = ||f^*||^2$, so $||f|| = ||f^*||$.

7.4 Lemma. Let A be a commutative B^*-algebra. If $f \in A$ satisfies $f^* = f^{-1}$, then $|\hat{f}| = 1$. If $g \in A$ satisfies $g^* = g$, then $\hat{g}$ is real.

Proof. Suppose $f^* = f^{-1}$. Then also $(f^{-1})^* = f$. So $1 = ||f^*f|| = ||f||^2$, and $1 = ||(f^{-1})^*f^{-1}|| = ||f^{-1}||^2$ It follows that both $\sigma(f)$ and $\sigma(f^{-1})$ are contained in the unit disc Δ. This only happens when $|\lambda| = 1$ for every $\lambda \in \sigma(f)$, that is, when $\hat{f}$ has unit modulus.

Now if h belongs to any Banach algebra, the series

$$\sum_{n=0}^{\infty} \frac{h^n}{n!}$$

converges to an element e^h satisfying $\widehat{e^h} = e^{\hat{h}}$. One computes that e^h is invertible, and has inverse e^{-h}.

In the case at hand, the involution $h \to h^*$ is continuous, by 7.3. Consequently

$$(e^h)^* = \sum_{n=0}^{\infty} \frac{(h^n)^*}{n!} = \sum_{n=0}^{\infty} \frac{(h^*)^n}{n!} = e^{h^*}, \qquad h \in A.$$

Now suppose $g^* = g$. Set $f = e^{ig}$. Then

$$f^* = e^{-ig^*} = e^{-ig} = f^{-1}.$$

By the first half of the lemma, $\sigma(f)$ is a subset of the unit circle. Hence $\sigma(g)$ must be real. That completes the proof.

Proof of 7.1. By 5.3 and 7.3, the Gelfand transform is an isometry of A onto a closed subalgebra $\hat{A}$ of $C(M_A)$.

If $f \in A$, set $g = (f + f^*)/2$ and $h = (f - f^*)/2i$. Then $f = g + ih$, $g = g^*$, and $h = h^*$. So $f^* = g^* - ih^*$. Using 7.4, we obtain

$$\widehat{f^*} = \widehat{g^*} - i\widehat{h^*} = \hat{g} - i\hat{h} = \bar{\hat{f}}.$$

This formula shows, in particular, that if $\hat{f} \in \hat{A}$, then the complex conjugate $\bar{\hat{f}}$ of $\hat{f}$ also belongs to $\hat{A}$. Since $\hat{A}$ contains the constants, and $\hat{A}$ separates the points of M_A, $\hat{A}$ must coincide with $C(M_A)$, by the Stone-Weierstrass theorem. This completes the proof.

8. Compactifications

Let S be a topological space. Let $CB(S)$ denote the algebra of bounded continuous complex-valued functions on S. The algebra $CB(S)$ is a commutative B^*-algebra, with the norm

$$\|f\|_S = \sup_{s \in S} |f(s)|$$

and the involution

$$f^* = \bar{f}.$$

A family $\mathscr{F}$ of functions on S is **self-adjoint** if $\bar{f} \in \mathscr{F}$ whenever $f \in \mathscr{F}$. The family $\mathscr{F}$ is **separating** if whenever $s \neq t$ belong to S, there is $f \in \mathscr{F}$ such that $f(s) \neq f(t)$.

A **compactification** of the topological space S is a compact Hausdorff space X and a continuous one-to-one map τ of S onto a dense subset $\tau(S)$ of X. It is often convenient to identify $s \in S$ with $\tau(s) \in X$.

Every compactification X of S determines a closed separating self-adjoint subalgebra A of $CB(S)$ which contains the constants, namely, the restriction of $C(X)$ to S.

Conversely, suppose that A is a closed separating self-adjoint subalgebra of $CB(S)$ which contains the constants. Then A is a B^*-subalgebra of $CB(S)$. By the Gelfand-Naimark theorem, A is isometrically isomorphic to $C(M_A)$.

Every $s \in S$ determines the evaluation homomorphism $\tau(s)$ at s, defined by

$$\hat{f}(\tau(s)) = f(s), \qquad f \in A.$$

The function τ from S into M_A is continuous, by the definition of the topology of M_A. Since A separates the points of S, τ is one-to-one.

If $g \in C(M_A)$, and g is zero on $\tau(S)$, then g is the Gelfand transform of the function identically zero on S, so $g = 0$. Hence $\tau(S)$ is dense in M_A.

It follows that M_A is a compactification of S. This proves the following theorem.

8.1 Theorem. Let S be a topological space. There is a bijective correspondence between compactifications X of S and closed separating self-adjoint subalgebras A of $CB(S)$ which contain the constants. The algebra A associated with the compactification X consists of the functions in $CB(S)$ which extend continuously to X. The compactification X associated with A is the maximal ideal space of A.

As an example, let R be the real line and let A be the algebra of uniform limits on R of trigonometric polynomials of the form

$$f(t) = \sum_{j=1}^{n} c_j e^{ia_j t},$$

where the a_j are real and the c_j complex. The algebra A is a commutative B^*-algebra. The resulting compactification is called the **Bohr compactification** of the real line. It can be given the structure of a compact topological group. The reals become embedded as a dense subgroup. But the embedding of R is not a homeomorphism.

Criteria that insure that the embedding τ be a homeomorphism are given by the following theorem.

8.2 Theorem. Let S be a topological space, and let A be a closed separating self-adjoint subalgebra of $CB(S)$ containing the constants. The canonical embedding $\tau: S \to M_A$ is a homeomorphism if and only if either of the following conditions is satisfied:

(i) For each closed subset F of S and $s \in S \backslash F$, there is a function $f \in A$ such that $f(s) = 0$ while $f|_F = 1$.

(ii) For each closed subset F of S and $s \in S \backslash F$, there is a function $f \in A$ such that $f(s) \notin \overline{f(F)}$ [the closure of $f(F)$].

Proof. In view of the fact that $A = C(M_A)$, (i) and (ii) each occur if and only if for all closed subsets F of S and $s \in S \backslash F$, $\tau(s)$ does not belong to the closure $\overline{\tau(F)}$ of $\tau(F)$ in M_A. This occurs if and only if for each closed subset F of S, $\tau(S \backslash F)$ is the intersection of $\tau(S)$ and the open subset $M_A \backslash \overline{\tau(F)}$ of M_A. And this occurs if and only if τ^{-1} is continuous. This proves the theorem.

A topological space S is **completely regular** if for every closed subset F of S and $s \in S \backslash F$, there is a continuous complex-valued function f on S such that $f(s) = 0$ while $f|_F = 1$. If S is a completely regular Hausdorff space, then 8.2 applies to the algebra $A = CB(S)$. This leads to the following theorem.

8.3 Theorem (Stone-Čech Theorem). Let S be a completely regular Hausdorff space. Then there is a compact Hausdorff space X such that S is homeomorphic to a dense subset of X, and every bounded continuous complex-valued function on S extends continuously to X.

9. The Algebra L^∞

Let S be a set, and let Σ be a σ-algebra of subsets of S. Let μ be a probability measure on (S, Σ), that is, μ is a positive measure on (S, Σ) such that $\mu(S) = 1$.

The Banach space $L^\infty(\mu)$ is a commutative B^*-algebra, with pointwise multiplication, and with the involution

$$f^* = \bar{f}, \qquad f \in L^\infty(\mu).$$

By the Gelfand-Naimark theorem, $L^\infty(\mu)$ is isometrically isomorphic to $C(Y)$, where Y is the maximal ideal space of $L^\infty(\mu)$.

The functional $\hat{f} \to \int f d\mu, f \in L^\infty(\mu)$, is a continuous linear functional on $C(Y)$. By the Riesz representation theorem, there is a regular Borel measure $\hat{\mu}$ on Y such that

$$\int f d\mu = \int \hat{f} d\hat{\mu}, \qquad f \in L^\infty(\mu).$$

We wish to show that $L^\infty(\hat{\mu}) = L^\infty(\mu)$, that is, that $L^\infty(\hat{\mu}) = C(Y)$.

Let χ_E denote the characteristic function of $E \in \Sigma$. Since $\chi_E^2 = \chi_E$, the Gelfand transform satisfies $\hat{\chi}_E^2 = \hat{\chi}_E$. Hence $\hat{\chi}_E$ is the characteristic function of a clopen (closed and open) subset U_E of Y.

Conversely, if U is a clopen subset of Y, its characteristic function is continuous, and so must be the transform of $\chi \in L^\infty(\mu)$. Since $\widehat{\chi^2} = \hat{\chi}$, we have $\chi^2 = \chi$, and $\chi = \chi_E$ for some $E \in \Sigma$. Hence every clopen subset U of Y is of the form U_E, for some $E \in \Sigma$.

The following lemma shows that Y is totally disconnected.

9.1 Lemma. The clopen subsets $\{U_E \colon E \in \Sigma\}$ form a basis for the topology of Y.

Proof. Since finite linear combinations of characteristic functions are dense in $L^\infty(\mu)$, the topology of Y has a basis of sets of the form

$$U = \{y \colon |\sum a_j \hat{\chi}_{E_j}(y)| < 1\},$$

where the a_j are complex and the E_j are disjoint. Since $\hat{\chi}_{E_j} \cdot \hat{\chi}_{E_k} = 0$ for $j \neq k$, each such set U coincides with U_E, where E is the intersection of those sets $S \backslash E_j$ for which $|a_j| > 1$. That proves the lemma.

Note that $\mu(E) = \hat{\mu}(U_E)$ for all $E \in \Sigma$. In particular, $\hat{\mu}(U_E) > 0$ unless χ_E is a null function, that is, unless U_E is empty. Hence we deduce the following corollary, stating that the closed support of $\hat{\mu}$ is Y.

9.2 Corollary. If U is an open non-empty subset of Y, then $\hat{\mu}(U) > 0$.

The next lemma shows that U is extremely disconnected, that is, the closure of every open subset of Y is open.

9.3 Lemma. Suppose U is open. Let

$$c = \sup\{\mu(D) \colon D \in \Sigma, U_D \subseteq U\}.$$

Suppose $E_n \in \Sigma$ is a sequence such that $U_{E_n} \subseteq U$ while $\mu(E_n) \to c$. Let $E = \bigcup_{n=1}^{\infty} E_n$. Then $\bar{U} = U_E$.

Proof. Replacing E_n by $E_1 \cup \ldots \cup E_n$, we can assume that the E_n increase to E.

Suppose $F \in \Sigma$ is such that U_F does not meet $\bar{U}$. Then U_F does not meet U_{E_n}, so $\mu(F \cap E_n) = 0$. Hence $\mu(F \cap E) = 0$, and U_F does not meet U_E. Since the complement of $\bar{U}$ is the union of such U_F, we have $U_E \subseteq \bar{U}$.

Now suppose $F \in \Sigma$ satisfies $U_F \subseteq U$. Then $\mu(E_n) \leq \mu(E_n \cup F) \leq c$, and so $\mu(E) = \mu(E \cup F) = c$. It follows that $U_F \subseteq U_E$. Hence $U \subseteq U_E$, and $\bar{U} \subseteq U_E$. This proves the lemma.

The next lemma states that $\hat{\mu}$ is a normal measure.

9.4 Lemma. If V is a Borel subset of Y, then

$$\hat{\mu}(\text{int } V) = \hat{\mu}(V) = \hat{\mu}(\bar{V}).$$

Proof. Suppose first that U is open. Let E_n and E be as in 9.3, and assume that the E_n are increasing. Then $U_{E \backslash E_n} \supseteq \bar{U} \backslash U$, and $\hat{\mu}(U_{E \backslash E_n}) = \mu(E \backslash E_n) \to 0$. So $\hat{\mu}(\bar{U} \backslash U) = 0$, or $\hat{\mu}(U) = \hat{\mu}(\bar{U})$.

Now if V is a Borel set, choose a decreasing sequence U_n of open sets such that $V \subseteq U_n$ and $\hat{\mu}(U_n) \to \hat{\mu}(V)$. Then $\hat{\mu}(\bar{V} \backslash V) \leq \hat{\mu}(\bar{U}_n \backslash V) = \hat{\mu}(\bar{U}_n) - \hat{\mu}(V) \to 0$.

Hence $\hat{\mu}(\bar{V}) = \hat{\mu}(V)$ for all Borel sets V. By taking complements, we obtain $\hat{\mu}(\text{int } V) = \hat{\mu}(V)$ for all Borel sets V.

9.5 Theorem. The natural injection $C(Y) \subseteq L^\infty(\hat{\mu})$ is an isometric isomorphism of $C(Y)$ and $L^\infty(\hat{\mu})$.

Proof. Since the closed support of $\hat{\mu}$ is Y, the injection of $C(Y)$ into $L^\infty(\hat{\mu})$ is an isometry. It suffices to show that the range is all of $L^\infty(\hat{\mu})$.

Let V be a Borel subset of Y. From 9.3 we see that the closure of int V coincides with U_F for some $F \in \Sigma$.

Now int $V \subseteq U_F \subseteq \bar{V}$. By 9.4, χ_V differs from the continuous function χ_{U_F} by a null function.

By taking finite linear combinations of characteristic functions of Borel sets, and then uniform limits, we see that every bounded Borel function on

Y differs from a function in $C(Y)$ by a null function. Hence the natural injection is surjective. This proves the theorem.

There is one further remark which is sometimes useful. Suppose $0 < p < \infty$, $f \in L^p(\mu)$, and $f \geq 0$. Then we can regard f as a lower semicontinuous function on Y. In fact, f is the increasing limit of the functions $f_n = \min(f, n)$, and each f_n can be regarded as a continuous function on Y.

10. Normal Operators on Hilbert Space

As another application of the Gelfand-Naimark theorem, we will discuss the spectral resolution of normal operators on Hilbert space.

By H we will denote a Hilbert space. The (non-commutative) Banach algebra of bounded operators on H will be denoted by $B(H)$. The adjoint operator of $T \in B(H)$ will be denoted by T^*. The correspondence $T \to T^*$ is an involution of $B(H)$, providing the definition of involution given in section 7 is appropriately modified to handle non-commutative algebras.

10.1 Lemma. If T is a bounded linear operator on a Hilbert space H, then $\|T^*T\| = \|T\|^2$.

Proof. $\|T^*T\| \leq \|T^*\|\|T\| = \|T\|^2 = \sup_{x \in H} (Tx, Tx)/\|x\|^2$
$= \sup_{x \in H} (T^*Tx, x)/\|x\|^2 \leq \|T^*T\|$.

A continuous linear operator N on Hilbert space H is **normal** if N commutes with its adjoint N^*. If N is normal, the closed subalgebra A of $B(H)$ generated by N and N^* is a commutative Banach algebra with involution. By 10.1, A is a B^*-algebra.

By the Gelfand-Naimark theorem, A is isometrically isomorphic to $C(M_A)$. Every $\phi \in M_A$ is completely determined by the value $\phi(N)$. Consequently M_A can be identified with the spectrum $\sigma_A(N)$ of the operator N as an element of A.

10.2 Lemma. The spectrum $\sigma_A(N)$ coincides with the set $\sigma(N)$ of all complex numbers λ such that $\lambda - N$ is not an invertible operator in $B(H)$.

Proof. If $\lambda - N$ has no inverse in $B(H)$, it certainly has no inverse in A. So $\sigma(N) \subseteq \sigma_A(N)$.

Now suppose $T = \lambda - N$ is invertible in $B(H)$. Then so is T^*, and so T^*T is invertible in $B(H)$. Since $T^{-1} = (T^*T)^{-1}T^*$, it suffices to show that $(T^*T)^{-1}$ belongs to A.

Since $\widehat{T^*T} = |\hat{T}|^2 \geq 0$ on M_A, $(t - T^*T)^{-1} \in A$ for $t < 0$. Since $(t - T^*T)^{-1}$ converges to $-(T^*T)^{-1}$ in $B(H)$ as $t \to 0-$, we obtain $(T^*T)^{-1} \in A$. This completes the proof.

Consequently A is isometrically isomorphic to $C(\sigma(N))$. If f is a continuous function on $\sigma(N)$, the operator corresponding to f will be denoted by $f(N)$.

10.3 Theorem. The isometric isomorphism $f \rightarrow f(N)$ of $C(\sigma(N))$ onto A extends to a homomorphism $g \rightarrow g(N)$ of the algebra of bounded Baire functions on $\sigma(N)$ into $B(H)$, such that for every bounded Baire function g,

$$\|g(N)\| \leq \|g\|_{\sigma(N)}$$

and

$$\bar{g}(N) = g(N)^*.$$

The extension is unique, subject to the following continuity condition: Whenever the sequence $\{g_n\}_{n=1}^{\infty}$ of bounded Baire functions on $\sigma(N)$ converges pointwise boundedly to g, then

$$\lim_{n\rightarrow\infty} (g_n(N)x, y) = (g(N)x, y), \qquad x, y \in H.$$

Proof. The uniqueness assertion is clear, since every bounded Baire function is a pointwise bounded limit of a sequence of continuous functions. We must prove the existence.

For $f \in C(\sigma(N))$ and $x, y \in H$, define

$$L(f, x, y) = (f(N)x, y).$$

The functional L is bilinear in f and x, and conjugate linear in y. Moreover,

$$|L(f, x, y)| \leq \|f\|\|x\|\|y\|.$$

For fixed x and y, there is a measure μ_{xy} on $\sigma(N)$ such that

$$L(f, x, y) = \int f d\mu_{xy}, \qquad f \in C(\sigma(N)).$$

This allows us to extend the definition of L to bounded Baire functions g, by defining

$$L(g, x, y) = \int g d\mu_{xy}.$$

It is easily seen that there is an operator $g(N)$ on H such that $L(g, x, y) = (g(N)x, y)$, $x, y \in H$. The correspondence $g \rightarrow g(N)$ is linear, and satisfies the three displayed assertions of 10.3.

If f and g are continuous on $\sigma(N)$, and $x, y \in H$, then

$$((fg)(N)x, y) = (f(N)g(N)x, y) = (g(N)x, f(N)^*y).$$

Fixing $g \in C(\sigma(N))$, we see that this equality must continue to hold for all bounded Baire functions f. Then fixing the bounded Baire function f, we see that it must continue to hold for all bounded Baire functions g. Consequently $(fg)(N) = f(N)g(N)$ for all bounded Baire functions f and g. This completes the proof.

For each Baire set $E \subseteq \sigma(N)$, define $P(E) = \chi_E(N)$, χ_E the characteristic function of E. Since $\hat{\chi}_E$ is real, and $\chi_E^2 = \chi_E$, $P(E)$ is a self-adjoint projection in $B(H)$.

Suppose g is continuous on $\sigma(N)$, and let ω be the modulus of continuity of g. Let E_j, $1 \leq j \leq n$, be disjoint Baire sets which cover $\sigma(N)$, such that the diameter of each E_j does not exceed δ. If $\lambda_j \in E_j$ is fixed, then $|g(\lambda) - g(\lambda_j)| \leq \omega(\delta)$ on E_j, so

$$||g - \sum_{j=1}^{n} g(\lambda_j)\chi_{E_j}|| \leq \omega(\delta).$$

By 10.3,

$$||g(N) - \sum_{j=1}^{n} g(\lambda_j)P(E_j)|| \leq \omega(\delta).$$

Letting $\delta \to 0$, we obtain

$$g(N) = \int_{\sigma(N)} g(\lambda)dP(\lambda), \quad g \in C(\sigma(N)).$$

In particular,

$$N = \int_{\sigma(N)} \lambda dP(\lambda).$$

This is the spectral theorem for bounded normal operators on Hilbert space.

NOTES

The basic theory of commutative Banach algebras is finding its way into an increasing number of elementary texts. The reader's favorite real variable textbook probably covers much of this material. Four outstanding expositions, of varying lucidity, are those of Naimark [1], Loomis [1], Bourbaki [1], and Dunford and Schwartz [2]. The reader who wishes to get the word from the proverbial horse's mouth will enjoy the accounts of Gelfand [1], and of Gelfand, Raikov, and Shilov [1], reflecting the early state of the field. For more on disconnected compact spaces, see Diximer [1]. In Segal and Kunze [1], one can find a detailed treatment of the spectral theory of families of commuting self-adjoint operators on Hilbert space.

EXERCISES

1. If A is generated by f, together with the elements $(\lambda - f)^{-1}$, $\lambda \notin \sigma(f)$, then

(a) M_A is homeomorphic to $\sigma(f)$.

(b) f generates A if and only if $\sigma(f)$ does not separate the plane.

2. Let e^A denote the subgroup of A^{-1} of exponentials e^g, $g \in A$.

(a) If $||1 - f|| < 1$, then $f \in e^A$.

(b) A^{-1} is an open subset of A. The map $f \to f^{-1}$ is an analytic homeomorphism of A^{-1}, whose Frechet derivative at $f \in A^{-1}$ is the linear operator $T_f(g) = -f^{-2}g$.

(c) The connected component of the identity in A^{-1} is e^A.

(d) $f \in e^A$ if and only if 0 belongs to the unbounded component of the resolvent of f.

3. If B is a subspace of A of codimension one which does not meet e^A, then B is a maximal ideal. *Hint:* For each fixed $f \in A$, apply the appropriate functional to the entire function $F(\lambda) = e^{\lambda f}$.

4. (Cohen's Factorization Theorem): A closed ideal J in A has an **approximate identity** if there is $c > 0$ such that whenever $\epsilon > 0$ and $h_1, \ldots, h_n \in J$, there is $f \in J$ satisfying $\| f \| \leq c$ and $\| fh_j - h_j \| < \epsilon$, $1 \leq j \leq n$. If J has an approximate identity, and $g \in J$, then $g = g_0 g_1$, where $g_0 \in J$ and $g_1 \in J$. *Hint:* Fix $a > 0$ so that $a(c + 1) < \frac{1}{2}$. Suppose $f_1, \ldots, f_{m-1} \in J$ are chosen so that if $1 \leq k \leq m - 1$, then (i) $\| f_k \| \leq c$, (ii) $h_k = \sum_{j=1}^{k} a(1 - a)^j f_j + (1 - a)^k$ belongs to A^{-1}, and (iii) $\| h_k^{-1} g - h_{k-1}^{-1} g \| \leq 1/2^k$. Choose $f_m \in J$ such that $\| f_m \| \leq c$, $\| f_m g - g \| \leq \epsilon_m$, and $\| f_m f_k - f_k \| \leq \epsilon_m$, $1 \leq k \leq m - 1$. Show that if ϵ_m is sufficiently small, then $af_m + 1 - a = p_m \in A^{-1}$, $\| p_m^{-1} f_k - f_k \|$ is small for $1 \leq k \leq m - 1$, $\sum_{k=1}^{m-1} a(1 - a)^{k-1} p_m^{-1} f_k + (1 - a)^{m-1} = p_m^{-1} h_m = q_m \in A^{-1}$, $\| q_m - h_{m-1} \|$ is small, $\| q_m^{-1} - h_{m-1}^{-1} \|$ is small, and $\| h_m^{-1} g - h_{m-1}^{-1} g \| < 1/2^m$.

5. Let B be a commutative Banach algebra. If $e \in B$ satisfies $eg = g$ for all $g \in B$, and $e \neq 0$, then $\| e \| \geq 1$. The operator norm

$$||| g ||| = \sup \{\| gh \| : h \in B, \| h \| \leq 1\}$$

defines an equivalent Banach algebra norm on B for which $||| e ||| = 1$ (so that B becomes a commutative Banach algebra with identity).

6. For a commutative Banach algebra B without identity, define M_B to be the family of non-zero complex-valued homomorphisms of B.

(a) Every $\phi \in M_B$ is continuous and has unit norm.

(b) B is isometrically isomorphic to a maximal ideal in a commutative Banach algebra A with identity.

(c) Endowed with the weak-star topology, M_B is homeomorphic to the locally compact topological space obtained by deleting the maximal ideal B from M_A.

(d) A maximal ideal J of B is the kernel of some $\phi \in M_B$ if and only if B/J is a field. This occurs if and only if J is a regular maximal ideal, i.e., there is $h \in B$ such that $hf - f \in J$ for all $f \in B$.

(e) The maximal ideals of B which are not regular coincide with the linear subspaces of codimension one in B which contain B^2.

7. A **point derivation** at $\phi \in M_A$ is a linear functional L on A which satisfies

$$L(fg) = \phi(f)L(g) + \phi(g)L(f), \qquad f, g \in A.$$

(a) The point derivations at ϕ coincide with the linear functionals L on A which satisfy $L(1) = 0$ and $L(A_\phi^2) = 0$. There exist non-trivial point derivations at ϕ if and only if $A_\phi^2 \neq A_\phi$. There exist non-trivial continuous point derivations at ϕ if and only if A_ϕ^2 is not dense in A_ϕ. (Note that A_ϕ^2 is the ideal generated by $A_\phi \cdot A_\phi$.)

(b) If J is a maximal ideal in A_ϕ, then either $J = A_\psi \cap A_\phi$ for some $\psi \in M_A$, or $J = B \cap A_\phi$, where B is the kernel of a point derivation at ϕ.

(c) The kernel B of a continuous point derivation is a closed subalgebra of A satisfying $M_B = M_A$.

(d) Suppose that B is a maximal proper closed subalgebra of A containing the identity, such that B is of finite codimension in A. Then B has codimension one in A. Either there exist $\phi, \psi \in M_A$ such that $B = \{f \in A: \phi(f) = \psi(f)\}$, or there exists a continuous point derivation at some point of M_A whose kernel is B. *Hint:* First note that $J = \{f \in B: fA \subseteq B\}$ is a maximal ideal in B which is also an ideal in A.

(e) If B is a closed subalgebra of A of finite codimension, then there is a chain of subalgebras

$$B = B_0 \subseteq B_1 \subseteq \ldots \subseteq B_n = A$$

such that B_{j-1} has codimension one in B_j, $1 \leq j \leq n$. Also, M_B is obtained from M_A by identifying a finite number of pairs of points.

8. The Volterra integral operator V on $L^2(0, 1)$ is defined by

$$(Vf)(s) = \int_0^s f(t)dt, \qquad f \in L^2(0, 1).$$

The spectral radius of V is 0. The Banach subalgebra A of $B(L^2(0, 1))$ generated by V has precisely one maximal ideal.

9. For $1 \leq p < \infty$, the Banach space l^p of sequences $a = \{a_j\}_{j=1}^\infty$ satisfying $\|a\|_p = \{\Sigma|a_j|^p\}^{1/p} < \infty$, together with pointwise multiplication, is a commutative Banach algebra without identity. Identify the regular maximal ideals (cf. exercise 6d) of l^p. Show that l^p has maximal ideals which are not regular.

10. Every closed ideal in $C(X)$, X a compact Hausdorff space, is of the form $\{f \in C(X): f|_F = 0\}$, for some closed subset F of X.

11. If A is the subalgebra of $C(\Delta)$ generated by $P(\Delta)$ and $|z|$, then $\partial_A = \Delta$ while $M_A \neq \Delta$.

12. If A is the subalgebra of $C(\Delta \times [0,1])$ of functions which are analytic on the slice $(\text{int } \Delta) \times \{0\}$, then $M_A = \partial_A = \Delta \times [0,1]$.

13. The Gelfand transform is bicontinuous if and only if there is $c > 0$ such that $\|f^2\| \geq c\|f\|^2$ for all $f \in A$.

14. **(Wiener-Levy Theorem)**: If $f \in C(b\Delta)$ has an absolutely convergent Fourier series, and if g is analytic in a neighborhood of $f(b\Delta)$, then $g \circ f$ has an absolutely convergent Fourier series.

15. A commutative Banach algebra B (with or without identity) is **regular** if for every closed subset E of M_B and every $\phi \in M_B \backslash E$, there is $f \in B$ such that $\phi(f) = 1$, while $\hat{f}$ vanishes on E.

(a) If B is regular, and B is embedded as a maximal ideal in A, then A is regular.

(b) If A is regular, and J is a closed ideal in A, then J and A/J are regular.

(c) If A is regular, then $M_A = \partial_A$.

(d) If A is regular, and $\{U_j\}_{j=1}^n$ is an open cover of M_A, then there are $f_j \in A$ such that the closed support of $\hat{f}_j$ is contained in U_j, and $\sum_{j=1}^n \hat{f}_j = 1$.

Hint: Proceed by induction on n.

(e) If A is regular, and if $h \in C(M_A)$ belongs locally to $\hat{A}$, then $h \in \hat{A}$.

16. Let $*$ be an involution on A.

(a) If $f \in A$, then $\sigma(f^*) = \overline{\sigma(f)}$.

(b) If $f \in A$, then $\|\hat{f}\|_{M_A} = \|\widehat{f^*}\|_{M_A}$.

(c) If $\|f^*f\| = \|f^*\|\|f\|$ for all $f \in A$, then A is a B^*-algebra. *Hint:* Note that $\|g\| = \|\hat{g}\|_{M_A}$ whenever $g = g^*$, then take $g = f^*f$.

(d) If $\|f\|^2 \le \|f^*f\|$ for all $f \in A$, or if $\|f\|\|f^*\| \le \|f^*f\|$ for all $f \in A$, then A is a B^*-algebra.

17. Let B be a maximal ideal in A.

(a) Every involution $*$ of B extends uniquely to an involution of A.

(b) Suppose $\|f^*f\| = \|f\|^2$ for all $f \in B$. Then $|||f||| = \|f\|$ for all $f \in B$, where $|||\cdot|||$ is the operator norm defined in exercise 5. With the norm $|||\cdot|||$, A becomes a commutative B^*-algebra.

18. The Bohr compactification of the real numbers (defined in section 8) is a compact topological group which contains the real numbers as a dense subgroup. The embedding of the reals into the Bohr group is not a homeomorphism.

19. The compact Hausdorff space X is extremely disconnected (the closure of every open set is open) if and only if $C_R(X)$ is a complete lattice (with the usual ordering).

20. Let $Y = M_{L^\infty(\mu)}$, where μ is a finite positive measure. The clopen subsets of Y, ordered by inclusion, form a complete lattice. Any linear chain of distinct elements of this lattice is at most countable. If $\mathscr{E}$ is any family of clopen subsets of Y, and E is the lattice supremum of $\mathscr{E}$, then E is the lattice supremum of a subset of $\mathscr{E}$ which is at most countable.

21. Let P be the spectral resolution associated with a normal operator N on a Hilbert space H.

(a) If $x \in H$ is fixed, then $E \to P(E)x$ is a countably-additive set function on the Borel subsets E of $\sigma(N)$.

(b) If g is analytic in a neighborhood of $\sigma(N)$, then the operator $g(N) = \int_{\sigma(N)} g(\lambda)dP(\lambda)$ coincides with $\frac{1}{2\pi i}\int_\Gamma g(\lambda)(\lambda - N)^{-1}d\lambda$, where Γ is an appropriate contour surrounding $\sigma(N)$.

CHAPTER II

UNIFORM ALGEBRAS

A **uniform algebra** on a compact Hausdorff space X is a uniformly closed subalgebra of $C(X)$ which contains the constants and separates the points of X. When endowed with the supremum norm

$$\|f\|_X = \sup_{x \in X} |f(x)|,$$

the uniform algebra A becomes a Banach algebra. Identifying each $x \in X$ with the evaluation homomorphism at x, we will regard X as a closed subset of the maximal ideal space M_A of A. The Shilov boundary ∂_A of A then becomes a closed subset of X.

This chapter is devoted to the study of uniform algebras. Particular attention is given to algebras of analytic functions on planar sets.

1. Algebras on Planar Sets

Let K be a compact subset of the complex plane. Associated with K there are three uniform algebras in which we shall be especially interested.

The algebra ***P*(*K*)** consists of the functions in $C(K)$ which can be approximated uniformly on K by polynomials in z.

The algebra ***R*(*K*)** consists of all functions in $C(K)$ which can be approximated uniformly on K by rational functions with poles off K.

The algebra ***A*(*K*)** consists of the functions in $C(K)$ which are analytic on the interior of K.

Evidently $P(K) \subseteq R(K) \subseteq A(K)$. Our main problem will be to decide when equality holds among either of these inclusions. We also wish to decide when a given function $f \in A(K)$ belongs to either $R(K)$ or $P(K)$.

It is easy to find sets K for which $P(K) \neq R(K)$—the unit circle will do. The Swiss cheeses provide examples of compact sets K with no interior, such that $R(K) \neq C(K) = A(K)$. These are constructed as follows.

A **Swiss cheese** is a compact set K obtained from the closed unit disc Δ by deleting a sequence $\{\Delta_j\}_{j=1}^{\infty}$ of open discs, such that the closures of the

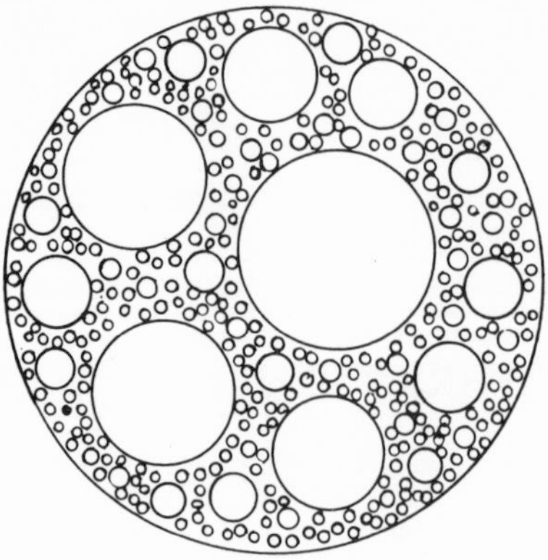

Figure 1. The Swiss cheese.

Δ_j are pairwise disjoint, and such that $K = \Delta \backslash (\bigcup_{j=1}^{\infty} \Delta_j)$ has no interior (cf. fig. 1). If f is a rational function with poles off K, then by Cauchy's theorem, $\int_{b\Delta} f(z)dz - \sum_{j=1}^{\infty} \int_{b\Delta_j} f(z)dz = 0$. If the radii r_j of the Δ_j are sufficiently small so that $\sum r_j < \infty$, then the measure ν which is dz on $b\Delta$ and $-dz$ on $b\Delta_j$, $1 \leq j < \infty$, is finite. And $\int f d\nu = 0$ for all $f \in R(K)$. Hence $R(K) \neq C(K)$ whenever $\sum r_j < \infty$.

Now we turn to some positive results. Recall that the differential operators $\partial/\partial z$ and $\partial/\partial \bar{z}$ are defined by

$$\frac{\partial}{\partial z} = \frac{1}{2}\left(\frac{\partial}{\partial x} - i\frac{\partial}{\partial y}\right)$$

$$\frac{\partial}{\partial \bar{z}} = \frac{1}{2}\left(\frac{\partial}{\partial x} + i\frac{\partial}{\partial y}\right)$$

A continuously differentiable function g is analytic on an open set D if and only if $\partial g/\partial \bar{z} = 0$ on D. If g is analytic on D, then the usual derivative of g coincides with $\partial g/\partial z$. If D is an open set whose boundary Γ consists of a finite number of smooth curves, and if f is continuously differentiable on $D \cup \Gamma$, then Green's formula for f assumes the form

$$f(\zeta) = \frac{1}{2\pi i}\int_{\Gamma} \frac{f(z)}{z - \zeta}dz - \frac{1}{\pi}\iint_D \frac{\partial f}{\partial \bar{z}} \frac{1}{z - \zeta}dxdy, \qquad \zeta \in D.$$

This reduces to Cauchy's formula when f is analytic in D.

1.1 Theorem. Let K be a compact subset of the complex plane. If $f \in C(K)$ extends to be continuously differentiable in a neighborhood of K, and if $\partial f/\partial \bar{z} = 0$ on K, then $f \in R(K)$.

Proof. We can extend f to be continuously differentiable on the complex plane, and to have compact support. By Green's formula,

$$f(\zeta) = -\frac{1}{\pi}\iint \frac{\partial f}{\partial \bar{z}} \frac{1}{z - \zeta}dxdy, \qquad \zeta \in C,$$

where the integral is extended over the complex plane C.

Let μ be a measure on K which is orthogonal to $R(K)$. By Fubini's theorem,

$$\int f d\mu = -\frac{1}{\pi}\iint \frac{\partial f}{\partial \bar{z}}\left[\int_K \frac{1}{z - \zeta}d\mu(\zeta)\right]dxdy.$$

Since the inner integral is zero when $z \in K^c$, and $\partial f/\partial \bar{z} = 0$ when $z \in K$, we obtain $\int f d\mu = 0$.

Hence every continuous linear functional on $C(K)$ which is orthogonal to $R(K)$ is also orthogonal to f. By the Hahn-Banach theorem, $f \in R(K)$.

1.2 Corollary. Let K be a compact subset of the complex plane. If $f \in C(K)$, and f extends to be analytic in a neighborhood of K, then $f \in R(K)$.

This corollary can also be proved by writing

$$f(\zeta) = \frac{1}{2\pi i} \int_\Gamma \frac{f(z)}{z - \zeta} dz,$$

where Γ is an appropriate contour surrounding K. The Riemann sums which approximate this integral are rational functions which approximate f uniformly on K.

1.3 Theorem. Let K be a compact subset of the complex plane. The maximal ideal space of $R(K)$ is K. The Shilov boundary of $R(K)$ is the topological boundary bK of K.

Proof. Every $\phi \in M_{R(K)}$ is determined by its value $\phi(z)$ at the coordinate function z. Since the function $z - \phi(z)$ is not invertible in $R(K)$, $\phi(z)$ must belong to K. Consequently ϕ must coincide with the evaluation homomorphism at $\phi(z)$. This proves the first half of the theorem.

Suppose $z_0 \in bK$. If $z_1 \in K^c$ is near z_0, the function $(z - z_1)^{-1} \in R(K)$ assumes its maximum modulus on K only near z_0. It follows that z_0 belongs to the Shilov boundary of $R(K)$, and $bK \subseteq \partial_{R(K)}$. On the other hand, each function in $R(K)$ is analytic on int (K), and so assumes its maximum modulus on bK. If follows that $bK = \partial_{R(K)}$. That completes the proof.

The union of K and the bounded components of K^c will be denoted by $\hat{K}$. Evidently $\hat{K}$ is compact, and $\hat{\hat{K}} = \hat{K}$. The sets K and $\hat{K}$ coincide if and only if K^c is connected. The topological boundary of $\hat{K}$ is the outer boundary of K, that is, the set of points in bK which are also in the boundary of the unbounded component of K^c.

If f is a polynomial, then $\|f\|_K = \|f\|_{\hat{K}}$, by the maximum modulus principle. Consequently any sequence of polynomials which converges uniformly on K to $g \in P(K)$ will also converge uniformly on $\hat{K}$ to an extension $\hat{g} \in P(\hat{K})$ of g, satisfying $\|\hat{g}\|_{\hat{K}} = \|g\|_K$. The isometric isomorphism $g \to \hat{g}$, which turns out to be the Gelfand transform, allows us to identify the algebras $P(K)$ and $P(\hat{K})$.

1.4 Theorem. Let K be a compact subset of the complex plane. The maximal ideal space of $P(K)$ is the union $\hat{K}$ of K and the bounded components of the complement of K. The Shilov boundary of $P(K)$ is the outer boundary of K, that is, the topological boundary of $\hat{K}$. Also,

$$P(K) = P(\hat{K}) = R(\hat{K}).$$

Proof. In view of 1.3, it suffices to show that $P(\hat{K}) = R(\hat{K})$. For this, it suffices to show that $(z - \lambda)^{-1} \in P(\hat{K})$ whenever $\lambda \notin \hat{K}$.

Replacing K by $\hat{K}$, we can assume that K^c is connected.

Let U be the set of all $\lambda \in K^c$ such that $(z - \lambda)^{-1} \in P(K)$. We must show that $U = K^c$.

If $\lambda_n \in U$, and $\lambda_n \to \lambda \in K^c$, then $(z - \lambda_n)^{-1}$ tends uniformly on K to $(z - \lambda)^{-1}$. Hence $\lambda \in U$, and U is a closed subset of K^c.

If $|\lambda|$ is large, the series

$$(z - \lambda)^{-1} = \sum_{j=0}^{\infty} \frac{(-1)^{j+1}}{\lambda^{j+1}} z^j$$

converges uniformly on K. So $\lambda \in U$ if $|\lambda|$ is large, and U is not empty.

Suppose $\lambda_0 \in U$. Then $1/(z - \lambda_0)^n \in P(K)$ for all $n \geq 1$. If λ is sufficiently near λ_0, the series

$$(z - \lambda)^{-1} = \sum_{n=0}^{\infty} \frac{(\lambda - \lambda_0)^n}{(z - \lambda_0)^{n+1}}$$

converges uniformly on K. Consequently $\lambda \in U$ when λ is near λ_0, and U is open.

Since K^c is connected, it must coincide with U. This proves the theorem.

1.5 Corollary (Runge's Theorem). Let K be a compact subset of the plane with connected complement. Then every function analytic in a neighborhood of K can be approximated uniformly on K by polynomials in z.

Proof. This follows from 1.2 and 1.4.

Now we turn our attention to $A(K)$. It is clear that the Shilov boundary of $A(K)$ is bK. It is true, but not at all clear, that the maximal ideal space of $A(K)$ is K. Before we prove this, let's become a little better acquainted with the algebra $A(K)$.

It will be convenient to extend the definition of $A(K)$ as follows. Let K be a compact subset of the Riemann sphere S^2, and let U be an open subset of S^2 which is contained in K. Then $\boldsymbol{A(K,U)}$ will denote the algebra of functions continuous on K which are analytic on U.

The algebra $A(K,U)$ is uniformly closed and contains the constants. However, if $K = S^2$, $A(K,U)$ may fail to separate the points of K.

1.6 Lemma. If $A(K,U)$ contains one non-constant function, then there are three functions in $A(K,U)$ which separate the points of K.

Proof. If $K \neq S^2$, then $(z - \lambda)^{-1}$ separates the points of K, for any $\lambda \in K^c$. If $K = S^2$ but U is not dense in K, one easily constructs two functions which separate the points of K.

Suppose U is dense in S^2. For convenience, we rotate the sphere so that

$\infty \in U$. Let g be a non-zero function in $A(K,U)$ such that $g(\infty) = 0$. Since g does not vanish identically on U, there is $z_0 \in U$ such that $g(z_0) \neq 0$. Then the three functions $g(z)$, $zg(z)$, and $[g(z) - g(z_0)]/(z - z_0)$ belong to $A(K,U)$ and separate the points of K. This proves the lemma.

One criterion that ensures that $A(K,U)$ contain non-constant functions is that the complement of U in S^2 have positive area. In fact, let E be a compact subset of the complex plane of positive planar measure, and let

$$f(\zeta) = \iint_E \frac{dxdy}{z - \zeta}.$$

Since f is the convolution of a bounded function with compact support and the locally integrable function $1/z$, f is continuous. Also, f is analytic off E, and f is analytic at ∞. Since $f'(\infty) = \lim_{\zeta \to \infty} \zeta f(\zeta) = \iint_E dxdy \neq 0$, f is not constant.

One interesting class of examples, first studied by Wermer, are the algebras $A(S^2, S^2 \backslash \Gamma)$, where Γ is an arc (= homeomorph of the unit interval) on S^2 of positive measure. Suppose $g_1, g_2, g_3 \in A(S^2, S^2 \backslash \Gamma)$ separate the points of S^2. The correspondence $z \to (g_1(z), g_2(z), g_3(z))$ embeds Γ as an arc in C^3 which is not polynomially convex (cf. chapter 3). Another interesting property of this algebra is that every value assumed by $f \in A(S^2, S^2 \backslash \Gamma)$ on S^2 is assumed by f on Γ. In fact, if $\lambda \notin f(\Gamma)$, then $f - \lambda$ has a continuous logarithm on Γ, and hence in a neighborhood of Γ. By the argument principle, $f - \lambda$ cannot vanish on $S^2 \backslash \Gamma$.

In order to deal with the maximal ideals of $A(K,U)$, we will consider integrals of the form

$$G(\zeta) = \frac{1}{\pi} \iint \frac{f(z) - f(\zeta)}{z - \zeta} h(z) dxdy,$$

where f is continuous on S^2, and h is a bounded Borel function on the complex plane with compact support. The convolutions $\iint f(z)h(z)(z - \zeta)^{-1} dxdy$ and $\iint h(z)(z - \zeta)^{-1} dxdy$ are continuous on S^2 and analytic at ∞. Hence G is continuous on S^2. And G is analytic at ∞ if f is.

If $t \neq 0$ is complex, then

$$\frac{G(\zeta + t) - G(\zeta)}{t} = \frac{1}{\pi} \iint \frac{(f(z) - f(\zeta))h(z)}{(z - \zeta)(z - \zeta - t)} dxdy$$
$$- \frac{1}{\pi} \frac{f(\zeta + t) - f(\zeta)}{t} \iint \frac{h(z)}{z - \zeta - t} dxdy.$$

Letting t tend to zero, and observing that these convolution integrals vary continuously with t, we deduce that G is analytic wherever f is.

We will be interested in these integrals when $h = \partial g / \partial \bar{z}$, where g is a continuously differentiable function with compact support. In this case, we obtain from Green's formula

$$G(\zeta) = \frac{1}{\pi}\iint \frac{f(z) - f(\zeta)}{z - \zeta}\frac{\partial g}{\partial \bar{z}}dxdy$$

$$(*) \qquad = f(\zeta)g(\zeta) + \frac{1}{\pi}\iint f(z)\frac{\partial g}{\partial \bar{z}}\frac{1}{z - \zeta}dxdy.$$

If f is continuously differentiable, then

$$G(\zeta) = -\frac{1}{\pi}\iint \frac{\partial f}{\partial \bar{z}}\frac{g(z)}{z - \zeta}dxdy.$$

This formula holds even for continuous f, if we interpret $\partial f/\partial \bar{z}$ as a distribution derivative.

From the formula $(*)$, it is clear that G is analytic off the closed support of g. Also, from $(*)$ it follows that $f - G$ is analytic on the interior of the set on which g attains the value 1.

1.7 Lemma. Let $f \in C(S^2)$, let g be a continuously differentiable function supported on the disc $\Delta(z_0; \delta)$, and let

$$G(\zeta) = \frac{1}{\pi}\iint \frac{f(z) - f(\zeta)}{z - \zeta}\frac{\partial g}{\partial \bar{z}}dxdy.$$

Then $G \in C(S^2)$, G is analytic wherever f is, G is analytic off the disc $\Delta(z_0; \delta)$, and $f - G$ is analytic on the interior of the level set $g^{-1}(1)$. Also,

$$\| G \|_{S^2} \leq 4\delta \left\| \frac{\partial g}{\partial \bar{z}} \right\|_\infty \sup_{|z - z_0| \leq \delta} |f(z) - f(z_0)|.$$

Proof. In view of the preceding discussion, it suffices to verify the estimate for $\| G \|_{S^2}$.

Suppose $|\zeta - z_0| \leq \delta$. A crude estimate yields

$$|G(\zeta)| \leq \frac{2}{\pi} \sup_{|z - z_0| \leq \delta} |f(z) - f(z_0)| \left\| \frac{\partial g}{\partial \bar{z}} \right\|_\infty \iint_{\Delta(z_0; \delta)} \frac{dxdy}{|z - \zeta|}.$$

The last integral is maximized when $\zeta = z_0$, in which case its value is computed, using polar coordinates, to be $2\pi\delta$. This yields the desired estimate for $G(\zeta)$, providing $\zeta \in \Delta(z_0; \delta)$. Since G is analytic off $\Delta(z_0; \delta)$ and at ∞, the same estimate obtains for all $\zeta \in S^2$.

1.8 Theorem. Let f be a continuous function on the Riemann sphere S^2, which is analytic on the open subset U of S^2. Let $z_0 \in S^2$. Then there is a sequence $\{f_n\}_{n=1}^\infty$ of continuous functions on S^2 such that f_n is analytic on U, f_n is analytic in a neighborhood of z_0, and $f_n \to f$ uniformly on S^2.

Proof. We can assume that $z_0 = 0$. Let $\{g_n\}_{n=1}^\infty$ be a sequence of continuously differentiable functions such that $g_n(z) = 0$ when $|z| \geq 2/n$, $g_n(z) = 1$ when $|z| \leq 1/n$, and $|\partial g_n/\partial \bar{z}| \leq 2n$. Set

$$G_n(\zeta) = \frac{1}{\pi}\iint \frac{f(z) - f(\zeta)}{z - \zeta}\frac{\partial g_n}{\partial \bar{z}}dxdy.$$

By 1.7,

$$\| G_n \|_{S^2} \leq 16 \sup_{|z| \leq 2/n} |f(z) - f(0)|.$$

Consequently G_n tends uniformly to zero on S^2. In view of 1.7, the functions $f_n = f - G_n$ do the trick.

The preceding theorem has been strengthened considerably by Vitushkin [1]. He shows that the statement of the theorem remains true if the point z_0 is replaced by a twice continuously differentiable curve on S^2. The Wermer arc shows that some smoothness assumption on the curve is necessary.

1.9 Theorem (Arens' Theorem). Let U be an open subset of the Riemann sphere S^2, and let K be a compact subset of S^2 containing U. Suppose the algebra $A(K,U)$ of functions in $C(K)$ which are analytic on U contains a non-constant function. Then $A(K,U)$ is a uniform algebra on K whose maximal ideal space is K.

Proof. That $A(K,U)$ is a uniform algebra follows from 1.6. It suffices to show that every $\phi \in M_A$ is evaluation at some point of K.

If U is empty, then $A(K,U) = C(K)$. This case has already been treated in I.3.1.

If U is not empty, we can suppose $\infty \in U$. Choose $\phi \in M_A$ which is not evaluation at ∞, and choose $g \in A(K,U)$ such that $g(\infty) = 0$ while $\phi(g) = 1$. Then $zg \in A(K,U)$. Set $z_0 = \phi(zg)$, so that $\phi((z - z_0)g) = 0$.

Suppose $f \in A(K,U)$. Extending f continuously to S^2 and applying 1.8, we find $f_n \in A(K,U)$ such that f_n extends analytically to a neighborhood of z_0, and $f_n \to f$ uniformly on K. Then $(f_n - f_n(z_0))/(z - z_0) \in A(K,U)$, and

$$\begin{aligned}\phi(f) - f(z_0) &= \lim\,(\phi(f_n) - f_n(z_0)) \\ &= \lim\,\phi((f_n - f_n(z_0))g) \\ &= \lim\,\phi\left(\frac{f_n - f_n(z_0)}{z - z_0}\right)\phi((z - z_0)g) \\ &= 0.\end{aligned}$$

Consequently ϕ evaluates every $f \in A(K,U)$ at z_0. Clearly $z_0 \in K$. That proves the theorem.

From 1.8 and 1.9 we obtain the following.

1.10 Corollary. Let K be a compact subset of the plane. The maximal ideal space of $A(K)$ is K. For every $z_0 \in K$, the ideal of functions in $A(K)$ which vanish at z_0 is the closed principal ideal generated by $z - z_0$.

2. Representing Measures

Let A be a uniform algebra on X, and let $\phi \in M_A$. A **representing measure for ϕ** is a positive measure μ on X such that

$$\phi(f) = \int f d\mu, \qquad f \in A.$$

The set of representing measures for ϕ will be denoted by M_ϕ. Each of the measures μ in M_ϕ satisfies $\|\mu\| = \int d\mu = 1$. The set M_ϕ is convex and weak-star compact.

The set M_ϕ of representing measures for ϕ coincides with the set of norm-preserving extensions of ϕ from A to $C(X)$. In fact, any $\mu \in M_\phi$ represents a norm-preserving extension of ϕ. Conversely, if ν represents a norm-preserving extension of ϕ, then $\int d|\nu| = 1 = \int d\nu$. So ν is positive, and $\nu \in M_\phi$.

The functional ϕ can be defined on Re (A) by setting

$$\phi(\text{Re}\,(f)) = \text{Re}\,\phi(f), \qquad f \in A.$$

Then $\phi(u) = \int u d\mu$ for any $u \in \text{Re}\,(A)$ and $\mu \in M_\phi$. This shows that ϕ is continuous on Re (A), and that $\phi(u) \geq 0$ if $u \geq 0$, $u \in \text{Re}\,(A)$.

If L is a monotone (positive) extension of ϕ from Re (A) to $C_R(X)$, then $L(v) \leq L(\|v\|_X) = \|v\|_X$ for every $v \in C_R(X)$. Consequently L is continuous, and L is represented by a measure $\mu \in M_\phi$. Hence M_ϕ coincides with the set of monotone extensions of ϕ from Re (A) to $C_R(X)$.

Let $v \in C_R(X)$, and let B be the subspace of $C_R(X)$ spanned by Re (A) and v. We can extend ϕ monotonically to B by setting $\phi(v) = c$, where c is any number satisfying

$$\sup\,\{\phi(u)\colon u \in \text{Re}\,(A),\, u \leq v\} \leq c \leq \inf\,\{\phi(u)\colon u \in \text{Re}\,(A),\, u \geq v\}.$$

By Zorn's lemma, we can extend this extension monotonically to $C_R(X)$. Hence, for any c as above, there exists $\mu \in M_\phi$ such that $\int v d\mu = c$.

2.1 Theorem. Let A be a uniform algebra on X, and let $\phi \in M_A$. If $v \in C_R(X)$, then

$$\sup\,\{\phi(u)\colon u \in \text{Re}\,(A),\, u \leq v\} = \inf\left\{\int v d\mu\colon \mu \in M_\phi\right\},$$

$$\inf\,\{\phi(u)\colon u \in \text{Re}\,(A),\, u \geq v\} = \sup\left\{\int v d\mu\colon \mu \in M_\phi\right\}.$$

Proof. We confine ourselves to the first equality.

By the preceding remark, the supremum of the $\phi(u)$ is equal to $\int v d\mu$ for some $\mu \in M_\phi$. On the other hand, if $\mu \in M_\phi$, $u \in \text{Re}\,(A)$, and $u \leq v$, then $\phi(u) = \int u d\mu \leq \int v d\mu$. So the supremum of the $\phi(u)$ never exceeds the infimum. That proves the theorem.

A **complex representing measure** on X for ϕ is any complex measure μ such that

$$\phi(f) = \int f d\mu, \qquad f \in A.$$

The complex representing measures for ϕ coincide with the continuous extensions of ϕ from A to $C(X)$.

2.2 Theorem. Let A be a uniform algebra, and let $\phi \in M_A$. Let μ be a complex representing measure for ϕ. There is a representing measure for ϕ which is absolutely continuous with respect to μ.

Proof. Let A_ϕ denote the kernel of ϕ. Let H^2 and H_ϕ^2 be the closures of A and A_ϕ respectively in $L^2(|\mu|)$. If $f \in A_\phi$, then $\int |1-f|^2 d|\mu| \geq \int (1-f)^2 d\mu = 1$. So $1 \notin H_\phi^2$, and $H_\phi^2 \neq H^2$. Also, $A_\phi H^2 \subseteq H_\phi^2$.

Choose $F \in H^2$ such that $\int |F|^2 d|\mu| = 1$, while F is orthogonal to H_ϕ^2 in $L^2(|\mu|)$. If $f \in A_\phi$, then $fF \in H_\phi^2$. So $fF \perp F$, and $\int f|F|^2 d|\mu| = 0$. Hence $|F|^2 d|\mu|$ is the required representing measure.

2.3 Theorem. Let A be a uniform algebra on X, and let $\phi \in M_A$. Then there is a representing measure on X for ϕ with minimal closed support. If $\mu \in M_\phi$ has minimal closed support, then the support of μ is either perfect or consists of one point. If $f \in A$ vanishes on an open subset of the support of μ, then $\phi(f) = 0$.

Proof. M_ϕ can be partially ordered, by defining $\nu \prec \eta$ if the closed support of ν is contained in the closed support of η. By Zorn's lemma, and the weak-star compactness of M_ϕ, there exists $\mu \in M_\phi$ which is minimal with respect to this ordering. Such a μ has minimal (but not necessarily minimum) closed support.

Suppose $f \in A$ satisfies $\phi(f) \neq 0$. Since $\phi(f)\phi(g) = \int fg d\mu$ for all $g \in A$, the measure $fd\mu/\phi(f)$ is a complex representing measure for ϕ. By 2.2 and the minimality of μ, f cannot vanish on an open subset of the support of μ.

In particular, if x is an isolated point of the support of μ, then $\phi(f)=0$ whenever $f \in A$ satisfies $f(x) = 0$. In this case μ must evidently be the point mass at x. This completes the proof.

A positive measure μ is a **Jensen measure** for ϕ if Jensen's inequality is valid:

$$\log|\phi(f)| \leq \int \log|f|\, d\mu, \qquad f \in A.$$

Applying Jensen's inequality to e^g and e^{-g}, $g \in A$, we obtain

$$\operatorname{Re}\phi(g) = \int \operatorname{Re}(g) d\mu, \qquad g \in A.$$

Hence every Jensen measure for ϕ is a representing measure for ϕ.

The properties mentioned in 2.3 enjoyed by minimal measures are shared by Jensen measures. But it is not immediately clear that Jensen measures exist.

2.4 Theorem. Let A be a uniform algebra on X, and let $\phi \in M_A$. There exists a Jensen measure on X for ϕ.

Proof. Let Q be the subset of $C_R(X)$ of functions u which have the following property: There exist $c > 0$ and $f \in A$ such that $\phi(f) = 1$ and $u > c \log |f|$.

If $u \in Q$ and $b > 0$, then $bu \in Q$. Also, Q contains all strictly positive functions in $C_R(X)$—take $f \equiv 1$ in the definition of Q.

Suppose that u_1 and u_2 belong to Q, say $u_1 > c_1 \log|f_1|$ and $u_2 > c_2 \log|f_2|$. We can assume that c_1 and c_2 are rational, say $c_j = p_j/q_j$ where p_j and q_j are integers. Then $u_1 + u_2 > (1/q_1q_2)\log|f_1^{p_1q_2}f_2^{p_2q_1}|$. So $u_1 + u_2 \in Q$, and Q is convex.

Now $0 \notin Q$. By the separation theorem for convex sets, there is a non-zero measure μ on X such that $\int u\,d\mu \geq 0$ for all $u \in Q$. Since the strictly positive functions in $C_R(X)$ belong to Q, μ is positive. Multiplying μ by a positive constant, we can assume that $\int d\mu = 1$.

Suppose $f \in A$ satisfies $\phi(f) = 1$. For each $\epsilon > 0$, $\log(|f| + \epsilon)$ belongs to Q, and $\int \log(|f| + \epsilon)d\mu \geq 0$. Letting ϵ tend to 0, we obtain

$$0 \leq \int \log|f|\,d\mu, \qquad f \in A,\ \phi(f) = 1.$$

Applying this inequality to $g/\phi(g)$, we obtain Jensen's inequality for all $g \in A$ such that $\phi(g) \neq 0$. The inequality is trivial if $\phi(g) = 0$.

3. Dirichlet Algebras

Let A be a uniform algebra on X. The algebra A is **dirichlet on X** if $\operatorname{Re}(A)$ is uniformly dense in $C_R(X)$. This occurs if and only if there are no non-zero real measures on X which are orthogonal to A.

If A is dirichlet on X, then X is the Shilov boundary of A. In fact, if $x \in X$, there are non-negative functions $\operatorname{Re}(f)$ in $\operatorname{Re}(A)$ whose zero sets lie arbitrarily close to $\{x\}$. The functions e^{-f} attain their maximum modulus only near x.

An algebra which is dirichlet on its Shilov boundary will be referred to simply as a **dirichlet algebra.**

The prototype of a dirichlet algebra is the algebra $P(\Delta)$ of uniform limits of polynomials on the closed unit disc Δ. The space $\operatorname{Re} P(\Delta)$ consists of the functions continuous on Δ which are harmonic on int Δ, and whose harmonic conjugates extend continuously to $b\Delta$. Consequently $\operatorname{Re} P(\Delta)$ contains the continuously differentiable functions on $b\Delta$, and these are dense in $C_R(b\Delta)$.

Let A be a dirichlet algebra on X, and let $\phi \in M_A$. The difference of any two representing measures for ϕ is a real measure orthogonal to A. Consequently, every homomorphism $\phi \in M_A$ has a unique representing measure on X. By 2.4, this representing measure must be a Jensen measure.

The representing measure for 0 on $P(\Delta)$ is the measure $d\theta/2\pi$ on $b\Delta$. Since $d\theta/2\pi$ must be a Jensen measure, we obtain

$$\log|f(0)| \leq \frac{1}{2\pi}\int_0^{2\pi} \log|f(e^{i\theta})|\,d\theta, \qquad f \in P(\Delta).$$

This is the classical Jensen's inequality.

3.1 Theorem. Suppose A is a dirichlet algebra on X. Then A contains the characteristic function of every open-closed subset of X. In particular, if X is disconnected, then M_A is disconnected.

Proof. Suppose X is the disjoint union of two closed disjoint subsets U and V. By dirichlicity, there is $f \in A$ such that $|\mathrm{Re}(f) - 1| \leq \frac{1}{4}$ on U and $|\mathrm{Re}(f)| \leq \frac{1}{4}$ on V. By I.4.2, the topological boundary of $f(M_A)$ is contained in $f(X)$. Hence $f(M_A)$ cannot meet the strip $\{\frac{1}{4} < \mathrm{Re}(z) < \frac{3}{4}\}$.

Let h be the characteristic function of the half-plane $\{\mathrm{Re}(z) > \frac{1}{2}\}$. Then h is analytic in a neighborhood of $f(M_A)$. By I.5.1, the composition $h \circ f$ belongs to A. The function $h \circ f$ is an idempotent which is 1 on U and 0 on V. This proves the theorem.

Theorem 3.1 shows that if K is a compact connected subset of the plane such that $R(K)$ is dirichlet (on bK), then bK is connected. The annulus algebra, for instance, cannot be dirichlet.

We now wish to prove that $P(K)$ is a dirichlet algebra for any compact subset K of the complex plane. In order to prove this, we must use the solvability of the Dirichlet problem for planar domains with smooth boundaries. The fact we need is the following.

Dirichlet's principle. Let D be a domain in the plane whose boundary Γ consists of a finite number of smooth closed curves. Let $u \in C_R(\Gamma)$ be continuously differentiable on Γ. Then u can be extended to be continuous on $D \cup \Gamma$ and harmonic on D. The harmonic extension of u minimizes the Dirichlet integral

$$\iint_D \left[\left(\frac{\partial v}{\partial x}\right)^2 + \left(\frac{\partial v}{\partial y}\right)^2\right] dx\,dy$$

among all continuously differentiable functions v on $D \cup \Gamma$ which coincide with u on Γ.

Let K be a compact subset of the plane. A point $z \in bK$ is said to satisfy **Lebesgue's condition** if $\int_S \frac{dr}{r} = +\infty$, where S consists of all r, $0 < r < 1$, such that the circle of radius r and center z meets the complement of K.

3.2 Lemma. Let K be a compact subset of the complex plane. Let K_n be compact subsets of the complex plane such that bK_n consists of a finite number of smooth closed curves, $K_{n+1} \subseteq \mathrm{int}(K_n)$, and $\bigcap_{n=1}^{\infty} K_n = K$. Suppose u is continuously differentiable on the complex plane, and u_n is the harmonic extension of $u|_{bK_n}$ to $\mathrm{int}(K_n)$. If every $z \in bK$ satisfies Lebesgue's condition, then u_n converges to u uniformly on bK.

Proof. The lemma will be proved by contradiction.

Suppose the u_n do not converge uniformly to u on bK. Passing to a subsequence, if necessary, we can choose $z_n \in bK$ and $\epsilon > 0$ such that $|u_n(z_n) - u(z_n)| > 3\epsilon$ for all n. Furthermore, we can assume that z_n converges to 0, $|z_n| > |z_{n+1}|$ for all n, and $u(0) = 0$.

Choose $\delta > 0$ so small that $|u(z)| < \epsilon$ for $|z| \leq \delta$. We can assume that $|z_n| < \delta$ for all n. Then $|u_n(z_n)| > 2\epsilon$ for all n.

Suppose there is a circle with center 0 and radius r, $0 < r \leq \delta$, which does not meet K. Then the circle does not meet K_n, for large n. Since $|u| \leq \epsilon$ on $b(K_n \cap \{|z| \leq r\})$, we obtain $|u_n| \leq \epsilon$ on $K_n \cap \{|z| \leq r\}$ for large n. This contradicts $|u_n(z_n)| > 2\epsilon$. It follows that every circle with center 0 and radius r, $0 < r \leq \delta$, meets K.

Let E_n be the set of r such that $|z_n| \leq r \leq \delta$, and such that the circle with center 0 and radius r meets K_n^c. Such circles must also meet bK_n.

The E_n form an increasing sequence of closed sets. The set $E = \bigcup_{n=1}^{\infty} E_n$ is the set of r, $0 < r \leq \delta$, such that the circle of radius r meets K^c. By hypothesis, $\int_E \frac{dr}{r} = +\infty$.

For fixed $r \in E_n$, we consider the set F of $z \in K_n$ such that $|z| \leq r$. The boundary of F consists of the part of bK_n inside $\{|z| < r\}$, together with certain arcs on $\{|z| = r\}$. Since u_n is harmonic on int (F), $|u_n(z_n)| \geq 2\epsilon$, and $|u_n| \leq \epsilon$ on the part of bF in $\{|z| < r\}$, there must be a point w on an arc of $bF \cap \{|z| = r\}$ for which $|u_n(w)| \geq 2\epsilon$. Since this arc must have an endpoint in bK_n, the variation of u_n along the arc is at least ϵ. Hence

$$\epsilon^2 \leq \left[\int \left|\frac{\partial u_n}{\partial \theta}\right| d\theta\right]^2 \leq 2\pi \int \left(\frac{\partial u_n}{\partial \theta}\right)^2 d\theta,$$

where the integration is performed on the set $K_n \cap \{|z| = r\}$. Using Dirichlet's principle, we obtain

$$\begin{aligned}\iint_{K_n} \left[\left(\frac{\partial u}{\partial x}\right)^2 + \left(\frac{\partial u}{\partial y}\right)^2\right] dxdy &= \iint_{K_n} \left[\left(\frac{\partial u_n}{\partial r}\right)^2 + \frac{1}{r^2}\left(\frac{\partial u_n}{\partial \theta}\right)^2\right] rdrd\theta \\ &\geq \int_{E_n} \int_{K_n \cap \{|z|=r\}} \frac{1}{r}\left(\frac{\partial u_n}{\partial \theta}\right)^2 d\theta dr \\ &\geq \epsilon^2 \int_{E_n} \frac{dr}{r}.\end{aligned}$$

The first integral decreases as $n \to \infty$, while the last integral tends to $\int_E \frac{dr}{r} = +\infty$ as $n \to \infty$. This contradiction establishes the lemma.

3.3 Theorem (Walsh-Lebesgue Theorem). Let K be a compact subset of the plane with connected complement. Then every continuous real-valued function on bK can be approximated uniformly on bK by harmonic polynomials in x and y.

Proof. Let $K = \bigcap_{n=1}^{\infty} K_n$, where K_n is compact, $K_{n+1} \subseteq \text{int}(K_n)$, bK_n consists of a finite number of smooth disjoint closed curves, and K_n^c is connected.

Let u be a real function on the complex plane which is continuously differentiable. Let u_n be the harmonic extension of $u|_{bK_n}$ to int (K_n). By 3.2, u_n converges uniformly to u on bK.

Since int (K_n) is simply connected, the harmonic conjugate $*u_n$ of u_n can be defined on int (K_n). Then $u_n + i{*u_n}$ is analytic in a neighborhood of K. By Runge's theorem, $u_n + i{*u_n}$ can be approximated uniformly on K by polynomials in z. The real parts of these polynomials will be harmonic polynomials in x and y which converge uniformly on bK to u_n. It follows that u, and hence any function in $C_R(bK)$, is uniformly approximable on bK by harmonic polynomials.

3.4 Corollary. Let K be a compact subset of the plane. Then $P(K)$ is a dirichlet algebra (on the outer boundary of K).

4. Logmodular Algebras

Let H^∞ be the algebra of bounded analytic functions on the open unit disc. By the theorem of Fatou, the radial limits $\lim_{r \to 1} f(re^{i\theta})$ for every $f \in H^\infty$ exist almost everywhere with respect to angular measure $d\theta$. The radial limits determine a function, also denoted by f, in $L^\infty(d\theta)$. The correspondence between a function and its boundary values is an isometry of H^∞ and a closed subalgebra of $L^\infty(d\theta)$. Every $f \in H^\infty$ can be recovered from its boundary values by the Poisson integral formula

$$f(re^{i\theta}) = \frac{1}{2\pi}\int_0^{2\pi} f(e^{i\varphi})P_r(\varphi - \theta)d\varphi,$$

where P_r is the Poisson kernel.

Let Y be the maximal ideal space of $L^\infty(d\theta)$. According to I.9, we can identify $L^\infty(d\theta)$ and $C(Y)$. So H^∞ becomes a closed subalgebra of $C(Y)$.

4.1 Lemma. The algebra H^∞ of bounded analytic functions on the interior of the unit disc is a uniform algebra on the maximal ideal space Y of $L^\infty(d\theta)$. The Shilov boundary of H^∞ is Y. If $u \in C_R(Y)$, there is an invertible function f in H^∞ such that $\log|f| = u$.

Proof. Suppose $u \in C_R(Y) = L^\infty_R(d\theta)$. Then

$$u(re^{i\theta}) = \frac{1}{2\pi}\int_0^{2\pi} u(e^{i\varphi})P_r(\varphi - \theta)d\varphi$$

extends u to be bounded and harmonic on int Δ. The extended function has radial boundary values almost everywhere, which agree with u. If $*u$ is the harmonic conjugate of u, and $f = e^{u+i{*u}}$, then f is an invertible function in H^∞ satisfying $\log|f| = u$.

It follows easily that H^∞ separates the points of Y, and that Y is the Shilov boundary of H^∞. This completes the proof.

If $f \in H^\infty$ satisfies $f^2 = f$, then either $f \equiv 0$ or $f \equiv 1$. Since Y is not connected, the proof of 3.1 shows that H^∞ is not a dirichlet algebra. On the other hand, 4.1 provides a property of H^∞ which is a useful substitute for dirichlicity.

A uniform algebra A on X is **logmodular on X** if $\log|A^{-1}|$ is uniformly dense in $C_R(X)$. If A is logmodular on X, then X is the Shilov boundary of A. By 4.1, H^∞ is logmodular on Y.

Dirichlet algebras are logmodular. In fact, if $f \in A$, then $\mathrm{Re}\,(f) = \log|e^f|$. So $\mathrm{Re}\,(A) \subseteq \log|A^{-1}|$.

4.2 Theorem. Let A be a logmodular algebra on X. Then every $\phi \in M_A$ has a unique representing measure on X.

Actually, we shall prove a more general result, which will be useful later.

4.3 Theorem (Hoffman's Uniqueness Theorem). Let A be a uniform algebra on X, and let $\phi \in M_A$. Suppose that $u \in C_R(X)$ is such that tu is in the closure of $\log|A^{-1}|$ for all real t. Then all representing measures for ϕ agree on u.

Proof. Let μ and ν be representing measures for ϕ. Suppose $v = \log|f|$, where $f \in A^{-1}$. Then $|\phi(f)| = |\int f d\mu| \leq \int e^v d\mu$, and $|\phi(1/f)| \leq \int e^{-v} d\nu$. Hence $1 = |\phi(f)\phi(1/f)| \leq \int e^v d\mu \int e^{-v} d\nu$. Approximating tu by such functions v, we obtain

$$h(t) = \int e^{tu} d\mu \int e^{-tu} d\nu \geq 1, \qquad -\infty < t < \infty.$$

From the expansion

$$h(t) = \left[1 + t\int u d\mu + \frac{t^2}{2}\int u^2 + \dots\right]\left[1 - t\int u d\nu + \frac{t^2}{2}\int u^2 d\nu + \dots\right],$$

we see that h is real analytic. Since $h(0) = 1 \leq h(t)$, we must have $0 = h'(0) = \int u d\mu - \int u d\nu$. That proves the theorem.

5. Maximal Subalgebras

A uniform algebra A on X is a **maximal subalgebra** of $C(X)$ if $A \neq C(X)$, while there is no uniform algebra B distinct from A and $C(X)$ satisfying $A \subseteq B \subseteq C(X)$. A uniform algebra A on X is a maximal subalgebra of $C(X)$ if and only if whenever $g \in C(X)$ does not belong to A, then every function in $C(X)$ can be approximated uniformly on X by polynomials in g with coefficients in A. Consequently, maximality theorems can be interpreted as approximation theorems.

5.1 Theorem (Wermer's Maximality Theorem). Let Δ be the closed unit disc in the complex plane. Then $P(\Delta)$ is a maximal subalgebra of $C(b\Delta)$.

Proof. Let B be a uniform algebra such that $P(\Delta) \subseteq B \subseteq C(b\Delta)$. There are two cases that can occur.

First, suppose there is no $\phi \in M_B$ such that $\phi(z) = 0$. Then $1/z \in B$. Since every $g \in C(b\Delta)$ can be approximated uniformly by polynomials in z and $1/z$, $B = C(b\Delta)$.

Second, suppose there is $\phi \in M_B$ such that $\phi(z) = 0$. The restriction of ϕ to $P(\Delta)$ is then evaluation at 0. Any representing measure on $b\Delta$ for ϕ must represent "evaluation at 0" on $P(\Delta)$, and so must coincide with $d\theta/2\pi$. Hence $d\theta/2\pi$ is multiplicative on B. If $f \in B$ and $n \geq 1$, then

$$\frac{1}{2\pi}\int_0^{2\pi} f(e^{i\theta})e^{in\theta}d\theta = \left(\frac{1}{2\pi}\int_0^{2\pi} f(e^{i\theta})d\theta\right)\left(\frac{1}{2\pi}\int_0^{2\pi} e^{in\theta}d\theta\right) = 0.$$

Since the negative Fourier coefficients of f vanish, f belongs to $P(\Delta)$ (cf. I.3.2). Hence $B = P(\Delta)$.

6. Hulls

Let A be a uniform algebra on X. The ***A*-convex hull** of a closed subset E of X is the set $\hat{E} \subseteq M_A$ of homomorphisms which extend continuously to the uniform closure of $A|_E$ in $C(E)$. Since every homomorphism of the closure of $A|_E$ automatically yields a homomorphism of A, we have the following.

6.1 Theorem. Let A be a uniform algebra on X. The A-convex hull $\hat{E}$ of a closed subset E of X is the maximal ideal space of the uniform closure of $A|_E$ in $C(E)$.

Now $\phi \in \hat{E}$ if and only if there is a constant $c > 0$ such that $|\phi(f)| \leq c\,\|f\|_E$ for all $f \in A$. Since ϕ determines a homomorphism of $A|_E$, this constant, when it exists, can be chosen to be one.

Also, $\phi \in \hat{E}$ if and only if there is a representing measure for ϕ supported on E.

6.2 Theorem (Glicksberg's Lemma). Let A be a uniform algebra on X, and let U be a non-empty open subset of X. Every function $f \in A$ which vanishes on U also vanishes on the complement of $\widehat{X\backslash U}$. If X is the Shilov boundary of A, then the complement of $\widehat{X\backslash U}$ is an open subset of M_A which meets U.

Proof. Let $\phi \in M_A$, $\phi \notin \widehat{X\backslash U}$. Let μ be a representing measure for ϕ with minimal closed support. By the remark preceding the theorem, μ cannot be supported on $X\backslash U$, so $\mu(U) > 0$. By 2.3, $\phi(f) = 0$ for every $f \in A$ which vanishes on U. This proves the first statement of the theorem.

Suppose X is the Shilov boundary of A. Since $X\backslash U$ is not a boundary for A, there exists $\phi \in U$ which has no representing measure supported on $X \backslash U$. This ϕ cannot belong to $\widehat{X \backslash U}$. That completes the proof.

As an application of 6.2 and Wermer's maximality theorem, we give a proof of Radó's theorem.

6.3 Theorem (Radó's Theorem). Let Δ be the closed unit disc in the complex plane, and let $f \in C(\Delta)$. If f is analytic at every point $z \in \text{int}\,\Delta$ for which $f(z) \neq 0$, then f is analytic on $\text{int}\,\Delta$.

Proof. Let B be the uniform algebra on Δ generated by $P(\Delta)$ and f. Since the functions in B are analytic where f is not zero, and they are analytic in the interior of $f^{-1}(0)$, we have $\partial_B \subseteq b\Delta \cup bf^{-1}(0)$.

Let $U = \partial_B \cap \text{int}\,\Delta$. Then U is an open subset of ∂_B.

Suppose that U is not empty. Since f vanishes on U, f must vanish on an open subset of M_B which meets U, by 6.2. But this contradicts $U \subseteq bf^{-1}(0)$.

Hence U is empty, and $\partial_B \subseteq b\Delta$. Consequently B can be considered a closed subalgebra of $C(b\Delta)$. By Wermer's maximality theorem 5.1, B coincides with $P(\Delta)$. That makes f analytic on $\text{int}\,\Delta$.

7. Decomposition of Orthogonal Measures

Let A be a uniform algebra on X, and let $\phi \in M_A$. Often we will find ourselves in the position of being able to verify a statement for each representing measure $m \in M_\phi$, and wanting to assert that the statement holds uniformly for all $m \in M_\phi$. We now prove a variation of von Neumann's minimax principle which will allow us to obtain uniform estimates from individual estimates. In the case of dirichlet algebras, or logmodular algebras, this technique is vacuous. The reader interested only in dirichlet algebras will find that the treatment offered in this section simplifies considerably in that special case.

7.1 Theorem (Minimax Theorem). Suppose that P is a convex subset of a vector space, and that M is a convex compact subset of a topological vector space. Let F be a non-negative function on $M \times P$ such that $m \to F(m,p)$ is concave and continuous, and $p \to F(m,p)$ is convex. Then

$$\sup_{m \in M} \inf_{p \in P} F(m,p) = \inf_{p \in P} \sup_{m \in M} F(m,p).$$

Proof. It is clear that

$$0 \leq \sup_{m \in M} \inf_{p \in P} F(m,p) \leq \inf_{p \in P} \sup_{m \in M} F(m,p) < \infty.$$

Suppose that the theorem is not true, so that

$$\sup_{m \in M} \inf_{p \in P} F(m,p) = c < c + \epsilon = \inf_{p \in P} \sup_{m \in M} F(m,p).$$

Let Q be the convex hull of the functions $F(\cdot,p)$ in $C_R(M), p \in P$. The functions f in Q are of the form $f(m) = \sum a_j F(m,p_j)$, where $a_j \geq 0$ and $\sum a_j = 1$. Since $p \to F(m,p)$ is convex,

$$\sup_{m \in M} f(m) \geq \sup_{m \in M} F(m, \sum a_j p_j) \geq c + \epsilon.$$

This shows that Q is at a positive distance from the convex set of $u \in C_R(X)$ satisfying $u \leq c$.

By the strict separation theorem for convex sets, there is a non-zero measure μ on M such that

$$\sup \left\{ \int u d\mu : u \in C_R(X),\, u \leq c \right\} < \inf \left\{ \int f d\mu : f \in Q \right\}.$$

In particular, $\int u d\mu \leq 0$ whenever $u \leq 0$. So $\mu \geq 0$. Multiplying by a positive constant, we can assume that $\int d\mu = 1$. The inequality then takes the form

$$c < \inf_{p \in P} \int f(m,p) d\mu.$$

We assert that there is a point $m_0 \in M$ such that

$$g(m_0) \geq \int g d\mu$$

for all concave functions $g \in C_R(M)$. In fact, let $\mu_\alpha = \sum_j a_{j\alpha} \delta_{m_{j\alpha}}$ be a net of convex combinations of point masses at points $m_{j\alpha} \in M$, such that μ_α converges weak-star to μ. Let $m_0 \in K$ be an adherent point of the net $\sum_j a_{j\alpha} m_{j\alpha}$. Then

$$\begin{aligned} g(m_0) &= \lim_\alpha g(\sum_j a_{j\alpha} m_{j\alpha}) \\ &\geq \lim_\alpha \sum_j a_{j\alpha} g(m_{j\alpha}) \\ &= \int g d\mu \end{aligned}$$

for all concave functions $g \in C_R(M)$. (The point m_0 is the barycenter of μ, usually written $\int m d\mu$. Barycenters crop up again in section 11.)

Since $m \to F(m,p)$ is concave, we obtain

$$\begin{aligned} c &< \inf_{p \in P} \int F(m,p) d\mu \leq \inf_{p \in P} F(m_0, p) \\ &\leq \sup_m \inf_{p \in P} F(m,p) = c. \end{aligned}$$

This contradiction establishes the theorem.

As the first application, we obtain a pointwise bounded approximation theorem for uniform algebras. The modification argument used in the proof is of especial importance.

7.2 Theorem (Hoffman-Wermer Theorem). Let A be a uniform algebra on X, and let $\phi \in M_A$. Suppose f is a continuous function on X which is in the closure of A in $L^1(m)$ for all $m \in M_\phi$. Then there is a sequence $f_n \in A$ such that $\|f_n\|_X \leq \|f\|_X$, and f_n converges to f M_ϕ-almost everywhere, that is, m-almost everywhere for all $m \in M_\phi$.

Proof. We apply the minimax theorem 7.1 to the function

$$F(m,g) = \int |f - g|\, dm, \qquad m \in M_\phi,\ g \in A.$$

The continuity of f guarantees that $m \rightarrow F(m, g)$ is continuous. By hypothesis,

$$\inf_{g \in A} \int |f - g|\, dm = 0, \qquad m \in M_\phi.$$

By the minimax theorem,

$$\inf_{g \in A} \sup_{m \in M_\phi} \int |f - g|\, dm = 0.$$

Consequently there are $g_n \in A$ such that

$$\sup_{m \in M_\phi} \int |f - g_n|\, dm < \frac{1}{n^2}.$$

For all $m \in M_\phi$, $\sum_{n=1}^{\infty} \int |f - g_n|\, dm < \infty$. Hence the set on which $g_n \rightarrow f$ has full m-measure, for all $m \in M_\phi$. That is, $g_n \rightarrow f$ M_ϕ-almost everywhere.

We can assume that $\|f\|_X = 1$. If a and b are complex numbers such that $|b| \leq 1$, then $\log^+ |a| \leq \max(0, |a| - 1) \leq |a - b|$. Hence

$$\sup_{m \in M_\phi} \int \log^+ |g_n|\, dm \leq \sup_{m \in M_\phi} \int |g_n - f|\, dm < \frac{1}{n^2}.$$

By 2.1 there are $h_n \in A$ such that $\log^+ |g_n| \leq \operatorname{Re}(h_n)$, while $\operatorname{Re} \phi(h_n) \leq 1/n^2$. We can assume that $\operatorname{Im} \phi(h_n) = 0$. For each $m \in M_\phi$,

$$\int e^{-h_n} dm = e^{-\phi(h_n)} \geq e^{-1/n^2} \geq 1 - \frac{1}{n^2}.$$

Since $|e^{-h_n}| \leq 1$, we obtain

$$\int |1 - e^{-h_n}|^2\, dm = 1 + \int |e^{-h_n}|^2\, dm - 2 \operatorname{Re} \int e^{-h_n} dm \leq \frac{4}{n^2}.$$

Hence $\sum_{n=1}^{\infty} \int |1 - e^{-h_n}|^2\, dm < \infty$ for all $m \in M_\phi$. And e^{-h_n} converges to 1 M_ϕ-almost everywhere. Now the sequence $f_n = g_n e^{-h_n}$ converges to f M_ϕ-almost everywhere, and $|f_n| \leq 1 = \|f\|_X$. That does it.

Combining the minimax principle and the modification argument used in 7.2, we also obtain the following.

7.3 Lemma (Forelli's Lemma). Let E be an F_σ-set such that $M_\phi(E) = 0$. Then there are $f_n \in A$ such that $\|f_n\|_X \leq 1$, $f_n(x) \to 0$ for all $x \in E$, and $f_n \to 1$ M_ϕ-almost everywhere.

Proof. Let $\{E_n\}$ be an increasing sequence of closed sets whose union is E. For each integer n, and for each $m \in M_\phi$, $m(E_n) = 0$. So

$$0 = \inf\left\{\int udm: u \in C(X), u \geq 0, u \geq n \text{ on } E_n\right\}.$$

By the minimax theorem, there exist $u_n \in C(X)$ such that $u_n \geq 0$, $u_n \geq n$ on E_n, and

$$\sup_{m \in M_\phi} \int u_n dm \leq \frac{1}{n^2}.$$

By 2.1, there are $h_n \in A$ such that $u_n \leq \operatorname{Re}(h_n)$, and $\operatorname{Re}\phi(h_n) < 2/n^2$. We can assume that $\operatorname{Im}\phi(h_n) = 0$.

Let $f_n = e^{-h_n}$. Then $|f_n| \leq 1$, and $f_n \to 0$ on E. As in 7.2, we obtain

$$\int |1 - f_n|^2\, dm \leq \frac{4}{n^2}.$$

Consequently $f_n \to 1$ M_ϕ-almost everywhere.

Now we wish to define a Lebesgue decomposition of a measure μ with respect to M_ϕ. Without defining them explicitly, we mention that there are two natural candidates for the Lebesgue decomposition with respect to a set of measures, and that the following lemma shows that these notions coincide in the case at hand.

7.4 Lemma. Let K be a weak-star compact convex set of positive measures on X. If a measure μ is singular to all measures in K, then there is an F_σ-set E such that $|\mu|(X \backslash E) = 0$ while $K(E) = 0$, that is, $m(E) = 0$ for all $m \in K$.

Proof. Let J be the convex set of $f \in C(X)$ such that $0 \leq f \leq 1$. For each $\nu \in K$,

$$\inf_{f \in J}\left\{\int (1-f)d\nu + \int fd|\mu|\right\} = 0.$$

By the minimax theorem,

$$\inf_{f \in J} \sup_{\nu \in K}\left\{\int (1-f)d\nu + \int fd|\mu|\right\} = 0.$$

Consequently there is a sequence $f_n \in J$ such that $\int f_n d|\mu| \leq 1/n^2$, while $\int (1 - f_n)d\nu \leq 1/n^2$ for all $\nu \in K$. The set of $x \in X$ such that $f_n(x) \to 1$ is a G_δ-set, whose complement can be taken to be the desired set E.

7.5 Corollary. Let K be a weak-star compact convex set of positive measures on X. Every complex measure μ on X has a unique decomposition

$$\mu = \mu_a + \mu_s,$$

where μ_a is absolutely continuous with respect to some $\nu \in K$, and μ_s is supported on a Borel set E such that $K(E) = 0$.

Proof. Let c be the supremum of $|\mu|(G)$ over all Borel sets G such that the restriction μ_G of μ to G is absolutely continuous with respect to some measure in K. Choose $\nu_n \in K$ and Borel sets F_n such that μ_{F_n} is absolutely continuous with respect to ν_n, and $|\mu|(F_n) \to c$. Let $F = \bigcup_{n=1}^{\infty} F_n$, and $\nu = \sum_{n=1}^{\infty} \nu_n/2^n \in K$. Then the restriction μ_a of μ to F is absolutely continuous with respect to ν. Evidently $\mu_s = \mu - \mu_a$ is singular to each measure in K. Now apply 7.4, to find the desired set E.

7.6 Theorem (Abstract F. and M. Riesz Theorem). Let A be a uniform algebra on X, and let $\phi \in M_A$. Let $\mu = \mu_a + \mu_s$ be the Lebesgue decomposition of a measure μ on X with respect to M_ϕ. If μ is orthogonal to A, then μ_a and μ_s are orthogonal to A.

Actually, we will prove a stronger result, which applies to ideals in A, or to subalgebras of $C(X)$ which contain A.

7.7 Theorem. Let A be a uniform algebra on X, and let B be a subspace of $C(X)$ which is an A-module. Let $\phi \in M_A$, and let $\mu = \mu_a + \mu_s$ be the Lebesgue decomposition of a measure μ on X with respect to M_ϕ. If μ is orthogonal to B, then μ_a and μ_s are orthogonal to B.

Proof. By 7.4, there is an F_σ-set E such that $M_\phi(E) = 0$ while $|\mu_s|(X\backslash E) = 0$. By the Forelli lemma 7.3, there are $f_n \in A$ such that $|f_n| \leq 1$, $f_n \to 0$ on E, and $f_n \to 1$ μ_a-almost everywhere. If $g \in B$, then also $gf_n \in B$, and

$$0 = \int g f_n \, d\mu \to \int g \, d\mu_a.$$

So μ_a and μ_s are orthogonal to B.

A representing measure $m \in M_\phi$ is **dominant** if every representing measure in M_ϕ is absolutely continuous with respect to m. If $\phi \in M_A$ has a unique representing measure, then that measure is certainly dominant.

7.8 Corollary (Ahern's Theorem). Let A be a uniform algebra on X. Suppose that $\phi \in M_A$ has a dominant representing measure m. Let $\mu = hdm + \mu_s$ be the Lebesgue decomposition of a measure μ on X with respect to m. If μ is orthogonal to A, then hdm and μ_s are orthogonal to A.

Another consequence of 7.6 is the following.

7.9 Corollary. Let $\mu = \mu_a + \mu_s$ be the Lebesgue decomposition of μ with respect to M_ϕ. If μ is orthogonal to the kernel A_ϕ of ϕ, then μ_a is orthogonal to A_ϕ, and μ_s is orthogonal to A.

Proof. Fix $m \in M_\phi$, and apply 7.6 to $\mu - cm$, where $c = \int d\mu$.

As an application of 7.9, we can prove a version of the usual F. and M. Riesz theorem.

7.10 Theorem (F. and M. Riesz Theorem). Let μ be a measure on the boundary $b\Delta$ of the closed unit disc Δ which is orthogonal to $P(\Delta)$. Then μ is absolutely continuous with respect to arc length.

Proof. The homomorphsim "evaluation at 0" of $P(\Delta)$ has a unique representing measure $d\theta/2\pi$ on $b\Delta$. By 7.6, the singular part μ_s of μ with respect to arc length $d\theta$ is orthogonal to $P(\Delta)$. That is, $\int z^n d\mu_s = 0$ for all $n \geq 0$. We must show that $\mu_s = 0$.

Suppose that n_0 is an integer such that $\int z^n d\mu_s = 0$ for $n > n_0$. Since $\int z^k z^{n_0} d\mu_s = 0$ for $k \geq 1$, the measure $z^{n_0}\mu_s$ is orthogonal to functions in $P(\Delta)$ which vanish at 0. By 7.9, $z^{n_0}\mu_s$ is orthogonal to $P(\Delta)$. In particular, $\int z^{n_0} d\mu_s = 0$.

This shows that there can be no "first" negative integer n_0 such that $\int z^{n_0} d\mu_s \neq 0$. It follows that $\int z^n d\mu_s = 0$ for all n, and $\mu_s = 0$.

As a final corollary of the abstract F. and M. Riesz theorem, there is the following decomposition theorem.

7.11 Theorem (Decomposition Theorem for Orthogonal Measures). Let A be a uniform algebra on X, and let μ be a measure on X which is orthogonal to A. Then there are $\phi_n \in M_A$ and measures μ_s and μ_n in $A^\perp$, $n = 1, 2, \ldots$, such that μ_s is singular with respect to M_ϕ for all $\phi \in M_A$, μ_n is absolutely continuous with respect to some representing measure for ϕ_n, the μ_n are pairwise mutually singular, and

$$\mu = \mu_s + \sum_n \mu_n,$$

the series converging in the total variation norm of measures.

Proof. Let c be the supremum of $\|\mu_a\|$, taken over the absolutely continuous part μ_a of μ with respect to M_ϕ for all $\phi \in M_A$. If $c = 0$, we set $\mu = \mu_s$, and we are done. If $c > 0$, choose $\phi_1 \in M_A$ such that the absolutely continuous part μ_1 of μ with respect to ϕ_1 satisfies $\|\mu_1\| \geq c/2$.

If we repeat this process, we arrive at a series $\sum \mu_n$, which may break off after a finite number of terms. At any rate, each μ_n is singular to $\mu - \sum_{j=1}^{n} \mu_j$. So μ_n is singular to μ_k for $k > n$, and the μ_n are pairwise mutually singular. It follows that $\sum \mu_n$ converges in the total variation norm of measures, and that $\|\sum \mu_n\| = \sum \|\mu_n\|$.

Let $\mu_s = \mu - \sum \mu_n$. For each $\phi \in M_A$ and each n, the norm of the absolutely continuous part of μ_s with respect to M_ϕ cannot exceed $2\|\mu_n\|$, by the choice of μ_n. Since $\|\mu_n\| \to 0$, μ_s is singular to M_ϕ, for all $\phi \in M_A$.

By the abstract F. and M. Riesz theorem, $\mu_n \in A^\perp$ for each n. Consequently $\mu_s \in A^\perp$. That proves the theorem.

8. Cauchy Transform

Let μ be a finite measure on the complex plane with compact support. The **Cauchy transform** of μ is defined by

$$\hat{\mu}(\zeta) = \int \frac{d\mu(z)}{z - \zeta}.$$

The Cauchy transform $\hat{\mu}$ is the convolution of μ and the locally integrable function $1/z$. So the integral defining $\hat{\mu}$ converges absolutely except for ζ belonging to a set of zero planar measure.

The relevance of the Cauchy transform is indicated by the following theorem.

8.1 Theorem. Let μ be a measure on the compact subset K of the complex plane. The Cauchy transform $\hat{\mu}$ vanishes off K if and only if μ is orthogonal to $R(K)$.

Proof. If $\mu \perp R(K)$, then clearly $\hat{\mu}$ vanishes off K.

Conversely, suppose that $\hat{\mu}$ vanishes off K. Let f be analytic in a neighborhood of K. Then

$$f(\zeta) = \frac{1}{2\pi i} \int_\Gamma \frac{f(z)}{z - \zeta} dz,$$

for an appropriate contour Γ surrounding K. Integrating with respect to μ, and interchanging the orders of integration in the double integral, we obtain $\int f d\mu = 0$. Consequently $\mu \perp R(K)$.

The Cauchy transform $\hat{\mu}$ is analytic off the closed support of μ. A converse of this statement is true.

8.2 Theorem. Let μ be a finite measure on the complex plane with compact support. Suppose U is an open set, and f is a function analytic on U, such that $f = \hat{\mu}$ $dxdy$-almost everywhere on U. Then $|\mu|(U) = 0$.

Proof. Let E be a rectangle in U such that $\int |z - \xi|^{-1} d|\mu|(z) < \infty$, and such that $f(\zeta) = \hat{\mu}(\zeta)$, for those $\zeta \in bE$ not belonging to a set of linear measure zero. By Fubini's theorem,

$$\begin{aligned}\mu(E) &= \int_E d\mu(z) = \frac{1}{2\pi i} \int_C \int_{bE} \frac{1}{\zeta - z} d\zeta d\mu(z) \\ &= \frac{1}{2\pi i} \int_{bE} \int_C \frac{d\mu(z)}{\zeta - z} d\zeta \\ &= -\frac{1}{2\pi i} \int_{bE} \hat{\mu}(\zeta) d\zeta = 0.\end{aligned}$$

Since such rectangles E generate the algebra of measurable subsets of U, μ must vanish identically on U.

8.3 Corollary. If $\hat{\mu} = 0$ $dxdy$-almost everywhere, then $\mu = 0$.

8.4 Corollary (Hartogs-Rosenthal Theorem). If K is a compact subset of the plane which has zero planar measure, then $R(K) = C(K)$.

Proof. If μ is a measure on K which is orthogonal to $R(K)$, then $\hat{\mu} = 0$ off K. By 8.3, $\mu = 0$.

8.5 Theorem (Wilken's Theorem). Let K be a compact subset of the plane. Let μ be a measure on bK which is orthogonal to $R(K)$. If μ is singular to the set of representing measures M_z for all $z \in K$, then $\mu = 0$.

Proof. Suppose $\mu \perp R(K)$, while $\mu \neq 0$. Choose z_0 so that $\int |z - z_0|^{-1} d|\mu|(z) < \infty$, and $\hat{\mu}(z_0) \neq 0$. Then $z_0 \in K$.

If f is a rational function with poles off K, then $(f(z) - f(z_0))/(z - z_0)$ is also a rational function with poles off K. So $\int (f(z) - f(z_0))/(z - z_0) d\mu(z) = 0$. Consequently,

$$f(z_0) = \frac{1}{\hat{\mu}(z_0)} \int \frac{f(z)}{z - z_0} d\mu(z)$$

for all $f \in R(K)$. Hence $d\mu(z)/[\hat{\mu}(z_0)(z - z_0)]$ is a complex representing measure for z_0.

By 2.2, there is a representing measure for z_0 which is absolutely continuous with respect to μ. Consequently μ cannot be singular to all representing measures. That proves the theorem.

The decomposition theorem 7.11 for measures orthogonal to $R(K)$ now reads as follows, in view of Wilken's theorem.

8.6 Theorem. Let K be a compact subset of the plane, and let μ be a measure on bK orthogonal to $R(K)$. Then there are points $z_j \in K$, representing measures m_j on bK for z_j, and functions $h_j \in L^1(m_j)$, such that $h_j m_j$ is orthogonal to $R(K)$, and

$$\mu = \sum h_j m_j,$$

the series converging in the total variation norm of measures.

Theorem 8.6 shows that if K is nowhere dense, and every point $z \in K$ has a unique representing measure (namely, the point mass at z), then $R(K) = C(K)$. This occurs, in particular, if $R(K)$ is a dirichlet algebra on K. Combining this observation with the Walsh-Lebesgue theorem 3.4, we obtain the following special case of Mergelyan's theorem.

8.7 Theorem (Lavrentiev's Theorem). Let K be a compact plane set. Then $P(K) = C(K)$ if and only if K is nowhere dense and the complement of K is connected.

9. Mergelyan's Theorem

The function algebraic techniques we have developed so far are sufficient to give a proof of Mergelyan's theorem on polynomial approximation. This proof is non-constructive, as opposed to the constructive proof we shall give in chapter VIII of a more general result on rational approximation. The proof rests on the dirichlicity of $P(K)$, together with the description of the measures orthogonal to $R(K)$ given in 8.6.

9.1 Theorem (Mergelyan's Theorem). Let K be a compact subset of the plane whose complement is connected. Then every function continuous on K and analytic on the interior of K can be approximated uniformly on K by polynomials.

Proof. It suffices to show that if μ is a measure on bK which is orthogonal to $P(K)$, then μ is orthogonal to $A(K)$. By 8.6, it suffices to show this for measures of the form $\mu = hdm$, where m is the representing measure for some point $z_0 \in K$.

Let $P_0(K)$ be the ideal of functions in $P(K)$ which vanish at z_0. Let H^2 and H_0^2 be the closures of $P(K)$ and $P_0(K)$ respectively in $L^2(m)$. Since $\int fgdm = 0$ for all $f \in P(K)$ and $g \in P_0(K)$, H^2 is orthogonal to $\overline{H_0^2}$ in $L^2(m)$. By the Walsh-Lebesgue theorem 3.4, $P(K)$ is a dirichlet algebra. Hence $P(K) + \overline{P_0(K)}$ is dense in $L^2(m)$, and

$$L^2(m) = H^2 \oplus \overline{H_0^2}.$$

Now evaluation at z_0 is also a homomorphism of $A(K)$. The only candidate for representing measure for z_0 on $A(K)$ is m. Hence m is multiplicative on $A(K)$.

Since $\int fgdm = 0$ for all $f \in A(K)$ and $g \in P_0(K)$, $A(K)$ is orthogonal to $\overline{H_0^2}$ in $L^2(m)$. Hence $A(K) \subseteq H^2$.

Now suppose $hdm \perp P(K)$, and $f \in A(K)$. Then f is in the closure of $P(K)$ in $L^1(dm)$. By the Hoffman-Wermer theorem 7.2, there is a bounded sequence $f_n \in P(K)$ such that $f_n \to f$ dm-almost everywhere. Then

$$\int fhdm = \lim_n \int f_n hdm = 0.$$

Consequently $hdm \perp A(K)$. That establishes the theorem.

The remainder of this section is devoted to several extensions of the idea of this proof. For this, we define a uniform algebra B to be **relatively maximal** if whenever A is a subalgebra of $C(M_B)$ such that $A \supseteq B$ and

$\partial_A = \partial_B$, then $A = B$. The proof of Mergelyan's theorem shows that $P(K)$ is relatively maximal. More generally, it shows the following.

9.2 Theorem. Let B be a dirichlet algebra on X. Suppose that the only measure in $B^\perp$ which is singular to all representing measures for homomorphisms $\phi \in M_A$ is the zero measure. Then B is relatively maximal.

9.3 Corollary. If $R(K)$ is a dirichlet algebra, then $R(K) = A(K)$.

With a little more effort, we can stretch the proof of 9.1 somewhat further.

9.4 Theorem. Let A and B be uniform algebras on X such that $A \supseteq B$. Suppose that the only measure in $B^\perp$ which is singular with respect to all representing measures for homomorphisms $\phi \in M_B$ is the zero measure. Suppose also that every representing measure for each homomorphism $\phi \in M_B$ is also multiplicative on A. Then $B = A$.

Proof. Let m be a representing measure for a fixed $\phi \in M_B$, and suppose that $h\,dm \in B^\perp$. By the decomposition theorem, it will suffice to show that $h\,dm \in A^\perp$.

Let $f \in A$, and suppose for convenience that $\int f\,dm = 0$. Then also $\int f^2 dm = 0$. So for any $\mu \in M_\phi$,

$$\frac{1}{2}\left(\int f d\mu\right)^2 = \frac{1}{2}\int f^2 d\mu = \int f^2 d\left(\frac{m+\mu}{2}\right)$$
$$= \left(\int f d\left(\frac{m+\mu}{2}\right)\right)^2 = \frac{1}{4}\left(\int f d\mu\right)^2.$$

Consequently $\int f d\mu = 0$ for all $\mu \in M_\phi$. By 2.1,

$$\inf\{\phi(u) : u \in \operatorname{Re}(B),\ u > \operatorname{Re}(f)\} = \sup_{\mu \in M_\phi} \int \operatorname{Re}(f) d\mu = 0.$$

By 2.1, there are $g_n \in B$ such that $\operatorname{Re}(g_n) > \operatorname{Re}(f)$ and $\phi(g_n) \to 0$.

Let $h_n = g_n - f \in A$. Then $\operatorname{Re}(h_n) > 0$ on X, and $\int h_n dm \to 0$. From I.4.3, it is clear that $\operatorname{Re}(h_n) > 0$ on M_A. By I.5.1, the function $\sqrt{h_n}$ belongs to A, where we take the principal branch of the square root. Now

$$\operatorname{Re}(\sqrt{h_n}) = |h_n|^{1/2} \cos\left(\frac{\arg h_n}{2}\right) \geq |h_n|^{1/2} \cos\frac{\pi}{4},$$

so

$$|h_n|^{1/2} \leq \sqrt{2}\operatorname{Re}(\sqrt{h_n}).$$

Hence

$$\int |h_n|^{1/2} dm \leq \sqrt{2}\operatorname{Re}\int \sqrt{h_n} dm = \sqrt{2}\operatorname{Re}\sqrt{\int h_n dm} \to 0.$$

Consequently $h_n \to 0$ in m-measure.

Now $e^{-h_n}e^{-f} = e^{-g_n}$ is a bounded sequence in B which tends to e^{-f} in

m-measure. Hence $\int e^{-f} h dm = \lim \int e^{-g_n} h dm = 0$. Similarly, $\int e^{tf} h dm = 0$ for all real t. Since $(e^{tf} - 1)/t$ converges uniformly to f as $t \to 0$, we obtain

$$\int f h dm = \lim_{t \to 0} \int \frac{e^{tf} - 1}{t} h dm = 0.$$

That proves the theorem.

9.5 Corollary. If $R(K)$ and $A(K)$ have the same representing measures on bK, then $R(K) = A(K)$.

9.6 Corollary. If $R(K)$ and $A(K)$ have the same real annihilating measures on bK, then $R(K) = A(K)$.

Note that 9.5 includes the other theorems in this section pertaining to $R(K)$.

10. Local Algebras

Let A be a uniform algebra on X. A function $f \in C(X)$ **belongs locally** on X to A if for every $x \in X$, there is a neighborhood U of x and a function $g \in A$ which coincides with f on U. The algebra A is **local on X** if every function which belongs locally on X to A is in A. The algebra A is **local** if it is local on its maximal ideal space.

An algebra which is not local on its Shilov boundary is $R(Y)$, where Y is the annulus $\{\frac{1}{2} \le |w| \le 1\}$. The function on bY which is 0 on one rim and 1 on the other belongs locally to $R(Y)$, while it does not belong to $R(Y)$.

From this algebra we construct a non-local algebra as follows. Let Δ be the unit disc, as usual, and let $X = \Delta \times bY$. Let A be the subalgebra of $f \in C(X)$ such that: for every fixed $w \in bY$, $f(\cdot, w) \in P(\Delta)$; $f(0, \cdot) \in R(Y)$; and $(\partial f/\partial z)(0, \cdot) \in R(Y)$. The functions in A can be written in the form

$$f(z,w) = f_0(w) + z f_1(w) + z^2 f_2(z,w),$$

where $f_0, f_1 \in R(Y)$ and $f_2(\cdot, w) \in P(\Delta)$ for each fixed $w \in bY$. The algebra is generated by the four functions w, $1/w$, z, and $z^2\bar{w}$. The maximal ideal space of A is obtained from $X \cup Y$ by identifying bY with the subset $\{0\} \times bY$ of X. The function which is z on $\Delta \times \{\frac{1}{2}\}$, and 0 elsewhere on M_A, belongs locally to A, although it does not belong to A.

We wish to show that $R(K)$ is local. We begin with the following lemma.

10.1 Lemma. Let $\hat{\mu}$ be the Cauchy transform of a finite measure μ with compact support, and let h be a continuously differentiable function with compact support. Then $h\hat{\mu}$ is the Cauchy transform of the measure $h\mu - (1/\pi)(\partial h/\partial \bar{z})\hat{\mu} dxdy$.

Proof. Set $\nu = (\partial h/\partial \bar{z})\hat{\mu} dxdy$. Then

$$\begin{aligned}\hat{\nu}(z_0) &= \iiint \frac{\partial h}{\partial \bar{z}} \frac{1}{(\zeta - z)(z - z_0)} d\mu(\zeta)dxdy \\ &= \iiint \frac{\partial h}{\partial \bar{z}} \left[\frac{1}{\zeta - z} + \frac{1}{z - z_0}\right] \frac{1}{\zeta - z_0} dxdyd\mu(\zeta) \\ &= \pi \int [h(\zeta) - h(z_0)] \frac{1}{\zeta - z_0} d\mu(\zeta) \\ &= \pi[\widehat{h\mu}(z_0) - h(z_0)\hat{\mu}(z_0)].\end{aligned}$$

That does it.

10.2 Theorem (Bishop Splitting Lemma). Let K be a compact subset of the plane. Let μ be a measure on K which is orthogonal to $R(K)$. Let $\{U_j\}_{j=1}^n$ be a finite open cover of K. Then there are measures μ_j such that $\mu = \sum_{j=1}^n \mu_j$, μ_j is orthogonal to $R(K)$, and the closed support of μ_j is contained in U_j.

Proof. Let $\{h_j\}_{j=1}^n$ be infinitely differentiable functions on the complex plane such that h_j has compact support which does not meet the closure of $K \backslash U_j$, and $\sum_{j=1}^n h_j = 1$ on K. By 10.1, there is a measure μ_j with compact support, such that $\hat{\mu}_j = h_j \hat{\mu}$. Since $\hat{\mu}$ vanishes off K, $\hat{\mu}_j$ vanishes off a compact subset of U_j. By 8.2, the closed support of μ_j is contained in U_j. By 8.1, μ_j is orthogonal to $R(\bar{U}_j)$, and, in particular, to $R(K)$.

10.3 Corollary (Localization Theorem). Let K be a compact subset of the plane. Suppose $f \in C(K)$ is such that every point $z \in K$ has a neighborhood $U(z)$ such that $f \in R(K \cap \overline{U(z)})$. Then $f \in R(K)$.

Proof. Let Δ_j, $1 \leq j \leq n$, be open discs which cover K, such that $f \in R(K \cap \bar{\Delta}_j)$, $1 \leq j \leq n$. By 10.2, every $\mu \perp R(K)$ can be written $\mu = \sum_{j=1}^n \mu_j$, where μ_j is supported on $K \cap \Delta_j$ and $\mu_j \perp R(K \cap \bar{\Delta}_j)$. In particular, $\int f d\mu = \sum \int f d\mu_j = 0$. Consequently $f \in R(K)$.

An an application, we can prove that $R(K) = A(K)$ for any compact set K whose complement has a finite number of components. In fact, the following more general theorem is true.

10.4 Theorem. Let K be a compact subset of the plane. Suppose that the diameters of the components of K^c are bounded away from zero. Then $R(K) = A(K)$.

Proof. Let d be the infimum of the diameters of the components of K^c. Let $\{\Delta_j\}_{j=1}^n$ be open discs of diameter $d/2$ which cover K. Then $K \cap \bar{\Delta}_j$ has a connected complement. By Mergelyan's theorem 9.1, together with

Runge's theorem, we have $R(K \cap \bar{\Delta}_j) = P(K \cap \bar{\Delta}_j) = A(K \cap \bar{\Delta}_j)$. By 10.3, $R(K) = A(K)$.

Using 1.8, we can improve the localization theorem to handle isolated bad points.

10.5 Theorem. Let K be a compact subset of the plane, and let $z_0 \in K$. Suppose that for every $z \neq z_0$ belonging to K, there is a closed disc $\Delta(z;\delta_z)$ such that $R(K \cap \Delta(z;\delta_z)) = A(K \cap \Delta(z;\delta_z))$. Then $R(K) = A(K)$.

Proof. By the localization theorem 10.3, every function in $A(K)$ which extends to be analytic in a neighborhood of z_0 belongs to $R(K)$. According to 1.8, such functions are dense in $A(K)$. Consequently $R(K) = A(K)$.

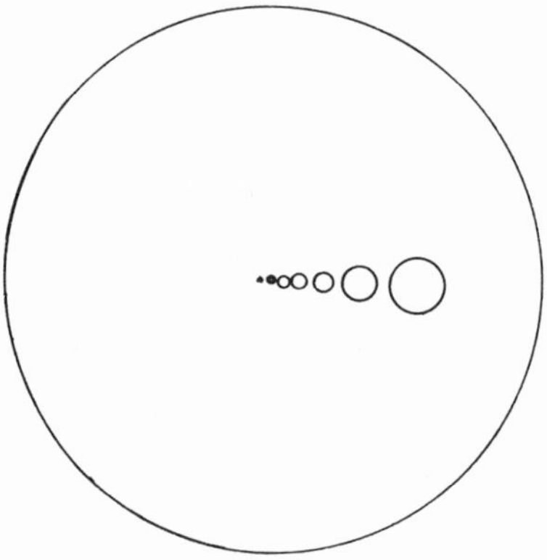

Figure 2.

As an example, let K be a compact set obtained from the closed unit disc Δ by deleting a sequence $\{\Delta_n\}_{n=1}^{\infty}$ of open discs whose radii tend to zero, and whose centers accumulate on a set E which is at most countable (cf. fig. 2, for a special configuration). Let F be the set of points in K which have no neighborhood U satisfying $R(K \cap \bar{U}) = A(K \cap \bar{U})$. Evidently F is closed, and $F \subseteq E$. In view of 10.5, F has no isolated points. By the Baire category theorem, F must be empty. By 10.5, $R(K) = A(K)$.

11. Peak Points

In this section, we will assume that X is a compact metric space, and that A is a uniform algebra on X.

A point $x \in X$ is a **peak point** if there is a function $f \in A$ such that $f(x) = 1$ while $|f(y)| < 1$ for $y \in X$, $y \neq x$. The function f which satisfies this condition is said to **peak at** x.

11.1 Theorem. Let A be a uniform algebra on the compact metric space X, and let $x \in X$. Suppose there are constants $0 < c < 1$ and $M \geq 1$ with the following property: For every neighborhood U of x, there exists $f \in A$ such that $f(x) = 1$, $\|f\|_X \leq M$, and $|f(y)| \leq c$ for $y \in X \backslash U$. Then x is a peak point of A.

Proof. Choose $0 < s < 1$ sufficiently close to 1 so that

$$M - 1 + s(c - M) < 0.$$

Choose a sequence $\{\epsilon_n\}_{n=0}^{\infty}$ decreasing to 0 so that

$$\epsilon_{n-1}(1 - s^n) + s^n(M - 1 + s(c - M)) < 0, \qquad n \geq 1.$$

Let $\{F_n\}_{n=0}^{\infty}$ be a sequence of closed subsets of X such that

$$\bigcup_{n=0}^{\infty} F_n = X \backslash \{x\}.$$

We choose a sequence of functions $\{h_n\}_{n=0}^{\infty}$ by induction, as follows. Take $h_0 = 1$. Supposing $h_0, \ldots, h_n$ have been chosen, let

$$W_n = \{y: \max_{1 \leq j \leq n} |h_j(y)| \geq 1 + \epsilon_n\}.$$

Take $h_{n+1} \in A$ such that $h_{n+1}(x) = 1$, $\|h_{n+1}\|_X \leq M$, and $\|h_{n+1}\|_{W_n \cup F_n} \leq c$. Define

$$h = (1 - s) \sum_{j=0}^{\infty} s^j h_j.$$

Then $h \in A$, while $h(x) = 1$.

Suppose $y \neq x$ does not belong to $\bigcup_{n=0}^{\infty} W_n$. Then $|h_j(y)| \leq 1$ for $0 \leq j < \infty$. Also, $y \in F_j$ for some j, so $|h_j(y)| < 1$ for at least one index j. Consequently $|h(y)| < 1$.

Now suppose $y \in \bigcup_{n=0}^{\infty} W_n$. Since the W_n are an increasing sequence of closed sets, there is an index m such that $y \in W_m$ while $y \notin W_{m-1}$. Then we have the estimates

$$|h_j(y)| \leq 1 + \epsilon_{m-1}, \qquad 1 \leq j \leq m - 1$$

$$|h_m(y)| \leq M$$

$$|h_j(y)| \leq c, \qquad j > m.$$

Consequently,

$$|h(y)| \leq (1 - s)\left\{(1 + \epsilon_{m-1}) \sum_{j=0}^{m-1} s^j + Ms^m + c \sum_{j=m+1}^{\infty} s^j\right\}$$
$$= 1 + \epsilon_{m-1}(1 - s^m) + s^m[M - 1 + s(c - M)] < 1.$$

Hence $|h(y)| < 1$ for all $y \in X$ such that $y \neq x$, and h peaks at x.

11.2 Corollary. If A is a uniform algebra on the compact metric space X, then the set of peak points of A is a G_δ-set.

Proof. Let d be a metric for X. Let U_n be the set of all $x \in X$ for which there exists a function $f \in A$ satisfying $\|f\| = 1$, $|f(x)| > \frac{1}{2}$, and $|f(y)| < \frac{1}{4}$ when $d(x,y) \geq 1/n$. Each U_n is open.

If x is a peak point, an appropriate power of the peaking function satisfies the above requirements. So $x \in U_n$ for all n.

Conversely, if $x \in U_n$ for all n, then appropriate multiples of the functions

f will satisfy the hypothesis of 11.1. Consequently $\bigcap_{n=1}^{\infty} U_n$ coincides with the set of peak points of A. That proves the corollary.

As an example, consider the Swiss cheese $K = \Delta \backslash \left(\bigcup_{j=1}^{\infty} \Delta_j \right)$ introduced in section 1. It is easy to see that every point belonging to $b\Delta$, or to the $b\Delta_j$, is a peak point for $R(K)$. However, $(b\Delta) \cup \left(\bigcup_{j=1}^{\infty} b\Delta_j \right)$ is not a G_δ-set, so there must be other peak points for $R(K)$ which lie deeper inside K.

11.3 Theorem. Let A be a uniform algebra on a compact metric space X. Then $x \in X$ is a peak point for A if and only if the point mass at x is the only representing measure for x. This occurs if and only if $\mu(\{x\}) = 1$ for every complex representing measure for x.

Proof. Suppose $f \in A$ peaks at x. Then f^n tends pointwise boundedly to the characteristic function of the singleton $\{x\}$ as $n \to \infty$. If μ is a complex representing measure for x, then

$$1 = f(x)^n = \int f^n d\mu \to \mu(\{x\}).$$

In turn, if $\mu(\{x\}) = 1$ for every complex representing measure for x, then the point mass at x is the only representing measure for x.

Now suppose that the point mass at x is the only representing measure for x. Let U be a neighborhood of x. Choose $v \in C_R(X)$ so that $v \leq 0$, $v(x) = 0$, and $v(y) < -2$ for $y \in X \backslash U$. By 2.1,

$$\sup \{u(x): u \in \operatorname{Re}(A), u \leq v\} = 0.$$

In particular, there is $g \in A$ such that $\operatorname{Re}(g) \leq 0$, $\operatorname{Re} g(x) > -1$, and $\operatorname{Re} g(y) < -2$ for $y \in X \backslash U$. Then $f = e^{g-g(x)} \in A$ satisfies $f(x) = 1$, $\|f\|_X \leq |e^{-g(x)}| \leq e$, and $|f(y)| \leq 1/e$ for $y \in X \backslash U$. Consequently the conditions of 11.1 are satisfied, with $c = 1/e$ and $M = e$. This shows that x is a peak point. That concludes the proof.

The importance of peak points for rational approximation is reflected by the following theorem. It shows that $R(K) = C(K)$ if and only if every point of K is a peak point of $R(K)$.

11.4 Theorem (Bishop's Peak Point Criterion for Rational Approximation). Let K be a compact subset of the plane, and let P be the set of peak points of $R(K)$. If $K \backslash P$ has zero planar measure, then $R(K) = C(K)$.

Proof. Let μ be a measure on K orthogonal to $R(K)$. Suppose z_0 is such that $\int |z - z_0|^{-1} d|\mu|(z) < \infty$, and $\hat{\mu}(z_0) \neq 0$. Then $z_0 \in K$. By the calculation performed in the proof of 8.5, $\mu/[\hat{\mu}(z_0)(z - z_0)]$ is a complex representing measure for z_0. Since $\mu(\{z_0\}) = 0$, this representing measure has no mass at z_0. By 11.3, z_0 is not a peak point of $R(K)$, that is, $z_0 \in K \backslash P$. Since $K \backslash P$ has zero area, $\hat{\mu}$ vanishes almost everywhere. By 8.3, $\mu = 0$. That proves the theorem.

The next two results, 11.5 and 11.6, depend on theorems concerning compact convex sets in locally convex linear topological vector spaces. The results will not be used later.

Suppose K is a compact convex subset of a locally convex topological vector space V. A point $x \in K$ is the **barycenter** of the probability measure μ on K if

$$L(x) = \int_K L(y)d\mu(y)$$

for all continuous linear functionals $L \in V^*$.

Every probability measure μ on K has a unique barycenter on K. It is easy to show that a point $x \in K$ is an extreme point of K if and only if the only probability measure on K with barycenter x is the point mass at x.

The content of the Krein-Milman theorem is that every $x \in K$ is the barycenter of a probability measure supported on the closure of the set of extreme points of K. This result can be sharpened, when K is metrizable.

Choquet's theorem. Let K be a compact convex metrizable subset of a locally convex topological vector space V. Then the set of extreme points E of K is a G_δ-set. Every $x \in K$ is the barycenter of a probability measure μ on K satisfying $\mu(E) = 1$.

Now we return to our uniform algebra A on the compact metric space X. Let Q be the set of linear functionals $L \in A^*$ such that $L(1) = 1 = ||L||$. The set Q is a convex subset of A^* which is compact and metrizable in the weak-star topology.

To each $x \in X$, there corresponds the evaluation functional $L_x \in Q$, defined by $L_x(f) = f(x)$. The correspondence $x \to L_x$ is a homeomorphism, which allows us to identify X with the weak-star closed subset of Q consisting of the evaluation functionals.

If $L \in Q$, there is a measure μ on X such that $||\mu|| = 1$, and $L(f) = \int f d\mu$ for all $f \in A$. In particular, $\int d\mu = 1 = \int d|\mu|$. So μ is a positive measure of unit mass, that is, μ is a probability measure.

Since every $L \in Q$ is the barycenter of a positive measure on X, the extreme points of Q are contained in X. The points of X which are extreme points of Q are precisely those which are the barycenter of a unique probability measure on X. That is, the extreme points of Q are the points of X which have a unique representing measure on X. By 11.3, these are the peak points of A.

11.5 Theorem. Let A be a uniform algebra on the compact metric space X. The extreme points of the convex set of functionals $L \in A^*$ satisfying $L(1) = 1 = ||L||$ are the functionals of the form $L_x(f) = f(x)$, $f \in A$, where x is a peak point of A.

From Choquet's theorem, we obtain the following.

11.6 Theorem. Let A be a uniform algebra on the compact metric space X, and let P be the set of peak points of A. Every $\phi \in M_A$ has a representing measure μ which is supported on P, that is, which satisfies $\mu(P) = 1$.

Recall that a subset E of X is a boundary for A if every function in A assumes its maximum modulus on E. From 11.6 it follows easily that the set P of peak points of A is a boundary for A. In fact, every $f \in A$ assumes its maximum modulus at some point $x_0 \in X$. Assume that $f(x_0) > 0$, and that μ is a representing measure on P for x_0. Then $\int (\|f\| - f)d\mu = \|f\| - f(x_0) = 0$. So f must coincide with $\|f\|$ on the closed support of μ. In particular, f assumes its maximum modulus on P.

Every boundary for A contains each peak point of A. It follows, assuming X is metrizable, that the set of peak points of A is a minimal boundary for A.

12. Peak Sets

Now we return to an arbitrary compact Hausdorff space X, and a uniform algebra A on X.

A closed subset E of X is a **peak set** if there is a function $f \in A$ such that $f(x) = 1$ for $x \in E$, and $|f(y)| < 1$ for $y \in X \backslash E$. The function f satisfying this condition is said to **peak on** E.

A countable intersection of peak sets is a peak set. In fact, if f_n peaks on E_n, then $\sum_{n=1}^{\infty} f_n/2^n$ peaks on $\bigcap_{n=1}^{\infty} E_n$. In particular, a finite intersection of peak sets is a peak set.

A closed subset E of X is a ***p*-set**, or **generalized peak set**, if it is the intersection of peak sets.

12.1 Lemma. A p-set is a peak set if and only if it is a G_δ-set.

Proof. If f peaks on E, then E is the intersection of the open sets $U_n = \{x: |f(x)| > 1 - 1/n\}$. So E is a G_δ-set.

The converse follows from the following more general assertion.

12.2 Lemma. Let E be a p-set, and let W be a G_δ-set containing E. Then there is a peak set F such that $E \subseteq F \subseteq W$.

Proof. Suppose $W = \bigcap_{n=1}^{\infty} W_n$, where the W_n are open. By compactness, we can find, for each fixed n, a finite number of peak sets $F_1, \ldots, F_m$ such that $E \subseteq F_1 \cap \ldots \cap F_m \subseteq W_n$. Consequently $E_n = F_1 \cap \ldots \cap F_m$ is a peak set, and $F = \bigcap_{n=1}^{\infty} E_n$ is the desired peak set.

From 12.1 it follows that the notions of p-set and peak set coincide when the underlying space X is metrizable.

We wish now to characterize the p-sets of A as the closed subsets E of

X such that $\mu_E \in A^\perp$ for all measures $\mu \in A^\perp$. Here μ_E is the restriction of μ to E. First we prove some lemmas, which depend only on linear structure.

Recall that if B_0 is a closed subspace of a Banach space B, then B/B_0 becomes a Banach space, when endowed with the quotient norm

$$\|x + B_0\| = \inf_{y \in B_0} \|x + y\|, \qquad x \in B.$$

A simple application of the Hahn-Banach theorem shows that the conjugate space B_0^* is isometrically isomorphic to $B^*/B_0^\perp$.

12.3 Lemma. Let B be a closed subspace of $C(X)$. Suppose E is a closed subset of X such that $\mu_E \in B^\perp$ for all measures $\mu \in B^\perp$. Then $B|_E$ is closed in $C(E)$. In fact, $B|_E$ is isometric to B/I_E, where I_E is the subspace of functions in B which vanish on E.

Proof. Here I_E is the kernel of the restriction operator $B \to B|_E$. So the restriction operator factors through the quotient B/I_E, yielding the linear operators

$$B \xrightarrow{S} B/I_E \xrightarrow{T} B|_E.$$

Both S and T are norm-decreasing, and S is surjective. We must show that T is an isometry.

The adjoint operators are given by

$$(B|_E)^* \xrightarrow{T^*} (B/I_E)^* \xrightarrow{S^*} B^*.$$

To show that T is an isometry, it suffices to show that T^* is an isometry. Since S^* is norm-decreasing, it suffices to show that $S^* \circ T^*$ is an isometry.

Representing B^* as $C(X)^*/B^\perp = M(X)/B^\perp$ and $(B|_E)^*$ as $M(E)/(B|_E)^\perp$, we can express $S^* \circ T^*$ explicitly in the form

$$(S^* \circ T^*)(\mu + (B|_E)^\perp) = \mu + B^\perp, \qquad \mu \in M(E).$$

Let $\mu \in M(E)$. If $\nu \in B^\perp$, then $\nu_E \in (B|_E)^\perp$, so $\|\mu + (B|_E)^\perp\| \leq \|\mu + \nu_E\| = \|\mu + \nu\| - \|\nu_{X\setminus E}\| \leq \|\mu + \nu\|$. Consequently $\|\mu + (B|_E)^\perp\| \leq \|\mu + B^\perp\|$. The reverse inequality is trivial, so $S^* \circ T^*$ is an isometry. This proves the lemma.

12.4 Lemma. Let B and E be as in 12.3. Let $f \in B|_E$, and let p be any positive continuous function on X such that $|f(y)| \leq p(y)$ for $y \in E$. Given $\epsilon > 0$, there is a function $g \in B$ such that $g|_E = f$, and $|g(x)| \leq p(x) + \epsilon$ for $x \in X$.

Proof. The fact that $B|_E$ is isometric to B/I_E means that given $f \in B|_E$ and $\epsilon > 0$, there exists $g \in B$ such that $g|_E = f$, and $\|g\|_X \leq \|f\|_E + \epsilon$. In other words, 12.4 holds in the special case $p = 1$. We will reduce the general case to this special case.

Let B/p be the subspace of functions of the form h/p, $h \in B$. A measure

ν is orthogonal to B/p if and only if ν/p is orthogonal to B. Consequently $\nu_E \in (B/p)^\perp$ whenever $\nu \in (B/p)^\perp$. Applying 12.3 to the subspace B/p, we find a function $g \in B$ such that $g|_E = f$ and $\|g/p\|_X \leq \|f/p\|_E + \epsilon/\|p\|_X$. Then $|g(x)| \leq p(x) + \epsilon$ for all $x \in X$.

12.5 Theorem. Let B be a closed subspace of $C(X)$. Let E be a closed subset of X such that $\mu_E \in B^\perp$ for all measures $\mu \in B^\perp$. Let $f \in B|_E$, and let p be a positive continuous function on X such that $|f(y)| \leq p(y)$ for $y \in E$. Then there is $g \in B$ such that $g|_E = f$ and $|g(x)| \leq p(x)$ for all $x \in X$.

Proof. Replacing the subspace B by B/p, as in the proof of 12.4, we see that it will suffice to prove the theorem in the case $p = 1$.

So we assume that $p = 1$, and, in particular, that $\|f\|_E \leq 1$. By 12.3, there is $g_1 \in B$ such that $g_1|_E = f$ and $\|g_1\|_X \leq \frac{5}{4}$.

Suppose $g_1, \ldots, g_{n-1}$ are chosen so that $g_j|_E = f$, $1 \leq j \leq n-1$. Let U_n be the set of $x \in X$ satisfying $|g_j(x)| < 1 + 1/2^{n+1}$, $1 \leq j \leq n-1$. Then U_n is an open set containing E. By 12.4, there is $g_n \in B$ such that $g_n|_E = f$, $\|g_n\|_X \leq 1 + 1/2^{n+1}$, and $|g_n(x)| \leq \frac{1}{2}$ when $x \notin U_n$.

Let $g = \sum_{n=1}^{\infty} g_n/2^n$. Then $g \in B$, and $g|_E = f$.

Suppose $x \in U_n$ while $x \notin U_{n+1}$. Then $|g_j(x)| \leq 1 + 1/2^{n+1}$ for $1 \leq j \leq n$, while $|g_j(x)| \leq \frac{1}{2}$ for $j > n$. Consequently

$$|g(x)| \leq \left(1 + \frac{1}{2^{n+1}}\right)\sum_{j=1}^{n}\frac{1}{2^j} + \frac{1}{2}\sum_{j=n+1}^{\infty}\frac{1}{2^j}$$
$$= 1 - \frac{1}{2^{n+1}}.$$

If $x \in U_n$ for all n, then $|g_n(x)| \leq 1$ for all n, and $|g(x)| \leq 1$. Consequently $|g(x)| \leq 1$ for all $x \in X$. This completes the proof.

If the hypothesis of 12.5 is strengthened to $\mu_E = 0$ for all measures $\mu \in B^\perp$, then $B|_E$ is dense in $C(E)$. Since $B|_E$ is closed, $B|_E = C(E)$. Hence the conclusion can be strengthened to assert that every function $f \in C(E)$ has an extension $g \in B$ which is appropriately dominated.

Applying this remark to the disc algebra, we obtain the following theorem as a consequence of the F. and M. Riesz theorem 7.10.

12.6 Theorem (Rudin-Carleson Theorem). Let Δ be the closed unit disc in the complex plane. Let E be a closed subset of $b\Delta$ of linear measure zero. For each $f \in C(E)$, there is a continuous function g on Δ which is analytic on the interior of Δ, such that $g|_E = f$ and $\|g\|_\Delta = \|f\|_E$.

12.7 Theorem (Glicksberg Peak Set Theorem). Let A be a uniform algebra on X, and let E be a closed subset of X. Then E is a p-set if and only if $\mu_E \in A^\perp$ for all measures $\mu \in A^\perp$.

Proof. The difficult half of the characterization is the backward implication. This follows from 12.5, by extending the function $1 \in A|_E$ to be dominated by appropriate positive functions.

For the forward implication, suppose first that F is a peak set, and that $f \in A$ peaks on F. If $\mu \in A^\perp$, then $f^n\mu \in A^\perp$ for all n. As $n \to \infty$, $f^n\mu$ tends weak-star to μ_F. So $\mu_F \in A^\perp$.

Suppose now that E is a p-set. Let $\mu \in A^\perp$. Let U be a G_δ-set such that $E \subseteq U$ while $|\mu|(U \backslash E) = 0$. Let F be a peak set such that $E \subseteq F \subseteq U$. By the preceding remark, $\mu_E = \mu_F \in A^\perp$. This proves the theorem.

Peak sets E satisfying $A|_E = C(E)$ are called **peak interpolation sets.** We have shown that a G_δ-set E is a peak interpolation set if and only if $|\mu|(E) = 0$ for all measures $\mu \in A^\perp$.

12.8 Corollary. If E and F are p-sets, then $E \cup F$ is a p-set. If E_j, $1 \leq j < \infty$, are p-sets, and $E = \bigcup_{j=1}^{\infty} E_j$ is closed, then E is a p-set.

Proof. For the first statement, note that $E \cap F$ is a p-set. So if $\mu \in A^\perp$, then $\mu_{E\cup F} = \mu_E + \mu_F - \mu_{E\cap F} \in A^\perp$. Hence $E \cup F$ is a p-set.

For the second statement, note that μ_E is the weak-star limit of $\mu_{E_1\cup\ldots\cup E_n}$ as $n \to \infty$. Hence $\mu \in A^\perp$ implies $\mu_E \in A^\perp$.

12.9 Corollary. If E is a p-set of A, and $F \subseteq E$ is a p-set of $A|_E$, then F is a p-set of A.

Proof. If $\mu \in A^\perp$, then $\mu_E \in A^\perp$. So $\mu_E \in (A|_E)^\perp$, and $\mu_F \in (A|_E)^\perp \subseteq A^\perp$.

A point $x \in X$ is a ***p*-point**, or **generalized peak point**, if the singleton $\{x\}$ is a p-set. Most of the theorems of section 11 are valid for uniform algebras on arbitrary compact spaces X, providing we replace "peak point" by "p-point."

For instance, suppose x satisfies the conditions of 11.1. The proof there shows that if U is a G_δ-set such that $x \in U$, then there is a peak set E satisfying $x \in E \subseteq U$. In particular, x is a p-point.

The proof of 11.2 fails. However, 11.3 and 11.5 carry over.

The analogue of theorem 11.6 is valid, providing the set P of p-points is a Borel set. The following more general theorem is valid, although we shall not prove it (for a proof, see Phelps [2]).

Theorem (Bishop-deLeeuw Theorem). Let A be a uniform algebra on the compact space X, and let P be the set of p-points of A. For each $x \in X$, there is a probability measure μ on the σ-algebra generated by the Borel sets and P, such that $\mu(P) = 1$, and

$$f(x) = \int f d\mu, \qquad f \in A.$$

The next theorem is a direct consequence of the Bishop-deLeeuw theorem. We give an alternative proof which hinges on 12.9.

12.10 Theorem. Let A be a uniform algebra on X. Every function $f \in A$ assumes its maximum modulus at some p-point $x \in X$.

Proof. There exists $y_0 \in X$ such that $|f(y_0)| = \|f\|$. Let E be the set of $y \in X$ such that $f(y) = f(y_0)$. We must show that E contains a p-point.

The set E is a peak set. In fact, $[1 + f/f(y_0)]/2$ peaks on E. By Zorn's lemma, there is a non-empty minimal p-set F contained in E. From 12.9, it follows that the algebra $A|_F$ has no non-trivial peak sets. Hence $A|_F$ has no non-constant functions, and F must consist of one point $\{x\}$. This proves the theorem.

The set of p-points of A is the **strong boundary** of A. If X is not metrizable, the strong boundary need not be a minimal boundary.

13. Antisymmetric Algebras

The uniform algebra A on X is **antisymmetric** if every real-valued function in A is constant. In this section, we consider the problem of expressing every uniform algebra as a continuous direct sum of antisymmetric algebras.

Let B be the real algebra consisting of the real-valued functions in a uniform algebra A. Let Y be the quotient space of X obtained by identifying two points x and y of X whenever $f(x) = f(y)$ for all $f \in B$. By the Stone-Weierstrass theorem, $B = C_R(Y)$.

Let $\{F_\alpha\}$ be the level sets of B, that is, the "points" of Y. Each F_α is evidently a p-set. Shilov showed that if $g \in C(X)$, and $g|_{F_\alpha} \in A|_{F_\alpha}$ for all F_α, then $g \in A$.

The algebras $A|_{F_\alpha}$ need not be antisymmetric. For instance, let $A = P(X)$, where X is the union of a line segment Y and a sequence of disjoint solid rectangles converging to Y. Every real-valued function in $P(X)$ is constant on each rectangle, and hence on Y. So the Shilov decomposition consists of the rectangles and Y. However, the restriction of $P(X)$ to Y is $C(Y)$.

By induction on the ordinals, it is possible to continue subdividing, until one arrives at antisymmetric algebras. The resulting theorem 13.1 is a generalization of the Stone-Weierstrass theorem.

An alternative proof based on the Krein-Milman theorem will be presented here. For this, we define a subset F of X to be a **set of antisymmetry** of A if every function in A which is real-valued on F must be constant on F.

13.1 Theorem (Bishop Antisymmetric Decomposition). Let A be a uniform algebra on X. Let $\{E_\alpha\}$ be the family of maximal sets of antisymmetry of A. The E_α are closed disjoint subsets of X whose union is X. Each E_α is a p-set. If $f \in C(X)$, and $f|_{E_\alpha} \in A|_{E_\alpha}$ for all E_α, then $f \in A$.

Proof. Every singleton $\{x\}$ is a set of antisymmetry. The union of any increasing family of sets of antisymmetry is a set of antisymmetry. By Zorn's lemma, every $x \in X$ is contained in a maximal set of antisymmetry.

If E and F are sets of antisymmetry, and $E \cap F$ is not empty, then $E \cup F$ is a set of antisymmetry. Consequently, the maximal sets of antisymmetry are disjoint.

The closure of any set of antisymmetry is a set of antisymmetry. So the maximal sets of antisymmetry are closed.

Let E be a maximal set of antisymmetry, and let F be the intersection of all peak sets containing E. We wish to show that $E = F$.

If $E \neq F$, there is a function $f \in A$ which is real-valued but not constant on F. Now f assumes some constant value c on E. The function $g = 1 - (f - c)^2/||f - c||^2$ satisfies $0 \leq g \leq 1$ on F. By 12.9, $g^{-1}(1) \cap F$ is a p-set, which is a proper subset of F containing E. This contradiction shows that $E = F$. Hence every maximal set of antisymmetry is a p-set.

Let μ be an extreme point of the unit ball of $A^\perp$, and let Y be the closed support of μ. We will show that Y is a set of antisymmetry.

Suppose $f \in A$ satisfies $0 < f(y) < 1$ for $y \in Y$. Let $a = ||f\mu||$ and $b = ||(1 - f)\mu||$. Then

$$a + b = \int |f|\, d|\mu| + \int |1 - f|\, d|\mu| = \int d|\mu| = 1.$$

Consequently,

$$\mu = a\left[\frac{f\mu}{a}\right] + b\left[\frac{(1 - f)\mu}{b}\right]$$

expresses μ as a convex combination of measures in the unit ball of $A^\perp$. Hence $f\mu/a = \mu$. So $f(y) = a$ for all $y \in Y$. It follows that Y is a set of antisymmetry. Hence every extreme point of the unit ball of $A^\perp$ is supported on some maximal set of antisymmetry. If $f \in C(X)$, and $f|_{E_\alpha} \in A|_{E_\alpha}$ for all maximal sets of antisymmetry E_α, then $\int f d\mu = 0$ for all extreme points μ of the unit ball of $A^\perp$. By the Krein-Milman theorem, $\int f d\nu = 0$ for all measures ν in the unit ball of $A^\perp$. Hence $f \in A$. That completes the proof.

13.2 Theorem. Let A be a uniform algebra on X, and let B be a closed subspace of $C(X)$ which is an A-module. If E is a p-set of A, then $\mu_E \in B^\perp$ for all measures $\mu \in B^\perp$. If $h \in C(X)$, and $h|_E \in B|_E$ for all maximal sets of antisymmetry E of A, then $h \in B$.

Proof. To prove the last statement, one shows that every extreme point of the unit ball of $B^\perp$ is supported on some maximal set of antisymmetry of A, just as in the proof of 13.1.

For the first statement, it suffices to prove that if F is a peak set of A, then $\mu_F \in B^\perp$ for all measures $\mu \in B^\perp$. For this, suppose $f \in A$ peaks on F. Then $f^n\mu \in B^\perp$, and $f^n\mu$ tends to μ_F in the weak-star topology. That does it.

Theorem 13.2 applies to ideals $I \subseteq A$. It shows, for instance, that every maximal ideal of A arises from a maximal ideal of $A|_E$, for some maximal set of antisymmetry E of A.

NOTES

The first Swiss cheese was introduced by (the Swiss mathematician) Alice Roth [1]. The Wermer arc algebras were introduced in Wermer [3], and further studied by Browder and Wermer [1, 2]. Arens' theorem is in Arens [3]. The existence of Jensen measures was proved by Bishop [8]. The discussion of the Walsh-Lebesgue theorem is based on Walsh [1] and Lebesgue [1]. Dirichlet algebras were introduced by Gleason [1], logmodular algebras by Hoffman [3]. Wermer's maximality theorem is in Wermer [1]. Glicksberg's lemma, together with the proof of Radó's theorem, is in Glicksberg [5].

The idea of introducing the minimax principle in this context is due to Glicksberg [7], who extended the Hoffman-Wermer theorem and the abstract F. and M. Riesz theorem to arbitrary uniform algebras. A related F. and M. Riesz theorem was later obtained by König and Seever [1]. The connection between the F. and M. Riesz theorems was noted by Rainwater [2], who proved 7.4. The original F. and M. Riesz theorem is in F. and M. Riesz [1]. A proof based on orthogonal projections in Hilbert space which made a large impact was given by Helson and Lowdenslager [1]. Later, Forelli [1] returned to the spirit of the original proof to give the primordial form of the Forelli lemma 7.3, valid for dirichlet algebras. Forelli's argument inspired Hoffman and Wermer to develop their modification technique appearing in 7.2, again in the context of dirichlet algebras (cf. Wermer [17]).

It was Bishop who introduced and systematically exploited the Cauchy transform for problems in rational approximation. In Bishop [1,4], the decomposition theorem for measures orthogonal to $P(K)$ is obtained, together with a precursor of the proof of Mergelyan's theorem presented in section 9. The "abstract" proof of Mergelyan's theorem was obtained by Glicksberg and Wermer [1]. An exposition of the proof, together with some simplifications, is in Carleson [3]. Wilken's theorem is in Wilken [3]. Corollary 9.6 is in Glicksberg [5], and the extension 9.4 comes from Garnett and Glicksberg [1].

The example of a non-local uniform algebra is due to Eva Kallin [1]. The Bishop splitting lemma is in Bishop [2]. This proof was shown me by A. Browder. Theorem 10.4 is due to Mergelyan [1].

The papers on peak points are Bishop [3] and Bishop and deLeeuw [1]. Theorem 11.1 is a variation of the Bishop-deLeeuw "$\frac{1}{4}$–$\frac{3}{4}$ criterion" from [3]. It is first stated explicitly, and used to obtain geometric criteria for peak points for $R(K)$, in Gonchar [1]. An equivalent version of 11.4 is in Mergelyan [1].

The Rudin-Carleson theorem is in Rudin [1] and Carleson [1]. It was extended to uniform algebras by Bishop [6]. The Glicksberg peak set theorem is in Glicksberg [1]. For the Shilov decomposition, see Rickart [1]. The antisymmetric decomposition is in Bishop [5]. The proof presented here is in Glicksberg [1], and is based on an idea of deBranges [1].

EXERCISES

1. Let U be an open subset of S^2, such that $A(S^2,U)$ contains non-constant functions.

(a) Every function in $A(S^2,U)$ is analytic off the Shilov boundary of $A(S^2,U)$.
(b) If U is connected, then every value assumed by $f \in A(S^2,U)$ on U is assumed by f on $S^2\backslash U$.
(c) The singular measure μ_s in the decomposition theorem 7.11 for measures orthogonal to $A(S^2,U)$ always vanishes.
(d) If $S^2\backslash U$ has Lebesgue density 1 at z_0, then z_0 is a peak point of $A(S^2,U)$. *Hint:* Apply 11.1 to functions of the form $\iint dxdy/(z - \zeta)$.
(e) The points of $S^2\backslash U$ which are not peak points of $A(S^2,U)$ have zero area.
(f) If $A(S^2,U)$ is dirichlet on $S^2\backslash U$, then $S^2\backslash U$ is connected.
(g) If $A(S^2,U)$ is dirichlet on $S^2\backslash U$, then $A(S^2,U)$ is a maximal subalgebra of $C(S^2\backslash U)$ if and only if U is connected.

2. If u is upper semi-continuous on X, then

$$\sup_{m \in M_\phi} \int u\,dm = \inf\{\phi(v): v \in \mathrm{Re}\,(A),\ v \geq u\}.$$

3. The algebra A is **normal** on X if whenever E and F are disjoint closed subsets of X, there is $f \in A$ such that $f = 0$ on E, and $f = 1$ on F. If A is normal on X, then $M_A = \partial_A = X$. Moreover, every $x \in X$ has a unique Jensen measure.

4. Suppose A and B are uniform algebras on X, such that every measure in $A^\perp$ is singular to every measure in $B^\perp$. Then $A \cap B$ is a uniform algebra on X, and $(A \cap B)^\perp = A^\perp + B^\perp$. If A and B are dirichlet, so is $A \cap B$. *Hint:* Use the Krein-Schmulian theorem to show that $A^\perp + B^\perp$ is weak-star closed.

5. For each complex number α of unit modulus, let E_α be the set of $\phi \in M_{H^\infty(\Delta)}$ such that $\phi(z) = \alpha$, and let $A_\alpha = H^\infty|_{E_\alpha}$.

(a) E_α is a peak set of H^∞.
(b) $M_{A_\alpha} = E_\alpha$, $\partial_{A_\alpha} = E_\alpha \cap \partial_{H^\infty}$.
(c) A_α is not antisymmetric.
(d) If $u \in C(\partial_{A_\alpha})$ satisfies $u \geq 0$, then there is $g \in A_\alpha$ such that $|g| = u$. *Hint:* Extend u so that $\log u \in L^1(d\theta)$, and take $f \in H^\infty$ such that $\log|f| = u$.
(e) A_α is not normal on ∂_{A_α}.
(f) Suppose Γ is a smooth curve in the complex plane forming part of the boundary of a domain D, such that D lies on one side of Γ. Let $z_0 \in \Gamma$, and let F be the set of $\phi \in M_{H^\infty(D)}$ such that $\phi(z) = z_0$. Then $H^\infty(D)|_F$ is isometrically isomorphic to A_α.
(g) The closed subalgebra of $L^\infty(d\theta)$ generated by H^∞ and $\bar{z}$ consists of all $f \in L^\infty$ such that $f|_{\partial A_\alpha} \in A_\alpha$ for all $\alpha \in b\Delta$.
(h) If $f \in C(b\Delta)$, then

$$\inf_{g \in P(\Delta)} \|f+g\|_{b\Delta} = \inf_{g \in H^\infty} \|f+g\|_\infty.$$

(i) $H^\infty + C(b\Delta)$ is closed in L^∞, and coincides with the algebra of part (g). *Hint:* Use (h).

6. Theorem 7.2 remains true if $H^1(m)$ is replaced by $H^p(m)$ for some fixed $p > 0$.

7. (a) If M_ϕ is norm-separable, then ϕ has a dominant representing measure.

(b) Let K be the compact subset of the plane obtained from the closed unit disc by deleting a sequence of open discs with disjoint closures, whose centers accumulate on a set whose intersection with int Δ is discrete. Then every $z \in K$ has a dominant representing measure for $R(K)$.

8. Show that $\partial\hat{\mu}/\partial\bar{z} = -\pi\mu$ in the sense of distributions.

9. Let g be a continuously differentiable function with compact support. Let T_g be the operator on $C(S^2)$ defined by

$$(T_g f)(\zeta) = \frac{1}{\pi}\iint \frac{f(z)-f(\zeta)}{z-\zeta}\frac{\partial g}{\partial \bar{z}}dxdy.$$

If $f_n \in C(S^2)$ converges uniformly on the support of g to f, then $T_g f_n \to T_g f$ uniformly on S^2. If μ is a measure on S^2, and T_g^* is the adjoint operator of T_g, then $T_g^*\mu$ is the measure whose Cauchy transform is $g\hat{\mu}$.

10. If $\{K_n\}_{n=1}^\infty$ are compact subsets of the complex plane such that $R(K_n) = C(K_n)$, and $K = \bigcup_{n=1}^\infty K_n$ is compact, then $R(K) = C(K)$.

11. Let A be the non-local algebra constructed in section 10.

(a) Verify that A has the properties claimed for it in section 10.

(b) A is not an integral domain.

(c) There are functions in A which vanish on an open subset of M_A without vanishing on ∂_A.

(d) A has non-local point derivations, that is, there is a point derivation L at some $\phi \in M_A$, and a function $f \in A$ which vanishes in a neighborhood of ϕ, such that $L(f) \neq 0$.

12. Let E be a closed A-convex subset of X, and let I_E be the ideal of functions in A which vanish on E.

(a) I_E has an approximate identity if and only if E is a p-set.

(b) If E is a p-set, and $f \in I_E$, then $f = f_1 f_2$, where $f_1, f_2 \in I_E$. (Cf. chapter I, exercise 4.)

(c) The only point derivation of A at a strong boundary point is the trivial one.

(d) If D is a derivation of A [i.e., a linear transformation satisfying $D(fg) = fD(g) + gD(f)$], then $D = 0$.

13. Let $X = \prod_{0 \leq a \leq 1} X_a$, where each X_a is the two-point space $\{0,1\}$. Then $C(X)$ has no peak points, and $C(X)$ has no "smallest" boundary. The strong boundary of $C(X)$ is X.

14. Let B be a closed subspace of $C(X)$, E a closed subset of X.

(a) $B|_E$ is closed in $C(E)$ is and only if there exists $c<1$ such that $\|\mu_E+(B|_E)^\perp\| \leq c\|\mu\|$ for all $\mu \in B^\perp$. This occurs if and only if there exists $d \geq 0$ such that $\|\mu_E+(B|_E)^\perp\| \leq d\|\mu_{X\setminus E}\|$ for all $\mu \in B^\perp$.

(b) The best possible constants c and d of (a) are related by $c = d/(1+d)$.

(c) $B|_E$ is isometric to B/I_E if and only if $\|\mu_E+(B|_E)^\perp\| \leq \|\mu\|/2$ for all $\mu \in B^\perp$.

(d) $B|_E = C(E)$ if and only if there exists $c<1$ such that $\|\mu_E\| \leq c\|\mu\|$ for all $\mu \in B^\perp$.

15. If A is logmodular on X, and E is a closed subset of X, then $A|_E$ is closed in $C(E)$ if and only if E is a p-set.

16. Suppose the unitary functions $\mathscr{U}$ in A separate the points of X.

(a) The functions in $\bar{\mathscr{U}}A$ are dense in $C(X)$. *Hint:* Apply the Stone-Weierstrass theorem.

(b) If E is a closed subset of X such that $A|_E = C(E)$, then E is a p-set.

17. Let $P(\Delta^n)$ be the uniform closure of the polynomials in $z_1, \ldots, z_n$ on the unit polycylinder Δ^n in C^n.

(a) $M_{P(\Delta^n)} = \Delta^n$ and $\partial_{P(\Delta^n)} = (b\Delta)^n$.

(b) $P(\Delta^n)$ is relatively maximal.

(c) If $f \in C(\Delta^n)$ is analytic off its zero set, then $f \in P(\Delta^n)$.

18. The following are equivalent, for a closed subset E of $(b\Delta)^n$.

(i) $P(\Delta^n)|_E = C(E)$.

(ii) E is a peak interpolation set.

(iii) There is $f \in P(\Delta^n)$ which vanishes precisely on E. *Hint:* Under the assumption (iii), f does not vanish on the interior of Δ^n. Compose $\log f$ with an appropriate fractional linear transformation.

19. If $\{F_\alpha\}$ are the maximal sets of antisymmetry of A, then $M_A = \bigcup_\alpha \hat{F}_\alpha$. The strong boundary of A is the union of the strong boundaries of the A_α. The closure of $\bigcup_\alpha \partial_{A|_{F_\alpha}}$ is ∂_A, although $\partial_A \neq \bigcup_\alpha \partial_{A|_{F_\alpha}}$ in general.

CHAPTER III

METHODS OF SEVERAL COMPLEX VARIABLES

In this chapter, we return to a commutative Banach algebra A with an identity. The algebra A can be expressed as a direct limit of its finitely generated subalgebras. And the study of the finitely generated subalgebras of A leads one to the study of analytic functions of several complex variables.

We assume the reader is familiar with the most elementary theory of analytic functions of several complex variables. The only non-elementary result used is the solvability of the Cousin problem for open polynomial polyhedra, and that result will not be needed until section 7.

This chapter is largely independent of the preceding chapter. Only section 9 relies decisively on results from chapter II.

1. Polynomial Convexity

The **polynomial convex hull** $\hat{K}$ of a bounded subset K of C^n consists of all $z \in C^n$ such that

$$|p(z)| \leq \sup_{w \in K} |p(w)|$$

for all polynomials p. A bounded subset K of C^n is **polynomially convex** if $\hat{K} = K$.

Polynomially convex subsets of C^n are closed. The polynomial convex hull of a bounded subset of C^n is polynomially convex.

A (**compact**) **polynomial polyhedron** is a subset of C^n of the form

$$\{z: |z_1| \leq M, \ldots, |z_n| \leq M, |p_1(z)| \leq M, \ldots, |p_m(z)| \leq M\},$$

where $M > 0$ is fixed, and $p_1, \ldots, p_m$ are fixed polynomials. The polynomial polyhedra are the simplest examples of polynomially convex subsets of C^n.

1.1 Lemma. If K is a polynomially convex subset of C^n, there are polynomial polyhedra $\{K_j\}_{j=1}^{\infty}$ such that $K_{j+1} \subseteq \operatorname{int}(K_j)$ and $\bigcap_{j=1}^{\infty} K_j = K$.

Proof. Let J be a polydisc $\{|z_j| \leq M, 1 \leq j \leq n\}$ such that $K \subseteq \text{int}(J)$. Let $\{U_j\}_{j=1}^{\infty}$ be open subsets of C^n such that $K \subseteq U_{j+1} \subseteq \bar{U}_{j+1} \subseteq U_j \subseteq J$, while $\bigcap_{j=1}^{\infty} U_j = K$. Suppose polynomial polyhedra $K_1, \ldots, K_m$ are chosen so that $K \subseteq \text{int}(K_j)$ and $K_j \subseteq U_{j-1} \cap \text{int}(K_{j-1})$, $1 \leq j \leq m$. For every $z \in J \backslash (U_m \cap \text{int}(K_m))$, there is a polynomial p such that $|p(z)| > M$ while $\|p\|_K < M$. Then $|p| > M$ in a neighborhood of z. By compactness of $J \backslash (U_m \cap \text{int}(K_m))$, there are a finite number of polynomials $p_1, \ldots, p_k$ such that $\|p_j\|_K < M$, $1 \leq j \leq k$, while $\max_{1 \leq j \leq k} |p_j(z)| > M$ for $z \in J \backslash (U_m \cap \text{int}(K_m))$. If K_{m+1} is the polynomial polyhedron determined by $p_1, \ldots, p_k$ and M, then $K \subseteq \text{int}(K_{m+1}) \subseteq U_m \cap \text{int}(K_m)$. Now $K \subseteq \bigcap_{j=1}^{\infty} K_j \subseteq \bigcap_{j=1}^{\infty} U_j = K$. So the K_j have the desired properties.

For K a compact subset of C^n, $\boldsymbol{P(K)}$ will denote the uniform closure on K of the algebra of polynomials in $z_1, \ldots, z_n$. Then $P(K)$ is a finitely generated uniform algebra on K.

Suppose $\{p_n\}_{n=1}^{\infty}$ is a sequence of polynomials converging uniformly on K to $f \in P(K)$. From the definition of polynomial convexity, we obtain

$$\|p_n - p_m\|_{\hat{K}} \leq \|p_n - p_m\|_K$$

for all n and m. Consequently $\{p_n\}_{n=1}^{\infty}$ converges uniformly on $\hat{K}$ to an extension $\hat{f}$ of f. Hence we can identify $\hat{K}$ with a subset of the maximal ideal space of $P(K)$, namely, the evaluation homomorphisms of $\hat{f}$ on $\hat{K}$.

1.2 Theorem. Let K be a compact subset of C^n. The maximal ideal space of $P(K)$ is the polynomial convex hull $\hat{K}$ of K.

Proof. Since every complex-valued homomorphism ϕ of the algebra of polynomials in $z_1, \ldots, z_n$ is evaluation at some point of C^n [namely, at $(\phi(z_1), \ldots, \phi(z_n))$], every homomorphism of $P(K)$ is evaluation at some point of C^n. Now "evaluation at z" extends continuously to $P(K)$ if and only if $|p(z)| \leq \|p\|_K$ for all polynomials p, that is, if and only if $z \in \hat{K}$. This proves the theorem.

Deciding which subsets of C^n are polynomially convex is not an easy task. Any compact convex subset of C^n is polynomially convex. The polydisc with center w and multiradius $r = (r_1, \ldots, r_n)$, defined by

$$\Delta(w; r) = \{z\colon |z_j - w_j| \leq r_j, 1 \leq j \leq n\},$$

is polynomially convex. The ball with center w and radius r, defined by

$$\left\{z\colon \sum_{j=1}^{n} |z_j - w_j|^2 \leq r^2\right\},$$

is also polynomially convex.

The union of two disjoint compact convex subsets of C^n is polynomially convex. The disjoint union of three balls in C^n is polynomially convex. However, the disjoint union of three polydiscs in C^n may not be polynomially convex. It is not known whether the disjoint union of four balls in C^n is polynomially convex.

1.3 Lemma. A compact subset K of C^1 is polynomially convex if and only if the complement of K is connected.

Proof. It was shown in II.1.4 that the maximal ideal space $\hat{K}$ of $P(K)$ is the union of K and the bounded components of the complement of K. Hence $\hat{K} = K$ if and only if K^c is connected. This proves the lemma.

Suppose $f_1, \ldots, f_n$ belong to the Banach algebra A. There is a natural projection π of M_A into C^n, defined by

$$\pi(x) = (\hat{f}_1(x), \ldots, \hat{f}_n(x)), \qquad x \in M_A.$$

The range of π is called the **joint spectrum** of $f_1, \ldots, f_n$, and is denoted by $\boldsymbol{\sigma}(f_1, \ldots, f_n)$. If $n = 1$, this definition reduces to the usual definition of the spectrum, by I.2.7.

Since π is continuous, $\sigma(f_1, \ldots, f_n)$ is a compact subset of C^n. A point $z = (z_1, \ldots, z_n)$ belongs to $\sigma(f_1, \ldots, f_n)$ if and only if $z_1 - f_1, \ldots, z_n - f_n$ belong to a common maximal ideal in A. This occurs if and only if there fail to exist $g_1, \ldots, g_n \in A$ such that $\sum_{j=1}^{n} (z_j - f_j)g_j = 1$.

The polynomial convex hull of $\sigma(f_1, \ldots, f_n)$ will be denoted by $\hat{\boldsymbol{\sigma}}(f_1, \ldots, f_n)$. In general, $\sigma(f_1, \ldots, f_n)$ is not polynomially convex: take A to be $R(\{1 \le |z| \le 2\})$, n to be 1, and f_1 to be the coordinate function z. However, the following theorem obtains.

1.4 Theorem. Let A be a finitely generated Banach algebra, generated by $f_1, \ldots, f_n$. Then the joint spectrum $\sigma(f_1, \ldots, f_n)$ of $f_1, \ldots, f_n$ is polynomially convex, and the natural projection π of M_A onto $\sigma(f_1, \ldots, f_n)$ is a homeomorphism.

Proof. Since every $x \in M_A$ is determined by its values $\hat{f}_1(x), \ldots, \hat{f}_n(x)$, the map π is one-to-one. Consequently the continuous map π is a homeomorphism.

Suppose $w \in \hat{\sigma}(f_1, \ldots, f_n)$. Then

$$\begin{aligned} |p(w)| &\le \sup\{|p(z)| : z \in \sigma(f_1, \ldots, f_n)\} \\ &= \sup\{|p(\hat{f}_1(x), \ldots, \hat{f}_n(x))| : x \in M_A\} \\ &\le \|p(f_1, \ldots, f_n)\| \end{aligned}$$

for all polynomials p. Hence

$$\phi(p(f_1, \ldots, f_n)) = p(w)$$

is a well-defined complex-valued homomorphism of the algebra of polynomials in $f_1, \ldots, f_n$, which extends continuously to A. Since $w = (\hat{f}_1(\phi), \ldots, \hat{f}_n(\phi))$, we obtain $w \in \sigma(f_1, \ldots, f_n)$. Hence $\hat{\sigma}(f_1, \ldots, f_n) = \sigma(f_1, \ldots, f_n)$.

1.5 Corollary. A uniform algebra A is finitely generated if and only if A is isometrically isomorphic to $P(K)$, for some compact polynomially convex subset K of C^n.

An example of a uniform algebra which is finitely generated is the algebra $R(K)$ of uniform limits of rational functions on a compact subset K of the complex plane. If $\{z_j\}_{j=1}^{\infty}$ is a sequence of points in K^c which meets every bounded component of K^c, then the functions z and $\sum_{j=1}^{\infty} c_j/(z - z_j)$ generate $R(K)$, providing the constants c_j are chosen appropriately.

2. Rational Convexity

The **rational convex hull** of a bounded subset K of C^n consists of all $z \in C^n$ such that $|f(z)| \leq \|f\|_K$ for all rational functions f which are analytic on K. A bounded subset K of C^n is **rationally convex** if it coincides with its rational convex hull.

The polynomial convex hull of a set K always contains the rational convex hull of K. So polynomially convex sets are rationally convex. A subset of C^2 which is rationally convex but not polynomially convex is the torus $\{(z, w): |z| = 1, |w| = 1\}$.

For a compact subset K of C^n, $\boldsymbol{R(K)}$ will denote the uniform closure on K of the algebra of rational functions which are analytic on K.

The following lemmas parallel results proved in section 1 for polynomial convexity. The proofs will be omitted.

2.1 Lemma. Let K be a compact subset of C^n. The maximal ideal space of $R(K)$ is the rational convex hull of K.

2.2 Lemma. Every compact subset of C^1 is rationally convex.

2.3 Lemma. Suppose the Banach algebra A is generated by $f_1, \ldots, f_n$, together with $(\lambda - f_j)^{-1}$, for λ belonging to the resolvent set of f_j and $1 \leq j \leq n$. Then the joint spectrum $\sigma(f_1, \ldots, f_n)$ of $f_1, \ldots, f_n$ is rationally convex, and the natural projection π of M_A onto $\sigma(f_1, \ldots, f_n)$ is a homeomorphism.

The following useful criterion relates rational convexity and polynomials.

2.4 Lemma. Let K be a compact subset of C^n. A point $z \in C^n$ belongs to the rational convex hull of K if and only if $p(z) \in p(K)$ for every polynomial p. This occurs if and only if every polynomial which vanishes at z also vanishes somewhere on K.

Proof. Suppose p is a polynomial which does not assume the value $p(z)$ on K. Then $1/(p - p(z))$ is a rational function which is analytic on K, but whose pole set contains z. Hence z cannot belong to the rational convex hull of K.

Conversely, suppose z_0 does not belong to the rational convex hull of K. Let f be a rational function which is analytic on K, and which satisfies $|f(z_0)| > \|f\|_K$. Write $1/(f - f(z_0)) = p/q$, where p and q are polynomials which are relatively prime. Then the pole set of $1/(f - f(z_0))$ coincides with the zero set of q. So q does not assume the value 0 on K, while $q(z_0) = 0$. That proves the lemma.

2.5 Theorem. There is a compact totally disconnected subset of C^2 which is not rationally convex (and, in particular, not polynomially convex).

Proof. Let E be a compact totally disconnected subset of the complex plane which has positive area. Let

$$f(\zeta) = \iint_E \frac{dx\,dy}{z - \zeta}.$$

Then f is a non-constant function on the Riemann sphere S^2, f is analytic on $S^2 \backslash E$, and $f(\infty) = 0$. This sort of function has already been discussed in II.1.

Let A be the closed subalgebra of continuous functions on S^2 generated by f and zf. Every function in A is analytic on $S^2 \backslash E$.

Let π be the map of the sphere into C^2, defined by $\pi(z) = (f(z), zf(z))$. Then π separates the points of S^2, except those lying in $f^{-1}(0)$. So $\pi(E)$ is homeomorphic to the quotient space of E obtained by identifying $f^{-1}(0) \cap E$ to a point. Evidently, $\pi(E)$ is totally disconnected. In fact, any quotient space of a compact totally disconnected space obtained by identifying a closed subset to a point is again totally disconnected.

If $g \in A$ vanishes somewhere on S^2, then g vanishes somewhere on E. This follows from the argument principle, as noted in II.1. Hence if $p(z_1, z_2)$ is any polynomial which vanishes somewhere on $\pi(S^2)$, then p vanishes somewhere on $\pi(E)$. By 2.4, $\pi(S^2)$ is contained in the rational convex hull of $\pi(E)$. And $\pi(S^2)$ properly contains $\pi(E)$. That proves the theorem.

By Antoine's theorem, every compact totally disconnected subset of a Euclidean space lies on an arc. By threading an arc through $\pi(E)$ so as to not cover all of $\pi(S^2)$, we obtain from 2.5 the following corollary.

2.6 Corollary. There is an arc in C^2 which is not rationally convex.

On the other hand, the following theorem shows that differentiable arcs in C^n are always rationally convex.

2.7 Theorem. Let K be a compact subset of C^n such that the projection of K onto each coordinate plane has zero area. Then $R(K) = C(K)$. In particular, K is rationally convex.

Proof. Let K_j be the projection of K onto the jth coordinate plane. By the Hartogs-Rosenthal theorem II. 8.4, every continuous function on K_j can be approximated uniformly on K_j by rational functions of z_j. Hence the real functions in $R(K)$ separate the points of K. By the Stone-Weierstrass theorem, $R(K) = C(K)$.

The theory of uniform approximation in C^n is a field which is largely undeveloped, and which, together with rational approximation on the plane, will be central in the development of uniform algebras in the years to come. Here we will give a sketch of some of the known results concerning uniform approximation on curves in C^n.

Chirka has shown that if K is an arc in C^n such that the projection of K into each coordinate plane has no interior, then every continuous function on K can be approximated uniformly on K by functions analytic in a neighborhood of K.

Suppose K is a continuously differentiable simple closed curve in C^n. Theorem 2.7 shows that $R(K) = C(K)$. It may happen that $P(K) \neq C(K)$— for instance, take K to be the unit circle in the complex plane. It can be shown, though, that either $P(K) = C(K)$, in which case K is polynomially convex, or $\hat{K}\backslash K$ is a one-dimensional analytic variety whose boundary is K. For a discussion of this type of result, see the paper of Stolzenberg [6].

These methods also show that if K is a continuously differentiable arc in C^n, then K is polynomially convex. From this fact, it is deduced that every continuous function on K can be approximated uniformly on K by polynomials.

A relevant example of Stolzenberg [2] is a compact subset K of C^2 such that $\hat{K} \neq K$, while the projection of $\hat{K}$ on both coordinate planes has zero area. This shows that $\hat{K}\backslash K$ cannot always be equipped with an analytic structure.

3. Circled Sets

A subset K of C^n is **circled** if whenever $(z_1, \ldots, z_n) \in K$ and $\lambda_1, \ldots, \lambda_n$ are complex numbers of modulus one, then $(\lambda_1 z_1, \ldots, \lambda_n z_n) \in K$. In this section we will extend Mergelyan's theorem to a class of circled sets in C^n. We will also determine the rational and polynomially convex hulls of compact connected circled sets in C^n.

Circles and annuli centered at the origin are circled subsets of C^1. One can visualize circled sets in C^2 by plotting $|z_1|$ and $|z_2|$ in the first quadrant of the plane. Each point not on the coordinate axes represents a torus in C^2.

Suppose f is a continuous function on the circle $\{|z| = r\}$, with Fourier series

$$f(z) \sim \sum_{k=-\infty}^{\infty} a_k z^k,$$

$$a_n = \frac{1}{2\pi r^n}\int_0^{2\pi} f(re^{i\theta})e^{-in\theta}\,d\theta = \frac{1}{2\pi i}\int_{|z|=r} f(z)z^{-n-1}dz.$$

The partial sums

$$S_m(z) = \sum_{k=-m}^{m} a_k z^k$$

need not converge to f uniformly. However, the Cesaro means of the partial sums

$$\sigma_m = \frac{1}{m+1}(S_0 + \cdots + S_m)$$

do converge uniformly to f. In fact,

$$\sigma_m(re^{i\theta}) = \frac{1}{2\pi}\int_{-\pi}^{\pi} f(re^{i\varphi})K_m(\theta - \varphi)d\varphi,$$

where K_m is the mth Fejer kernel. The Fejer kernels have the following properties:

(i) $K_m \geq 0$

(ii) $\frac{1}{2\pi}\int_{-\pi}^{\pi} K_m(\theta)d\theta = 1$

(iii) $\lim_{m\to\infty}\int_{\delta\leq|\theta|\leq\pi} K_m(\theta)d\theta = 0$, for each fixed $\delta > 0$.

From this, it follows that

$$|f(re^{i\theta}) - \sigma_m(re^{i\theta})| = \left|\frac{1}{2\pi}\int_{-\pi}^{\pi}[f(re^{i\theta}) - f(re^{i\varphi})]K_m(\theta - \varphi)d\varphi\right|$$

$$\leq \omega(f, r\delta) + 4\,||f||_\infty \int_\delta^\pi K_m(\varphi)d\varphi,$$

where ω is the modulus of continuity,

$$\omega(f, a) = \sup\{|f(z_1) - f(z_2)| : |z_1 - z_2| \leq a\}.$$

Now $\omega(f, r\delta)$ tends to zero with δ. By first choosing $\delta > 0$ small, and then choosing m large, we can make $||f - \sigma_m||_\infty$ arbitrarily small. So σ_m converges uniformly to f. This is Fejer's theorem.

Now suppose that f is a continuous function on the multitorus $T = \{|z_j| = r_j,\ 1 \leq j \leq n\}$ in C^n. Then f has a Fourier series expansion

$$f \sim \sum_{-\infty}^{\infty} a_{j_1\cdots j_n} z_1^{j_1} \ldots z_n^{j_n},$$

where

$$(*) \qquad a_{j_1\cdots j_n} = \frac{1}{(2\pi i)^n}\int_T f(z_1, \ldots, z_n) z_1^{-j_1-1} \ldots z_n^{-j_n-1}\,dz_1 \ldots dz_n.$$

Set $f_m(z)$ equal to

$$\frac{1}{(2\pi)^n}\int f(r_1e^{i\varphi_1}, \dots, r_ne^{i\varphi_n})K_{m-1}(\theta_1 - \varphi_1) \dots K_{m-1}(\theta_n - \varphi_n)d\varphi_1 \dots d\varphi_n.$$

Then

$$(**) \qquad f_m(z) = \sum_{|j_1|<m,\dots,|j_n|<m} \frac{(m-|j_1|)\dots(m-|j_n|)}{m^n} a_{j_1\dots j_n} z_1^{j_1}\dots z_n^{j_n}.$$

The same estimate used in the case of one variable shows that

$$\|f_m - f\|_T \leq \omega(f, r\delta) + \gamma(m, \delta)\,\|f\|_T,$$

where $r = \max\{r_1, \dots, r_n\}$, and $\lim_{m\to\infty} \gamma(m, \delta) = 0$, for each fixed $\delta > 0$. In particular, f_m tends uniformly to f as $m \to \infty$.

With these preliminaries, we can prove easily the following lemma.

3.1 Lemma. Let D be a bounded open subset of C^n which is connected and circled. Let f be an analytic function on D which is uniformly continuous. There is a sequence of rational functions $\{f_m\}_{m=1}^{\infty}$ whose pole sets do not meet D, such that each f_m is a polynomial in $z_1, \dots, z_n, z_1^{-1}, \dots, z_n^{-1}$, and f_m converges uniformly to f on D.

Proof. Let E be the subset of D of points not contained in any of the coordinate axes $\{z_j = 0\}$, $1 \leq j \leq n$. Then E is a connected, open, circled subset of C^n which is dense in D.

On each multitorus $\{|z_j| = r_j, 1 \leq j \leq n\}$ contained in E, the function f has a Fourier series

$$f \sim \sum a_{j_1\dots j_n}(r_1, \dots r_n)\, z_1^{j_1} \dots z_n^{j_n}.$$

The formula (*), and the analyticity of f, show that $a_{j_1\dots j_n}(r_1, \dots, r_n) = a_{j_1\dots j_n}(s_1, \dots s_n)$ if $(r_1, \dots, r_n)$ is near to $(s_1, \dots, s_n)$. Since E is connected, the Fourier coefficients of f are independent of $(r_1, \dots, r_n)$.

If f_m is defined by the formula (**), then

$$\|f_m - f\|_E \leq \omega(f, R\delta) + \gamma(m, \delta)\,\|f\|_E,$$

where R is chosen large enough so that E is contained in the polydisc $\{|z_j| \leq R, 1 \leq j \leq n\}$. Hence f_m converges uniformly on E to f.

Since the functions f_m are bounded on E, their pole sets cannot meet D. This can be seen either directly, or by showing from the formula (*) and the analyticity of f that $a_{j_1\dots j_n} = 0$ whenever D meets the coordinate axis $\{z_k = 0\}$ and $j_k < 0$. It follows that f_m tends uniformly to f on D. This proves the lemma.

The pole set of the approximating rational functions may meet the boundary of D. This occurs, for instance, when $D = \{|z_2| < |z_1| < 1\}$, and $f(z_1, z_2) = z_2^2/z_1$.

For a compact subset K of C^n, $A(K)$ will denote the uniform algebra of functions which are continuous on K and analytic on the interior of K. As an immediate consequence of 3.1, we obtain the following theorem.

3.2 Theorem. Let K be a compact circled subset of C^n, such that int(K) is connected and dense in K. Suppose also that int(K) meets every coordinate axis $\{z_j = 0\}$ which K meets. Then every $f \in A(K)$ can be approximated uniformly on K by polynomials in $z_1, \ldots, z_n, z_1^{-1}, \ldots, z_n^{-1}$ which are analytic in a neighborhood of K.

In particular, under the hypotheses of 3.2 we have $R(K) = A(K)$. If K meets the coordinate axis $\{z_j = 0\}$, then powers of z_j^{-1} cannot appear in the approximating polynomials. So $P(K) = A(K)$ if K meets every coordinate axis.

If K is a circled subset of C^n, let $\boldsymbol{L(K)}$ be the set of points $(s_1, \ldots, s_n) \in R^n$ such that $(e^{s_1}, \ldots, e^{s_n}) \in K$. If E is a subset of R^n, let $\boldsymbol{L^{-1}(E)}$ be the circled subset of C^n of points $z = (z_1, \ldots, z_n)$ such that $(\log|z_1|, \ldots, \log|z_n|) \in E$. If K is circled, then $L^{-1}(L(K))$ is that part of K which does not meet the coordinate axes.

Armed with the approximation theorem 3.2, we can compute the rational convex hull of the circled sets appearing there.

3.3 Theorem. Under the hypotheses of 3.2, the rational convex hull of K is the closure of $L^{-1}(\text{co } L(K))$, where "co" indicates "linear convex hull."

Proof. First we show that $L^{-1}(\text{co } L(K))$ is contained in the rational convex hull of K.

Suppose that $r, s \in R^n$ are such that the multitori $T_0 = \{|z_j| = e^{r_j}, 1 \le j \le n\}$ and $T_1 = \{|z_j| = e^{s_j}, 1 \le j \le n\}$ belong to the rational convex hull of K. We must show that the multitori

$$T_a = \{|z_j| = e^{(1-a)r_j + as_j}, 1 \le j \le n\}$$

belong to the rational convex hull of K, for $0 < a < 1$.

The rational convex hull of a circled set is clearly circled. So it will suffice to show that each T_a meets the rational convex hull of K, $0 < a < 1$.

For $1 \le j \le n$, and ζ complex, define

$$F_j(\zeta) = \exp[(1 + i\zeta)r_j - i\zeta s_j],$$

$$F(\zeta) = (F_1(\zeta), \ldots, F_n(\zeta)).$$

Then F maps the line $\{\text{Im}(\zeta) = a\}$ into T_a, $0 \le a \le 1$.

If $f(z) = z_1^{k_1} \ldots z_n^{k_n}$ is analytic in a neighborhood of K, where the k_j are integers, then

$$(f \circ F)(\zeta) = \prod_{j=1}^{n} \exp[(1 + i\zeta)k_j r_j - i\zeta k_j s_j]$$

is analytic on the strip $0 \le \text{Im}(\zeta) \le 1$.

Applying 3.2, we see that if f is any rational function which is analytic in a neighborhood of K, then $f \circ F$ is continuous on the strip $0 \leq \text{Im}(\zeta) \leq 1$, and is analytic and bounded on the strip $0 < \text{Im}(\zeta) < 1$. By a Phragmén-Lindelöf theorem,

$$|f(z)| \leq \|f\|_{T_0 \cup T_1} \leq \|f\|_K$$

for all $f \in R(K)$ and all $z \in F(\{0 < \text{Im}(\zeta) < 1\})$. So the rational convex hull of K meets each T_a, $0 < a < 1$. This completes the first half of the proof.

Now suppose that $z^0 = (z_1^0, \ldots, z_n^0)$ does not belong to the closure of $L^{-1}(\text{co}\, L(K))$. We must show that z^0 does not belong to the rational convex hull of K.

Suppose first that $z_j^0 = 0$ for some index j, while K does not meet the coordinate axis $\{z_j = 0\}$. Then $1/z_j$ is analytic on K and large at z^0. So z^0 cannot belong to the rational convex hull of K.

Next suppose that K meets every coordinate axis $\{z_j = 0\}$ for which $z_j^0 = 0$. Then the same is true of int(K), by our hypothesis.

Let U_ϵ be the set of $z \in C^n$ such that $||z_j| - |z_j^0|| \leq \epsilon$, $1 \leq j \leq n$. Then $L(U_\epsilon)$ is convex, and $L(U_\epsilon)$ is disjoint from co $L(K)$ for $\epsilon > 0$ small. By the separation theorem for convex sets, there are real numbers $c_1, \ldots, c_n$ not all zero, such that

$$\sum_{j=1}^{n} c_j \log |z_j| \geq \sum_{j=1}^{n} c_j \log |w_j|$$

for all $z \in U_\epsilon$ and $w \in K$ which do not meet the coordinate axes.

Since $\sum_{j=1}^{n} c_j \log |w_j|$ is bounded above, for all $w \in K$ not belonging to a coordinate axis, we see that $c_j \geq 0$ whenever int(K) meets $\{z_j = 0\}$. That is, $c_j \geq 0$ whenever K meets the coordinate axis $\{z_j = 0\}$.

Similarly, $c_j \leq 0$ whenever $z_j^0 = 0$. Consequently $c_j = 0$ whenever $z_j^0 = 0$.

Let $j_1 < \cdots < j_p$ be the indices for which $z_j^0 \neq 0$. If $\epsilon > 0$ is chosen so small that the j_qth coordinates of the points in U_ϵ are bounded away from zero, $1 \leq q \leq p$, then we obtain strict separation:

$$|z_{j_1}^0|^{c_{j_1}} \ldots |z_{j_p}^0|^{c_{j_p}} > \inf_{z \in U_\epsilon} |z_{j_1}|^{c_{j_1}} \ldots |z_{j_p}|^{c_{j_p}}$$
$$\geq \sup_{w \in K} |w_{j_1}|^{c_{j_1}} \ldots |w_{j_p}|^{c_{j_p}}.$$

By altering c_{j_q}, $1 \leq q \leq p$, so that they become rational and then multiplying by an integer, we can find integers m_{j_q}, $1 \leq q \leq p$, such that $f(z) = z_{j_1}^{m_{j_1}} \ldots z_{j_p}^{m_{j_p}}$ satisfies $|f(z^0)| > \|f\|_K$. Since the rational function f is analytic in a neighborhood of K, z^0 does not belong to the rational convex hull of K.

Theorem 3.3 can be used to determine the rational convex hull of any compact connected circled subset K of C^n. In fact, such a K can be written as a decreasing intersection of sets K_n which satisfy the hypotheses of 3.2.

The rational convex hull of K is then the intersection of the rational convex hulls of the K_n. From 3.3, and its proof, we deduce the following.

3.4 Theorem. Let K be a connected compact circled subset of C^n. If w does not belong to the rational convex hull of K, there are integers $k_1, \ldots, k_n$ such that the rational monomial $f(z) = z_1^{k_1} \ldots z_n^{k_n}$ is analytic in a neighborhood of K, and satisfies $|f(w)| > \|f\|_K$.

A circled set K is **complete** if whenever $(z_1, \ldots, z_n) \in K$ and $\lambda_1, \ldots, \lambda_n$ are complex numbers satisfying $|\lambda_j| \leq 1$, $1 \leq j \leq n$, then $(\lambda_1 z_1, \ldots, \lambda_n z_n) \in K$. A complete circled set K is connected, and it contains 0. Unless K is contained in the coordinate axes, 0 belongs to the interior of K.

Every function analytic on the polydisc $\{|z_1| \leq r_1, \ldots, |z_n| \leq r_n\}$ attains its maximum modulus over the polydisc on the multitorus $\{|z_1| = r_1, \ldots, |z_n| = r_n\}$. Consequently the polynomial convex hull $\hat{K}$ of a circled set K is complete.

In view of theorems 3.2 through 3.4, and the preceding remarks, we obtain the following theorem for polynomials.

3.5 Theorem. Let K be a compact circled subset of C^n. Then $\hat{K}$ is a complete circled set. If $w \notin \hat{K}$, there are non-negative integers $k_1, \ldots, k_n$ such that the monomial $f(z) = z_1^{k_1} \ldots z_n^{k_n}$ satisfies $|f(w)| > \|f\|_K$. Moreover, if int(K) is dense in K, then $P(K) = A(\hat{K})$. In this case, $\hat{K}$ is the smallest closed, complete circled set containing K whose image under L is convex.

If $K = \{|z_2| \leq |z_1| \leq 1\}$, then K is compact, $L(K)$ is convex and int (K) is connected and dense in K. However, K is not rationally convex. The rational convex hull of K coincides with $\hat{K} = \{|z_1| \leq 1, |z_2| \leq 1\}$. The function z_1^2/z_2 extends continuously from int (K) to K, and so belongs to $A(K)$. However, z_1^2/z_2 does not extend analytically to int $(\hat{K})$, and so does not belong to $P(K) = R(K) = A(\hat{K})$.

4. The Functional Calculus

We return now to a commutative Banach algebra A with an identity. In I.5, we showed that if F is analytic in a neighborhood of the spectrum $\sigma(f)$ of $f \in A$, then there is $g \in A$ such that $\hat{g} = F \circ \hat{f}$. In this section, we will generalize this result to analytic functions of several complex variables. We will show that if $f_1, \ldots, f_n \in A$, and if F is analytic in a neighborhood of the joint spectrum $\sigma(f_1, \ldots, f_n)$, then there exists $g \in A$ such that $\hat{g} = F(\hat{f}_1, \ldots, \hat{f}_n)$. For this, we will develop a functional calculus for elements of a Banach algebra.

In theorem 4.1, the functional calculus is axiomatized. In 4.2 through

4.6, other useful properties of the functional calculus are enumerated for easy reference. The remainder of this section, and the next, are devoted to defining the functional calculus and establishing these properties.

4.1 Theorem. There exists a unique rule assigning to every n-tuple $(f_1, \ldots, f_n)$ of elements in A, and to every function F analytic in a neighborhood of $\sigma(f_1, \ldots, f_n)$, an element $F(f_1, \ldots, f_n) \in A$, satisfying the three following conditions:

(i) If $F(z_1, \ldots, z_n) = \sum a_{i_1 \ldots i_n} z_1^{i_1} \ldots z_n^{i_n}$ is a polynomial, then

$$F(f_1, \ldots, f_n) = \sum a_{i_1 \ldots i_n} f_1^{i_1} \ldots f_n^{i_n}.$$

(ii) If F is analytic in a neighborhood U of $\sigma(f_1, \ldots, f_n)$, if $f_{n+1}, \ldots, f_m \in A$, and if $\tilde{F}$ is the extension of F to $U \times C^{m-n}$ defined by

$$\tilde{F}(z_1, \ldots, z_m) = F(z_1, \ldots, z_n),$$

then

$$\tilde{F}(f_1, \ldots, f_m) = F(f_1, \ldots, f_n).$$

(iii) If $\{F_k\}_{k=1}^{\infty}$ is a sequence of functions analytic on some neighborhood U of $\sigma(f_1, \ldots, f_n)$ which tends uniformly to F on U, then $F_k(f_1, \ldots, f_n)$ converges to $F(f_1, \ldots, f_n)$ in A.

The linearity and multiplicativity of the functional calculus on polynomials will force it to be linear and multiplicative on all analytic functions.

4.2 Theorem. If $f_1, \ldots, f_n \in A$, and F and G are analytic in a neighborhood of $\sigma(f_1, \ldots, f_n)$, then

(i) $(F + G)(f_1, \ldots, f_n) = F(f_1, \ldots, f_n) + G(f_1, \ldots, f_n)$.
(ii) $(FG)(f_1, \ldots, f_n) = F(f_1, \ldots, f_n)G(f_1, \ldots, f_n)$.

The next theorem shows that the element $F(f)$ given by the functional calculus coincides with that introduced in I.5.

4.3 Theorem. If $f \in A$, if $\sigma(f)$ is contained in a domain D with smooth boundary bD, and if F is analytic on $\bar{D}$, then

$$F(f) = \frac{1}{2\pi i} \int_{bD} F(z)(z - f)^{-1} dz.$$

4.4 Theorem (Product Theorem). Suppose F is analytic in a neighborhood of $\sigma(f_1, \ldots, f_n)$, and G is analytic in a neighborhood of $\sigma(g_1, \ldots, g_m)$. Then

$$H(z_1, \ldots, z_{n+m}) = F(z_1, \ldots, z_n)G(z_{n+1}, \ldots, z_{n+m})$$

defines a function H analytic in a neighborhood of $\sigma(f_1, \ldots, f_n, g_1, \ldots, g_m)$, and

$$H(f_1, \ldots, f_n, g_1, \ldots, g_m) = F(f_1, \ldots, f_n)G(g_1, \ldots, g_m).$$

4.5 Theorem (Composition Theorem). If F is analytic in a neighborhood of $\sigma(f_1, \ldots, f_n)$, then

$$\widehat{F(f_1, \ldots, f_n)} = F(\hat{f}_1, \ldots, \hat{f}_n).$$

The final theorem 4.6 will show, for instance, that if F is analytic in a neighborhood of $\sigma(f)$, and if G is analytic in a neighborhood of $F(\sigma(f)) = \sigma(F(f))$, then $(G \circ F)(f) = G(F(f))$.

4.6 Theorem (Substitution Theorem). Suppose $F_1, \ldots, F_m$ are analytic in a neighborhood of $\sigma(f_1, \ldots, f_n)$. Suppose G is analytic in a neighborhood of $\sigma(F_1(f_1, \ldots, f_n), \ldots, F_m(f_1, \ldots, f_n))$. Then the function

$$H(z_1, \ldots, z_n) = G(F_1(z_1, \ldots, z_n), \ldots, F_m(z_1, \ldots, z_n))$$

is analytic in a neighborhood of $\sigma(f_1, \ldots, f_n)$, and

$$H(f_1, \ldots, f_n) = G(F_1(f_1, \ldots, f_n), \ldots, F_m(f_1, \ldots, f_n)).$$

We showed in I.5 that the integral $\frac{1}{2\pi i}\int_{bD} f(z)(z - f)^{-1}dz$, given by 4.3, defines an element $g \in A$ satisfying $\hat{g} = F \circ \hat{f}$. This formula does not generalize immediately to several variables. The following lemma provides an alternate definition which does generalize.

4.7 Lemma. Suppose $f \in A$, $\sigma(f)$ is contained in the open set D with smooth boundary bD, and F is analytic on $\bar{D}$. Suppose w is an infinitely differentiable function on $\bar{D}$ such that $w = 1$ in a neighborhood of $\sigma(f)$, and $w = 0$ on bD. Then the A-valued function u on $\bar{D}$, defined by

$$u(z) = \begin{cases} (1 - w(z))(z - f)^{-1}, & z \in \bar{D}\backslash\sigma(f) \\ 0, & z \in \sigma(f), \end{cases}$$

is infinitely differentiable. Also,

$$(*) \qquad \frac{1}{2\pi i}\int_{bD} F(z)(z - f)^{-1}\, dz = \frac{1}{2\pi i}\iint_D F(z)du(z)dz.$$

The formula (*) defines an element $g \in A$ such that $\hat{g} = F \circ \hat{f}$. If $F(z) = \sum_{k=0}^{N} a_k z^k$ is a polynomial, then $g = \sum_{k=0}^{N} a_k f^k$.

Proof. Using Stokes' theorem to convert the boundary integral to a double integral, we have

$$\int_{bD} F(z)(z - f)^{-1}dz = \int_{bD} Fudz = \iint_D d(Fu)dz = \iint_D \frac{\partial}{\partial \bar{z}}(Fu)d\bar{z}dz$$
$$= \iint_D F\frac{\partial u}{\partial \bar{z}}d\bar{z}dz = \iint_D F\, dudz.$$

This proves (*).

By approximating the contour integral in (*) by finite Riemann sums and evaluating at points of M_A, we obtain $\hat{g} = F \circ \hat{f}$, as in I.5.

Note that the integrals in (*) depend neither on the domain D nor on the function w, subject to the conditions placed on them in the lemma. In particular if $F(z) = \sum_{k=0}^{N} a_k z^k$ is a polynomial, we can take D to be a large disc $\{|z| < R\}$. Then $(z - f)^{-1} = \sum_{k=0}^{\infty} f^k/z^{k+1}$ converges uniformly on bD. So

$$\frac{1}{2\pi i}\int_{bD} F(z)(z - f)^{-1}dz = \sum_{k=0}^{\infty} \frac{f^k}{2\pi i}\int_{bD} F(z)z^{-k-1}dz$$
$$= \sum_{k=0}^{N} a_k f^k.$$

This completes the proof.

Now we wish to find an analogue of the integral $\frac{1}{2\pi i}\iint_D F du dz$ which will be suitable for functions of several complex variables. First we establish some lemmas.

4.8 Lemma. Suppose $f_1, \ldots, f_n \in A$. There exist infinitely differentiable A-valued functions $v_1, \ldots, v_n$, defined on $C^n \backslash \sigma(f_1, \ldots, f_n)$, such that

$$\sum_{j=1}^{n} v_j(z)(z_j - f_j) = 1.$$

Proof. Suppose $z^0 = (z_1^0, \ldots, z_n^0) \in \sigma(f_1, \ldots, f_n)^c$. As remarked in section 1, there exist $g_1, \ldots, g_n \in A$ such that $\sum g_j(z_j^0 - f_j) = 1$. Let $U(z^0)$ be an open set containing z^0, on which $\sum g_j(z_j - f_j)$ is invertible. Define

$$h_k(z) = g_k\left[\sum_{j=1}^{n} g_j(z_j - f_j)\right]^{-1}, \qquad z \in U(z^0).$$

Then h_k is infinitely differentiable on $U(z^0)$, and

$$\sum h_k(z)(z_k - f_k) = 1, \qquad z \in U(z^0).$$

Now we will use a partition of unity to patch together the locally defined functions h_k to obtain globally defined functions v_k. This standard tool allows us to assert the following: There exists a locally finite open cover $\{W_k\}_{k=1}^{\infty}$ of $\sigma(f_1, \ldots, f_n)^c$ and infinitely differentiable real-valued functions $\{w_k\}_{k=1}^{\infty}$ on C^n such that $0 \leq w_k \leq 1$, w_k is supported on W_k, and $\sum_{k=1}^{\infty} w_k = 1$ on $\sigma(f_1, \ldots, f_n)^c$. Moreover, the cover $\{W_k\}$ can be chosen subordinate to the cover $\{U(z^0): z^0 \in \sigma(f_1, \ldots, f_n)^c\}$, that is, it can be chosen so that for every $1 \leq k < \infty$, there is a point $z^k \in \sigma(f_1, \ldots, f_n)^c$ such that $\bar{W}_k \subseteq U(z^k)$.

Suppose that $h_{1k}, \ldots, h_{nk}$ are the infinitely differentiable A-valued functions defined on $U(z^k)$, such that $\sum_{j=1}^{n} h_{jk}(z_j - f_j) = 1$ on $U(z^k)$. We extend h_{jk} arbitrarily to the complement of $U(z^k)$, and define

$$v_j(z) = \sum_{k=1}^{\infty} w_k(z) h_{jk}(z), \qquad z \in \sigma(f_1, \ldots, f_n)^c, 1 \leq j \leq n.$$

All but a finite number of terms of this series vanish in a neighborhood of any fixed point in $\sigma(f_1, \ldots, f_n)^c$. So the functions $v_1, \ldots, v_n$ are infinitely differentiable. Evidently they satisfy $\sum v_j(z)(z_j - f_j) = 1$ on $\sigma(f_1, \ldots, f_n)^c$.

4.9 Lemma. Let $f_1, \ldots, f_n \in A$, and let U be a neighborhood of $\sigma(f_1, \ldots, f_n)$. If w is an infinitely differentiable A-valued function supported on U which is 1 on a neighborhood of $\sigma(f_1, \ldots, f_n)$, then there are infinitely differentiable A-valued functions $u_1, \ldots, u_n$ defined on C^n, such that

$$\sum_{j=1}^{n} u_j(z)(z_j - f_j) = 1 - w(z), \qquad z \in C^n.$$

If $u_1, \ldots, u_n$ satisfy these conditions, then $du_1dz_1 \ldots du_ndz_n$ is supported on U. If w' and $u'_1, \ldots, u'_n$ are other infinitely differentiable A-valued functions which satisfy the same conditions, then there is an $(n-1)$-form τ supported on U, such that

$$du_1dz_1 \ldots du_ndz_n - du'_1dz_1 \ldots du'_ndz_n = d\tau dz_1 \ldots dz_n.$$

Proof. Let $v_1, \ldots, v_n$ be as in 4.8. Set $u_j = (1-w)v_j$ on $\sigma(f_1, \ldots, f_n)^c$, and extend u_j to be 0 on $\sigma(f_1, \ldots, f_n)$. Then $\sum u_j(z)(z_j - f_j) = 1 - w(z)$, $z \in C^n$.

Now $\sum u_j(z)(z_j - f_j) = 1$ in a neighborhood V of U^c. Taking the total differential, we obtain the following identity on V:

$$\sum (z_j - f_j)du_j + \sum u_j dz_j = 0.$$

If we then multiply by $du_1 \ldots du_{k-1}du_{k+1} \ldots du_ndz_1 \ldots dz_n$, we obtain $(z_k - f_k)du_1dz_1 \ldots du_ndz_n = 0$ on V. Multiplying by u_k and summing, we obtain $du_1dz_1 \ldots du_ndz_n = 0$ on V. Hence the closed support of $du_1dz_1 \ldots du_ndz_n$ is contained in U.

Next we will verify the final assertion of the theorem in two special cases. Later we will reduce the general case to these special cases.

Suppose q is an infinitely differentiable A-valued function supported on U, and suppose

$$\begin{aligned} u'_1 &= u_1 + q, \\ u'_j &= u_j, \qquad\qquad 2 \leq j \leq n, \\ w' &= w - (z_1 - f_1)q. \end{aligned}$$

Then $\sum (z_j - f_j)u'_j = 1 - w'$. Also

$$\begin{aligned} du'_1dz_1 \ldots du'_ndz_n - du_1dz_1 \ldots du_ndz_n &= dqdz_1du_2dz_2 \ldots du_ndz_n \\ &= d\tau dz_1 \ldots dz_n, \end{aligned}$$

where τ is a multiple of $qdu_2 \ldots du_n$. This proves the final assertion in this special case.

Suppose now that p is an infinitely differentiable A-valued function on C^n, and suppose

$$
\begin{aligned}
u_1' &= u_1 + (z_2 - f_2)p, \\
u_2' &= u_2 - (z_1 - f_1)p, \\
u_j' &= u_j, \qquad\qquad 3 \le j \le n, \\
w' &= w.
\end{aligned}
$$

Again we have $\sum (z_j - f_j)u_j' = 1 - w'$. From the identities

$$
\begin{aligned}
&du_1'du_2'dz_1dz_2 \\
&\qquad = du_1du_2dz_1dz_2 - (z_1 - f_1)du_1dpdz_1dz_2 - (z_2 - f_2)du_2dpdz_1dz_2, \\
&dw = -\sum (z_j - f_j)du_j - \sum u_jdz_j,
\end{aligned}
$$

we obtain

$$
\begin{aligned}
du_1'dz_1 \ldots du_n'dz_n - du_1dz_1 \ldots du_ndz_n &= -dwdpdz_1dz_2du_3dz_3 \ldots du_ndz_n \\
&= d\tau dz_1 \ldots dz_n,
\end{aligned}
$$

where τ is a multiple of $pdwdu_3 \ldots du_n$. This proves the final assertion also in this special case.

Now let w' and $u_1', \ldots, u_n'$ be as in the lemma. Then

$$
\begin{aligned}
u_k' - u_k &= \left[\sum_{j=1}^{n} u_j(z_j - f_j) + w\right]u_k' - \left[\sum_{j=1}^{n} u_j'(z_j - f_j) + w'\right]u_k \\
&= \sum_{j=1}^{n} (z_j - f_j)p_{jk} + q_k,
\end{aligned}
$$

where

$$
\begin{aligned}
p_{jk} &= -p_{kj} = u_ju_k' - u_j'u_k, \\
q_k &= wu_k' - w'u_k.
\end{aligned}
$$

Also,

$$
\begin{aligned}
w' - w &= -\sum (z_k - f_k)(u_k' - u_k) \\
&= -\sum_{j,k} (z_k - f_k)(z_j - f_j)p_{jk} - \sum_k (z_k - f_k)q_k \\
&= -\sum_k (z_k - f_k)q_k.
\end{aligned}
$$

Hence $u_1', \ldots, u_n'$ and w' can be obtained from $u_1, \ldots, u_n$ and w by a chain of substitutions of the forms we have already considered. This proves the lemma.

Now we are in a position to define the functional calculus. Suppose $f_1, \ldots, f_n \in A$, and F is analytic in a neighborhood U of $\sigma(f_1, \ldots, f_n)$. Let $u_1, \ldots, u_n$ and w be functions as in lemma 4.9. We define

$$
F(f_1, \ldots, f_n) = \frac{n!}{(2\pi i)^n} \int F du_1dz_1 \ldots du_ndz_n.
$$

The integral is well-defined, since $du_1dz_1 \ldots du_ndz_n$ is supported on U.

If τ is an $(n-1)$-form supported on U, and F is analytic on U, then

$$
d(F\tau) = Fd\tau + \sum \frac{\partial F}{\partial z_j} dz_j\tau.
$$

So

$$\int_U F d\tau dz_1 \ldots dz_n = \int_U d(F\tau dz_1, \ldots, dz_n) = 0$$

by Stokes' theorem, since $F\tau dz_1 \ldots dz_n$ vanishes in a neighborhood of bU. If $u_1', \ldots, u_n'$ and w' is another set of functions as in 4.9, then

$$\int F du_1 \ldots dz_n - \int F du_1' \ldots dz_n' = \int F d\tau dz_1 \ldots dz_n = 0.$$

Consequently, the definition of $F(f_1, \ldots, f_n)$ does not depend on the choice of $u_1, \ldots, u_n$ and w, subject to the conditions of 4.9.

If $n = 1$, then necessarily $u_1(z) = (1 - w(z))(z - f_1)^{-1}$. Consequently theorem 4.3 follows from lemma 4.7 and the definition of the functional calculus. Two other facts that we deduce immediately from the definition of the functional calculus are 4.1(iii) and 4.2(i), which are the continuity and linearity of the functional calculus respectively. Next we prove the product theorem.

Proof of 4.4. Suppose F is analytic in a neighborhood U of $\sigma(f_1, \ldots, f_n)$, and G is analytic in a neighborhood V of $\sigma(g_1, \ldots, g_m)$. Then

$$H(z_1, \ldots, z_n, \zeta_1, \ldots, \zeta_m) = F(z_1, \ldots, z_n)G(\zeta_1, \ldots, \zeta_m)$$

defines a function H analytic on the neighborhood $U \times V$ of $\sigma(f_1, \ldots, f_n, g_1, \ldots, g_m)$ in C^{n+m}. We must show that $H(f_1, \ldots, g_m) = F(f_1, \ldots, f_n)\, G(g_1, \ldots, g_m)$.

Choose $u_1(z), \ldots, u_n(z), w(z)$ and $v_1(\zeta), \ldots, v_m(\zeta), w'(\zeta)$ as in lemma 4.9, so that w is supported on U, w' is supported on V, $\sum u_j(z)(z_j - f_j) = 1 - w(z)$, and $\sum v_j(\zeta)(\zeta_j - g_j) = 1 - w'(\zeta)$. Then ww' is supported on $U \times V$, and

$$\sum_{j=1}^{n} u_j(z)(z_j - f_j) + w(z) \sum_{j=1}^{m} v_j(\zeta)(\zeta_j - g_j) = 1 - w(z)w'(\zeta).$$

Hence $H(f_1, \ldots, g_m)$ is given by the integral

$$\frac{(n+m)!}{(2\pi i)^{n+m}} \int F(z)G(\zeta) du_1 dz_1 \ldots du_n dz_n d(wv_1) d\zeta_1 \ldots d(wv_m) d\zeta_m.$$

We must show that this is equal to $F(f_1, \ldots f_n)G(g_1, \ldots, g_m)$, that is, to the integral

$$\frac{n!m!}{(2\pi i)^{n+m}} \int F(z)G(\zeta) du_1 \ldots dz_n dv_1 d\zeta_1 \ldots dv_m d\zeta_m.$$

Since w does not depend on ζ,

$$du_1 dz_1 \ldots du_n dz_n d(wv_1) d\zeta_1 \ldots d(wv_m) d\zeta_m = w^m du_1 dz_1 \ldots dv_m d\zeta_m.$$

Applying Stokes' theorem as before, we see that it suffices to find an $(n-1)$-form τ supported on U, such that

$$(n+m)!\, w^m du_1 dz_1 \ldots du_n dz_n - n!m! du_1 dz_1 \ldots du_n dz_n = d\tau dz_1 \ldots dz_n.$$

For this, it suffices to find $(n-1)$-forms $\tau_1, \ldots, \tau_{m-1}$ supported on U, such that

$$[(n+k)w^k - kw^{k-1}]du_1dz_1 \ldots du_ndz_n = d\tau_kdz_1 \ldots dz_n.$$

The desired form τ is then

$$\tau = \sum_{k=1}^{m} \frac{(n+k-1)!m!}{k!}\tau_k.$$

For $1 \leq k \leq (m-1)$, let τ_k be the obvious $(n-1)$-form which satisfies

$$\tau_kdz_1 \ldots dz_n = w^k \sum_{j=1}^{n} u_jdu_1dz_1 \ldots \widehat{du_j} \ldots du_ndz_n,$$

where the "hat" indicates that that symbol has been deleted from the expression. With the aid of the identity

$$dw = -\sum (z_j - f_j)du_j - \sum u_jdz_j,$$

we obtain

$$\begin{aligned} d\tau_kdz_1 \ldots dz_n &= nw^kdu_1 \ldots dz_n + kw^{k-1}dw \sum_{j=1}^{n} u_jdu_1 \ldots \widehat{du_j} \ldots dz_n \\ &= [nw^k - kw^{k-1} \sum (z_j - f_j)u_j]du_1 \ldots dz_n \\ &= [(n+k)w^k - kw^{k-1}]du_1dz_1 \ldots du_ndz_n, \end{aligned}$$

as desired. This completes the proof.

Suppose now that $f_1, \ldots, f_n \in A$, and that F is the monomial $F(z) = z_1^{i_1} \ldots z_n^{i_n}$. By repeated application of the product theorem, and the fact that the functional calculus behaves correctly on polynomials of one complex variable, we obtain $F(f_1, \ldots, f_n) = f_1^{i_1} \ldots f_n^{i_n}$. From the linearity of the functional calculus, we obtain property 4.1(i).

Property 4.1 (ii) now becomes a special case of the product theorem, obtained by taking $G = 1$. That completes the proof of 4.1 except for the uniqueness assertion, which will be proved in the next section.

Before proving 4.5, we establish the following lemma.

4.10 Lemma. Suppose $a = (a_1, \ldots, a_n) \in C^n$, and F is analytic in a neighborhood U of a. Suppose $u_1, \ldots, u_n$ and w are infinitely differentiable complex-valued functions on C^n, such that w is supported on U, and

$$\sum_{j=1}^{n} (z_j - a_j)u_j(z) = 1 - w(z), \qquad z \in C^n.$$

Then

$$F(a_1, \ldots, a_n) = \int_U F du_1dz_1 \ldots du_ndz_n.$$

Proof. Let A be the Banach algebra $C(\{a\})$ of continuous functions on the singleton $\{a\}$. The joint spectrum $\sigma(z_1, \ldots, z_n)$ of the coordinate functions is precisely the singleton $\{a\}$. By 4.1(i), the formula

$$P(a_1, \ldots, a_n) = \int P du_1 dz_1 \ldots du_n dz_n$$

holds for every polynomial P.

The function F can be approximated uniformly by polynomials on any compact polydisc contained in U with center a. By 4.1(iii), the element $\int F du_1 dz_1 \ldots du_n dz_n$ obtained from the functional calculus by applying F to the coordinate functions $z_1, \ldots, z_n$ must coincide with the constant $F(a_1, \ldots, a_n)$. This proves the lemma.

Proof of 4.5. Suppose $f_1, \ldots, f_n \in A$, and F is analytic in a neighborhood of $\sigma(f_1, \ldots, f_n)$. Choose $u_1, \ldots, u_n$ and w as in lemma 4.9. Suppose $\phi \in M_A$, and set $a = (\phi(f_1), \ldots, \phi(f_n))$. Applying 4.10 to the functions $\phi \circ u_1, \ldots, \phi \circ u_n$ and $\phi \circ w$, we obtain

$$\begin{aligned} \phi(F(f_1, \ldots, f_n)) &= \phi\left(\int F du_1 dz_1 \ldots du_n dz_n\right) \\ &= \int F d(\phi \circ u_1) dz_1 \ldots d(\phi \circ u_n) dz_n \\ &= F(\phi(f_1), \ldots, \phi(f_n)). \end{aligned}$$

This proves the theorem.

Now we have proved everything except for the uniqueness assertion of 4.1, the multiplicativity of the functional calculus 4.2(ii), and the substitution theorem 4.6. The proofs of these will be deferred to the next section.

5. Polynomial Approximation

The following important approximation theorem is a consequence of the properties of the functional calculus which we have already verified. It will be used to verify the remaining properties.

5.1 Theorem (Oka-Weil Approximation Theorem). Let K be a compact polynomially convex subset of C^n. Then every function analytic in a neighborhood of K can be approximated uniformly on K by polynomials.

Proof. The set K is the joint spectrum $\sigma(z_1, \ldots, z_n)$ of the coordinate functions $z_1, \ldots, z_n$, considered as elements of $P(K)$. If F is analytic in a neighborhood of K, the functional calculus provides an element $F(z_1, \ldots, z_n) \in P(K)$ which, by 4.5, must be the restriction of F to K. That is, $F \in P(K)$. That proves the theorem.

Recall that the joint spectrum $\sigma(f_1, \ldots, f_n)$ of $f_1, \ldots, f_n \in A$ need not be polynomially convex, and so functions analytic in a neighborhood of $\sigma(f_1, \ldots, f_n)$ need not be approximable uniformly on $\sigma(f_1, \ldots, f_n)$ by polynomials. The following lemma allows us to pass to a situation where polynomial approximation is possible.

5.2 Lemma (Arens-Calderón Lemma). Let U be a neighborhood in C^n of $\sigma(f_1, \ldots, f_n)$, where $f_1, \ldots, f_n$ belong to the Banach algebra A. Then there exist $f_{n+1}, \ldots, f_m \in A$ such that

$$\pi(\hat{\sigma}(f_1, \ldots, f_m)) \subseteq U,$$

where π is the projection of C^m onto C^n.

Proof. Here $\hat{\sigma}(f_1, \ldots, f_m)$ is the polynomial convex hull of $\sigma(f_1, \ldots, f_m)$. So $\pi(\hat{\sigma}(f_1, \ldots, f_m))$ is the joint spectrum of $f_1, \ldots, f_n$ with respect to the closed subalgebra of A generated by $f_1, \ldots, f_m$. Consequently, the more elements f_j we throw in, the smaller $\pi(\hat{\sigma}(f_1, \ldots, f_m))$ becomes.

If $(z_1, \ldots, z_n) \notin \sigma(f_1, \ldots, f_n)$, then there exist $f_{n+1}, \ldots, f_{2n} \in A$ such that $\sum_{j=1}^{n} (z_j - f_j)f_{n+j} = 1$. Hence $z_1 - f_1, \ldots, z_n - f_n$ belong to no common maximal ideal of the algebra generated by $f_1, \ldots, f_{2n}$. So $(z_1, \ldots z_n) \notin \pi(\hat{\sigma}(f_1, \ldots, f_{2n}))$. Consequently, there is a neighborhood of $(z_1, \ldots, z_n)$ which does not meet $\pi(\hat{\sigma}(f_1, \ldots, f_{2n}))$.

Now by a standard compactness argument, we can find $f_{n+1}, \ldots, f_m \in A$ such that U^c does not meet $\pi(\hat{\sigma}(f_1, \ldots, f_m))$. That proves the lemma.

The technique for applying the Arens-Calderón lemma is illustrated in the proof of the uniqueness of the functional calculus, which follows.

Proof of 4.1, concluded. Suppose that $F \to \tilde{F}(f_1, \ldots, f_n)$ is another functional calculus satisfying the requirements of 4.1. Suppose $f_1, \ldots, f_n \in A$, and F is analytic in a neighborhood U of $\sigma(f_1, \ldots, f_n)$. Choose $f_{n+1}, \ldots, f_m$ as in 5.2. Then $G(z_1, \ldots, z_m) = F(z_1, \ldots, z_n)$ is analytic in a neighborhood V of $\hat{\sigma}(f_1, \ldots, f_m)$. Let K be a compact polynomially convex set in C^m such that $K \subseteq V$ while $\hat{\sigma}(f_1, \ldots, f_m) \subseteq \operatorname{int}(K)$. By the Oka-Weil approximation theorem 5.1, there is a sequence of polynomials $\{p_k\}$ which converges uniformly to G on K. From the hypotheses, we obtain $\tilde{F}(f_1, \ldots, f_n) = \tilde{G}(f_1, \ldots, f_m) = \lim_{k \to \infty} p_k(f_1, \ldots, f_m) = G(f_1, \ldots, f_m) = F(f_1, \ldots, f_n)$. This proves the uniqueness.

Proof of 4.2(ii). Suppose that F and G are analytic in a neighborhood U of $\sigma(f_1, \ldots, f_n)$. By the Arens-Calderón lemma, there are $f_{n+1}, \ldots, f_m \in A$ such that $\pi(\hat{\sigma}(f_1, \ldots, f_m)) \subseteq U$, where π is the projection of C^m onto C^n. Let $\tilde{F}$ and $\tilde{G}$ be the extensions of F and G to $U \times C^{m-n}$ which are independent of the last $m - n$ coordinates. Choose a polynomially convex set J in C^m such that $\hat{\sigma}(f_1, \ldots, f_m) \subseteq \operatorname{int} J$, while $J \subseteq U \times C^{m-n}$. Then there are sequences $\{F_j\}$ and $\{G_j\}$ of polynomials on C^m such that F_j and G_j tend respectively to $\tilde{F}$ and $\tilde{G}$ uniformly on J. Then

$$\begin{aligned}(FG)(f_1,\ldots,f_n) = (\tilde{F}\tilde{G})(f_1,\ldots,f_m) &= \lim_{j\to\infty}(F_jG_j)(f_1,\ldots,f_m)\\ &= \lim_{j\to\infty} F_j(f_1,\ldots,f_m)G_j(f_1,\ldots,f_m)\\ &= \tilde{F}(f_1,\ldots,f_m)\tilde{G}(f_1,\ldots,f_m)\\ &= F(f_1,\ldots,f_n)G(f_1,\ldots,f_n).\end{aligned}$$

This concludes the proof of 4.2.

Proof of 4.6. We use the notation of 4.6, and set $g_j = F_j(f_1,\ldots,f_n)$, $1 \le j \le m$. The function

$$R(z,\zeta) = G(\zeta) - H(z)$$

is analytic in a neighborhood of $\sigma(f_1,\ldots,f_n, g_1,\ldots,g_m)$ in C^{n+m}, and $R(z_1,\ldots,z_n, F_1(z),\ldots,F_m(z)) = 0$. Hence we can write

$$R(z,\zeta) = \sum_{j=1}^{n}(\zeta_j - F_j(z))R_j(z,\zeta),$$

where the function $R_j(z,\zeta)$, defined to be equal to

$$\frac{R(z_1,\ldots,z_n,\zeta_1,\ldots,\zeta_j,F_{j+1}(z),\ldots,F_m(z)) - R(z_1,\ldots,\zeta_{j-1},F_j(z),\ldots,F_m(z))}{\zeta_j - F_j(z)},$$

is analytic in a neighborhood of $\sigma(f_1,\ldots,f_n, g_1,\ldots,g_m)$. By 4.2, we have $R(f_1,\ldots,f_n, g_1,\ldots g_m) = 0$. Hence $G(g_1,\ldots,g_m) = H(f_1,\ldots,f_n)$. This proves the theorem.

6. Implicit Function Theorem

As an application of the functional calculus, we prove the following theorem.

6.1 Theorem (Implicit Function Theorem for Banach Algebras). Let $f_0,\ldots,f_n$ belong to the Banach algebra A. Let $h \in C(M_A)$, and let $\sigma(h, f_0,\ldots,f_n)$ be the set of $(n+2)$-tuples $(h(x),\hat{f}_0(x),\ldots,\hat{f}_n(x))$, $x \in M_A$. Let $F(w, z_0,\ldots,z_n)$ be a function analytic in a neighborhood of $\sigma(h, f_0,\ldots,f_n)$ such that $F(h,\hat{f}_0,\ldots,\hat{f}_n) = 0$, while $\partial F/\partial w$ does not vanish on $\sigma(h, f_0,\ldots,f_n)$. Then there exists a unique element $g \in A$ such that $\hat{g} = h$ and $F(g, f_0,\ldots,f_n) = 0$.

Proof: Suppose $x_0 \in M_A$. Choose $\epsilon > 0$ and $\delta > 0$ so that whenever $|z_k - \hat{f}_k(x_0)| < \delta$, $0 \le k \le n$, then $F(w, z_0,\ldots,z_n) = 0$ has precisely one root in the disc $\{|w - h(x_0)| < \epsilon\}$. Let U be a neighborhood of x_0 such that $|h(x) - h(x_0)| < \epsilon$ and $|\hat{f}_k(x) - \hat{f}_k(x_0)| < \delta$, $0 \le k \le n$, whenever $x \in U$. Then $F(w,\hat{f}_0(x),\ldots,\hat{f}_n(x)) = 0$ has precisely one solution satisfying $|w - h(x_0)| < \epsilon$, namely, $h(x)$. So if $x, y \in U$ satisfy $\hat{f}_k(x) = \hat{f}_k(y)$, $0 \le k \le n$, then $h(x) = h(y)$. It follows that there is a neighborhood V of the diagonal in $M_A \times M_A$, such that whenever $(x,y) \in V$ satisfies $\hat{f}_k(x) = \hat{f}_k(y)$, $0 \le k \le n$, then $h(x) = h(y)$.

For every $(x, y) \in (M_A \times M_A)\backslash V$, there is $f \in A$ such that $\hat{f}(x) \neq \hat{f}(y)$. Then $\hat{f}$ separates a neighborhood of x from a neighborhood of y. By the compactness of $(M_A \times M_A)\backslash V$, there exists a finite set of elements $f_{n+1}, \ldots, f_m \in A$ such that if $x,y \in M_A$ satisfy $\hat{f}_k(x) = \hat{f}_k(y)$, $0 \le k \le m$, then $h(x) = h(y)$. This allows us to define a function H on $\sigma(f_0, \ldots, f_m)$ such that $h(x) = H(\hat{f}_0(x), \ldots, \hat{f}_m(x))$, $x \in M_A$. Evidently H must be continuous.

We extend F to a neighborhood of $\sigma(h, f_0, \ldots, f_n) \times C^{m-n}$ to be independent of the last $m - n$ coordinates. Then $F(H(z_0, \ldots, z_m), z_0, \ldots, z_m) = 0$ on $\sigma(f_0, \ldots, f_m)$. For each $(z_0^0, \ldots, z_m^0) \in \sigma(f_0, \ldots, f_m)$, the identity $F(w, z_0, \ldots, z_m) = 0$ defines a unique analytic function $w = G(z_0, \ldots, z_m)$ in a neighborhood of $(z_0^0, \ldots, z_m^0)$, subject to the condition $G(z_0^0, \ldots, z_m^0) = H(z_0^0, \ldots, z_m^0)$. Hence the analytic function G must agree with H on the intersection of $\sigma(f_0, \ldots, f_m)$ and some ball with center at $(z_0^0, \ldots, z_m^0)$. If two such balls overlap on a set which meets $\sigma(f_0, \ldots, f_m)$, then the two determinations of G agree on at least one point of the overlap, and so they coincide on the overlap.

Hence there is an extension $G(z_0, \ldots, z_m)$ of H which is analytic in a neighborhood of $\sigma(f_0, \ldots, f_m)$. Note that $F(G(z_0, \ldots, z_m), z_0, \ldots, z_m) = 0$ in a neighborhood of $\sigma(f_0, \ldots, f_m)$.

Let $g = G(f_0, \ldots, f_m)$. Then $g \in A$ satisfies $\hat{g} = h$. By the substitution theorem 4.6, $F(g, f_0, \ldots, f_m) = 0$. Since F is independent of the last $m - n$ coordinates, $F(g, f_0, \ldots, f_n) = 0$. That proves the existence of g.

To prove the uniqueness of g, suppose $g + r \in A$ is such that $\hat{g} + \hat{r} = h$ while $F(g + r, f_0, \ldots, f_n) = 0$. Then $\hat{r} = 0$. We must show that $r = 0$. The function $R(a, w, z_0, \ldots, z_n)$, defined to be equal to

$$\frac{1}{a^2}\left[F(w + a, z_0, \ldots, z_n) - F(w, z_0, \ldots, z_n) - a\frac{\partial F}{\partial w}(w, z_0, \ldots, z_n)\right],$$

is analytic in a neighborhood of $\sigma(r, g, f_0, \ldots, f_n)$. Multiplying R by a^2, and using the substitution theorem together with our hypotheses, we are led to the identity

$$r\left[\frac{\partial F}{\partial w}(g, f_0, \ldots, f_n) + rR(r, g, f_0, \ldots, f_n)\right] = 0.$$

Since $\partial F/\partial w$ does not vanish on $\sigma(g, f_0, \ldots, f_n)$, $\partial F/\partial w(\hat{g}, \hat{f}_0, \ldots, \hat{f}_n) + \hat{r}R(\hat{r}, \hat{g}, \hat{f}_0, \ldots, \hat{f}_n)$ does not vanish on M_A. Consequently $\partial F/\partial w(g, f_0, \ldots, f_n) + rR(r, g, f_0, \ldots, f_n)$ is invertible in A. It follows that $r = 0$. This proves the uniqueness of g, and the theorem is established.

If we apply the implicit function theorem to the function $F(w, z) = e^w - z$, we obtain the following corollary.

6.2 Corollary. If f belongs to the Banach algebra A, and $\hat{f}$ has a continuous logarithm on M_A, then $f = e^g$ for some $g \in A$.

Applying the theorem to functions of the form $F(w, z_0, \ldots, z_n) = \sum_{k=0}^{n} z_k w^k$, we obtain the following.

6.3 Corollary. Suppose $f_0, \ldots, f_n \in A$ and $h \in C(M_A)$ satisfy $\sum_{k=0}^{n} \hat{f}_k h^k = 0$, while $\sum_{k=1}^{n} k\hat{f}_k h^{k-1}$ does not vanish on M_A. Then there is a unique element $g \in A$ such that $\hat{g} = h$ while $\sum_{k=0}^{n} f_k g^k = 0$.

6.4 Corollary. Suppose f is an invertible element of A. If there is $h \in C(M_A)$ such that $h^n = \hat{f}$, then there is $g \in A$ such that $g^n = f$.

6.5 Corollary (Shilov Idempotent Theorem). Let E be an open-closed subset of the maximal ideal space M_A of A. Then there is a unique element f of A such that $f^2 = f$, while $\hat{f}$ is the characteristic function of E.

Proof. The characteristic function h of E satisfies the relation $h^2 - h = 0$ on M_A, while $2h - 1$ does not vanish on M_A. The theorem now follows from 6.3.

The Shilov idempotent theorem can also be obtained immediately from the functional calculus as follows.

By compactness of E and of $M_A \backslash E$, one can find elements $f_1, \ldots, f_n \in A$ such that $\pi(E)$ is disjoint from $\pi(M_A \backslash E)$, where π is the natural projection of M_A onto $\sigma(f_1, \ldots, f_n)$. Let $F(z_1, \ldots, z_n)$ be a function which is 1 in a neighborhood of $\pi(E)$, and 0 in a neighborhood of $\pi(M_A \backslash E)$. Then F is analytic in a neighborhood of $\sigma(f_1, \ldots, f_n)$. The element $f = F(f_1, \ldots, f_n)$ satisfies $f^2 = f$, while $\hat{f}$ is the characteristic function of E. This proves the existence.

Suppose $f + r \in A$ also has the property that $(f + r)^2 = f + r$, while $\hat{f} + \hat{r}$ is the characteristic function of E. In other words, $(2f + r - 1)r = 0$, while $\hat{r} = 0$. Since $2\hat{f} + \hat{r} - 1$ does not vanish on M_A, $2f + r - 1$ is invertible. We conclude that $r = 0$, and the idempotent f is unique.

7. Cohomology of the Maximal Ideal Space

Let $\{U_\alpha\}$ be an open cover of an open subset U of C^n. By **Cousin data** for the cover $\{U_\alpha\}$ we mean a family of functions $\{h_{\alpha\beta}\}$ analytic on $U_\alpha \cap U_\beta$, which satisfy

$$h_{\alpha\beta} + h_{\beta\gamma} + h_{\gamma\alpha} = 0$$

on $U_\alpha \cap U_\beta \cap U_\gamma$. In particular, $h_{\alpha\alpha} = 0$, and $h_{\alpha\beta} = -h_{\beta\alpha}$. The (additive) **Cousin problem** is to find functions h_α analytic on U_α, such that $h_\alpha - h_\beta = h_{\alpha\beta}$ on $U_\alpha \cap U_\beta$.

An **open polynomial polyhedron** in C^n is a subset of C^n of the form

$$U = \{z : |p_j(z)| < 1, 1 \leq j \leq m\},$$

where $p_1, \ldots, p_m$ are fixed polynomials. In this section, and in the next section, we wish to use the solvability of the Cousin problem for open polynomial polyhedra. This theorem, proved in some special cases by Cousin, is due to Oka.

7.1 Theorem. Let $\{h_{\alpha\beta}\}$ be Cousin data for the open cover $\{U_\alpha\}$ of an open polynomial polyhedron U. Then there exist functions h_α analytic on U_α, such that

$$h_\alpha - h_\beta = h_{\alpha\beta}$$

on $U_\alpha \cap U_\beta$.

We shall not prove this theorem. The reader interested in more details concerning the Cousin problem should consult the first chapter of Gunning and Rossi [1].

7.2 Theorem (Arens-Royden Theorem). Suppose $f \in C(M_A)$ does not vanish on M_A. Then there exists $g \in A^{-1}$ such that $f/\hat{g}$ has a continuous logarithm on M_A.

Proof: By the Stone-Weierstrass theorem, finite linear combinations of functions of the form $\hat{h}\bar{\hat{k}}$ are dense in $C(M_A)$, where h and k range over all elements of A. Hence there are $h_1, \ldots, h_{2n} \in A$ such that

$$\left\| 1 - \sum_{j=1}^{n} \frac{\hat{h}_j \bar{\hat{h}}_{n+j}}{f} \right\|_{M_A} < 1.$$

In particular, the values of $\sum_{j=1}^{n} \hat{h}_j \bar{\hat{h}}_{n+j}/f$ lie in the right half-plane. So $\sum_{j=1}^{n} \hat{h}_j \bar{\hat{h}}_{n+j}/f$ has a continuous logarithm on M_A—the principal branch will do. It suffices now to find $g \in A^{-1}$ such that $\sum_{j=1}^{n} \hat{h}_j \bar{\hat{h}}_{n+j}/\hat{g}$ has a continuous logarithm on M_A.

Now define

$$F(z) = \sum_{j=1}^{n} z_j \overline{z_{n+j}}\,.$$

Since $\sum_{j=1}^{n} \hat{h}_j \bar{\hat{h}}_{n+j}$ does not vanish on M_A, F does not vanish on $\sigma(h_1, \ldots, h_{2n})$. Let V be a neighborhood of $\sigma(h_1, \ldots, h_{2n})$ on which F does not vanish. By the Arens-Calderón lemma 5.2, there exist $h_{2n+1}, \ldots, h_m \in A$ and an open polynomial polyhedron U containing $\sigma(h_1, \ldots, h_m)$, such that $\pi(U) \subseteq V$, where π is the projection of C^m onto C^{2n}.

The function $F(z) = \sum_{j=1}^{n} z_j \bar{z}_{n+j}$ remains defined on C^m, and F does not vanish on U. Let $\{U_j\}_{j=1}^{\infty}$ be open balls which cover U, such that F has a continuous logarithm F_j on U_j. That is, F_j is a continuous function on U_j satisfying $e^{F_j} = F$ on U_j. On $U_j \cap U_k$ we define the function $F_{jk} = F_j - F_k$. Then $F_{jk}/2\pi i$ is an integer. Also, on $U_j \cap U_k \cap U_p$, we have

$$F_{jk} + F_{kp} + F_{pj} = 0.$$

Hence $\{F_{jk}\}$ are Cousin data for the cover $\{U_j\}$ of U.

By Oka's theorem 7.1, there are analytic functions G_j on U_j such that $G_j - G_k = F_{jk}$ on $U_j \cap U_k$. Then $e^{G_j} = e^{G_k}$ on $U_j \cap U_k$. So the function G on V, defined to be equal to e^{G_j} on U_j, is well-defined and analytic.

Now $F_j - G_j = F_k - G_k$ on $U_j \cap U_k$. Hence the function H on U which is equal to $F_j - G_j$ on U_j is well-defined, and it satisfies $e^H = F/G$.

Let $g = G(h_1, \ldots, h_m) \in A$. Then the function on M_A defined by

$$\tilde{H}(x) = H(h_1(x), \ldots, h_m(x)), \qquad x \in M_A,$$

is a continuous logarithm of $\sum \hat{h}_j \bar{\hat{h}}_{n+j}/\hat{g}$. This proves the theorem.

The exponentials e^A of elements in A form a subgroup of the multiplicative group A^{-1} of invertible elements of A. Theorems 6.2 and 7.2 show that the quotient group A^{-1}/e^A depends only on the topological space M_A. More precisely, the following corollary obtains.

7.3 Corollary. The correspondence $f \to \hat{f}e^{C(M_A)}$ is a homomorphism of the group A^{-1} onto the quotient group $C(M_A)^{-1}/e^{C(M_A)}$. The kernel of the homomorphism is e^A.

There are some other useful descriptions of the quotient group $C(X)^{-1}/e^{C(X)}$, X a compact Hausdorff space.

It is easy to see that $C(X)^{-1}/e^{C(X)}$ is isomorphic to the group of homotopy classes of maps of X into the circle group. The coset $fe^{C(X)}$ of the function $f \in C(X)^{-1}$ corresponds to the homotopy class of the map $f/|f|$ of X into the circle group.

Another description proceeds as follows. Suppose $f \in C(X)^{-1}$. Let $\{U_j\}$ be a finite open cover of X, so that $\log f$ has a continuous determination f_j on each U_j. The U_j and the f_j can be chosen so that $f_j - f_k$ is constant on $U_j \cap U_k$. Then $(f_j - f_k)/2\pi i$ is an integer n_{jk}. The integers $\{n_{jk}\}$ form a one cocycle on the nerve of the cover $\{U_j\}$. Hence they determine an element $[f]$ of $H^1(X, Z)$, the first Čech cohomology group of X with integer coefficients. The theorem of Bruschlinsky states that the cohomology class of the $\{n_{jk}\}$ depends only on f, and not on the determinations of the logarithms. Moreover, the correspondence $f \to [f]$ is a homomorphism of $C(X)^{-1}$ onto $H^1(X, Z)$, whose kernel is precisely $e^{C(X)}$. Hence we obtain the following corollary.

7.4 Corollary. The group A^{-1}/e^A is isomorphic to $H^1(M_A, Z)$, the first Čech cohomology group of M_A with integer coefficients.

Now suppose that A is a uniform algebra. Then $\log|A^{-1}|$ is an additive group, which contains $\mathrm{Re}\,(A)$ as a subgroup. The homomorphism $f \to \log|f|$ of A^{-1} onto $\log|A^{-1}|$ maps e^A onto $\mathrm{Re}\,(A)$. The induced quotient map is a homomorphism of A^{-1}/e^A onto the quotient group $\log|A^{-1}|/\mathrm{Re}(A)$.

Suppose fe^A belongs to the kernel of this homomorphism. Choose $h \in A$ such that $\log|f| = \mathrm{Re}\,(h)$. Then $g = fe^{-h}$ belongs to the coset fe^A, and $|g| = 1$. So the kernel of the homomorphism consists precisely of the cosets of the functions in A^{-1} which have unit modulus.

7.5 Theorem. If A is an antisymmetric uniform algebra, then the Čech cohomology group $H^1(M_A, Z)$ is isomorphic to the additive quotient group $\log|A^{-1}|/\mathrm{Re}\,(A)$.

Proof. Recall that a uniform algebra A is antisymmetric if the only real-valued functions in A are the constants (cf. II.13). In this case, if $g \in A^{-1}$ has unit modulus, then $g + 1/g = g + \bar{g}$ and $g - 1/g = g - \bar{g}$ must be constant. So g is constant, and $g \in e^A$. In view of the remarks preceding the theorem, the kernel of the homomorphism of $H^1(M_A,Z) \cong A^{-1}/e^A$ onto $\log|A^{-1}|/\mathrm{Re}\,(A)$ reduces to the identity. That does it.

8. Local Maximum Modulus Principle

A closed subset E of the maximal ideal space M_A of the Banach algebra A is a **peak set** if there is an element $f \in A$ such that $\hat{f} = 1$ on E, while $|\hat{f}| < 1$ off E. Peak sets of uniform algebras were investigated in II.12.

A closed subset E of M_A is a **local peak set** if there is $f \in A$ and a neighborhood U of E in M_A such that $\hat{f} = 1$ on E, while $|f| < 1$ on $U \backslash E$.

8.1 Theorem (Rossi's Local Peak Set Theorem). Local peak sets are peak sets.

Proof. Let E be a local peak set. Choose $f \in A$ and a neighborhood U of E such that $\hat{f} = 1$ on E and $|\hat{f}| < 1$ on $\bar{U}\backslash E$. Let $f_1 = f - 1$. Then $\mathrm{Re}\,(\hat{f}_1) < 0$ on $\bar{U}\backslash E$.

By the compactness of E, we can assume that U is the union of the basic neighborhoods

$$U_j = \{x : |f_k(x)| < 1, n_{j-1} < k \leq n_j\}, \qquad 1 \leq j \leq r,$$

where $1 = n_0 < \cdots < n_r$ and $f_2, \ldots, f_{n_r} \in A$. Let $V_j = \{z \in C^{n_r} : |z_k| < 1, n_{j-1} < k \leq n_j\}$, and set $V = \bigcup_{j=1}^{r} V_j$. Let $W = \bar{V}^c \cup \{\mathrm{Re}\,(z_1) < 0\}$. Then $V \cap W \subseteq \{\mathrm{Re}\,(z_1) < 0\}$. We wish to show that the open sets V and W cover $\sigma(f_1, \ldots, f_{n_r})$.

If $x \in M_A \backslash \bar{U}_j$, then $(\hat{f}_1(x), \ldots, \hat{f}_{n_r}(x)) \notin \bar{V}_j$. Hence if $x \in M_A \backslash \bar{U}$, then $(\hat{f}_1(x), \ldots, \hat{f}_{n_r}(x)) \notin \bar{V}_1 \cup \cdots \cup \bar{V}_r = \bar{V}$, and $(\hat{f}_1(x), \ldots, \hat{f}_{n_r}(x)) \in W$. If $x \in \bar{U} \backslash E$, then $\operatorname{Re} \hat{f}_1(x) < 0$, so $(\hat{f}_1(x), \ldots, \hat{f}_{n_r}(x)) \in W$. Finally, if $x \in U$, then $(\hat{f}_1(x), \ldots, \hat{f}_{n_r}(x)) \in V$. And so $(\hat{f}_1(x), \ldots, \hat{f}_{n_r}(x)) \in V \cup W$ for all $x \in M_A$. That is, $\sigma(f_1, \ldots, f_{n_r}) \subseteq V \cup W$.

Choose $f_{n_r+1}, \ldots, f_m \in A$ and an open polynomial polyhedron K in C^m such that $\sigma(f_1, \ldots, f_m) \subseteq K$ and $\pi(K) \subseteq V \cup W$, where π is the projection of C^m onto C^{n_r}. Let $K_1 = \pi^{-1}(V) \cap K$ and $K_2 = \pi^{-1}(W) \cap K$. Then $K_1 \cap K_2 \subseteq \{\operatorname{Re}(z_1) < 0\}$. So we can define a branch of $\log z_1$ on $K_1 \cap K_2$. The function $(\log z_1)/z_1$ is Cousin data for the cover $\{K_1, K_2\}$ of K. By 7.1, there are analytic functions g_1 on K_1 and g_2 on K_2 such that $g_2 - g_1 = (\log z_1)/z_1$ on $K_1 \cap K_2$. Define

$$G(z) = \begin{cases} z_1 e^{z_1 g_1(z)}, & z \in K_1 \\ e^{z_1 g_2(z)}, & z \in K_2. \end{cases}$$

Then G is well-defined and analytic on K. Set $g = G(f_1, \ldots, f_m) \in A$. Then $\hat{g}$ vanishes precisely on the set E.

Now $1/G$ is bounded on compact subsets of K_2. On K_1, we have

$$\frac{1}{G} = \frac{1}{z_1} + \frac{e^{-z_1 g_1} - 1}{z_1},$$

where $\operatorname{Re}(1/z_1) \leq 0$, and $(e^{-z_1 g_1} - 1)/z_1$ is bounded on each compact subset of K_1. Hence the range of $1/\hat{g}$ omits a half-plane of the form $\{\operatorname{Re}(w) > M\}$. It follows that the range of $\hat{g}$ omits the disc $\{|w - 1/M| < 1/M\}$. Hence for $\epsilon = 1/2M$, the function $\epsilon/(\epsilon - \hat{g})$ peaks on E. This proves the theorem.

The preceding proof simplifies somewhat when E consists of a single point. In this case, the theorem states that local peak points are peak points. The general case can be deduced from this special case by considering the algebra of those $g \in A$ such that $\hat{g}$ is constant on E.

8.2 Theorem (Local Maximum Modulus Principle). If U is an open subset of M_A, then for all $f \in A$,

$$\|\hat{f}\|_{\bar{U}} = \|\hat{f}\|_{(\partial_A \cap U) \cup bU}.$$

Proof. Let E be the set of $x \in \bar{U}$ such that $|\hat{f}(x)| = \|f\|_{\bar{U}}$. We must show that E meets either bU or ∂_A.

Suppose E does not meet bU. Modifying f by a constant multiple, we can assume that $\|f\|_{\bar{U}} = 1$, and that $f(x) = 1$ for some $x \in E$. Let F be the set of $y \in U$ such that $\hat{f}(y) = 1$. Then F is a non-empty closed subset of E. Also, $(1 + \hat{f})/2$ is 1 on F, while $|(1 + \hat{f})/2| < 1$ on $U \backslash F$. By 8.1, F is a peak set. Hence F meets ∂_A. This proves the theorem.

Try proving the following corollary without the local maximum modulus principle.

8.3 Corollary. If M_A is homeomorphic to the unit interval [0, 1], then $M_A = \partial_A$.

Proof. Otherwise there is a non-empty interval (a, b) contained in $M_A \backslash \partial_A$. By 8.2, every f in A which satisfies $\hat{f}(a) = \hat{f}(b) = 0$ must vanish identically on $[a, b]$. This is absurd.

9. Extensions of Uniform Algebras

Let A be a uniform algebra. A continuous function f on M_A is ***A*-holomorphic** at a point $x \in M_A$ if there is a neighborhood U of x such that f can be approximated uniformly on U by functions in A. A function f is A-holomorphic on a subset S of M_A if f is A-holomorphic at every point of S.

For f a continuous function on M_A, $[A, f]$ will denote the uniform algebra generated by A and f. Note that $\partial_{[A,f]} \subseteq M_A \subseteq M_{[A,f]}$.

The following lemma embodies one of the principal applications of the local maximum modulus principle.

9.1 Lemma. Let A be a uniform algebra, let U be an open subset of M_A, and let f be a continuous complex-valued function on M_A which is A-holomorphic on U. Then $\partial_{[A,f]} \subseteq \partial_A \cup U^c$.

Proof. Suppose $x \in U \backslash \partial_A$. There is an open set V containing x such that V does not meet ∂_A, while f can be approximated uniformly on V by functions in A. Hence the closure of $[A, f]$ in $C(\bar{V})$ coincides with the closure $\overline{A|_V}$ of A in $C(\bar{V})$.

By the local maximum modulus principle 8.2, the Shilov boundary of $\overline{A|_V}$ is contained in bV. So $\partial_{[A,f]}$ cannot meet V. Hence $\partial_{[A,f]}$ does not meet $U \backslash \partial_A$. That proves the lemma.

Now we turn to an abstract version of Radó's theorem II.6.3.

9.2 Theorem. Let A be a uniform algebra. Let f be a continuous function on M_A such that f is A-holomorphic on $M_A \backslash f^{-1}(0)$. Then $M_{[A,f]} = M_A$ and $\partial_{[A,f]} = \partial_A$.

Proof. Let $B = [A, f]$. By 9.1, $\partial_B \subseteq \partial_A \cup f^{-1}(0)$.

Suppose $\partial_B \neq \partial_A$. Then $\partial_B \backslash \partial_A$ is a non-empty relatively open subset of ∂_B on which f vanishes. By Glicksberg's lemma II.6.2, f vanishes on an open subset U of M_A which meets $\partial_B \backslash \partial_A$. But then $A|_U = B|_U$, so

$$U \cap (\partial_B \backslash \partial_A) \subseteq \partial_{\overline{B|_U}} = \partial_{\overline{A|_U}} \subseteq bU \cup \partial_A.$$

This contradiction shows that $\partial_B = \partial_A$.

Now let π be the projection of M_B onto M_A. If $x \in M_B$, then $\pi(x)$ is the restriction of the homomorphism x to A. We must show that every fiber $\pi^{-1}(y)$ consists of exactly one point, that is, that every $y \in M_A$ extends uniquely to B.

Let $F = f \circ \pi \in C(M_B)$. Then F is constant on every fiber $\pi^{-1}(y)$. Now F is B-holomorphic on $M_B \backslash F^{-1}(0)$. By the half of the theorem we have already proved, $\partial_{[B,F]} = \partial_B = \partial_A$. However, F coincides with f on M_A, and, in particular, on ∂_A. So $F = f$, and f is constant on every fiber $\pi^{-1}(y)$. It follows that $\pi^{-1}(y)$ consists of no more than one point. So π is in fact a homeomorphism, and $M_B = M_A$.

9.3 Corollary. Let A be a uniform algebra. Let B be a uniform algebra on M_A such that $A \subseteq B$, and such that every $g \in B$ is A-holomorphic on M_A. Then $M_B = M_A$ and $\partial_B = \partial_A$.

Proof. By 9.1, every function in B attains its maximum modulus on ∂_A. So $\partial_B = \partial_A$. By 9.2, every $x \in M_A$ extends uniquely to each $g \in B$. Hence $M_A = M_B$.

9.4 Theorem. Let A be a uniform algebra. Suppose $f \in C(M_A)$ satisfies a relation of the form

$$f^n + g_{n-1}f^{n-1} + \cdots + g_1 f + g_0 = 0,$$

where $g_0, \ldots, g_{n-1} \in A$. Then $M_{[A,f]} = M_A$ and $\partial_{[A,f]} = \partial_A$.

Proof. We can assume, by induction, that the theorem is true for all uniform algebras and all continuous roots of monic polynomials of degree $n - 1$.

Consider the formal derivative

$$h = nf^{n-1} + (n-1)g_{n-1}f^{n-2} + \cdots + g_1.$$

In a neighborhood of a point x at which $h(x) \neq 0$, the function f can be expressed as a convergent power series in the coefficients $g_0, \ldots, g_{n-1}$. Hence f is A-holomorphic off the set $h^{-1}(0)$. It follows that h is A-holomorphic off the set $h^{-1}(0)$. By 9.2, $\partial_{[A,h]} = \partial_A$ and $M_{[A,h]} = M_A$.

Now f satisfies a monic polynomial of degree $n - 1$ with coefficients in $[A, h]$, namely

$$f^{n-1} + \frac{n-1}{n} g_{n-1} f^{n-2} + \cdots + \frac{g_1 - h}{n} = 0.$$

By the induction assumption, $M_{[A,h,f]} = M_A$ and $\partial_{[A,h,f]} = \partial_A$. Since $[A, h, f] = [A, f]$, we are done.

9.5 Corollary. Let A be a uniform algebra, and let $f \in C(M_A)$. Let S_j, $1 \leq j \leq n$, be subsets of M_A such that $f|_{S_j} \in A|_{S_j}$, $1 \leq j \leq n$, while $\bigcup_{j=1}^{n} S_j = M_A$. Then $M_{[A,f]} = M_A$ and $\partial_{[A,f]} = \partial_A$.

Proof. If $g_j \in A$ coincides with f on S_j, then $\prod_{j=1}^{n} (f - g_j) = 0$. Now apply 9.4.

NOTES

For polynomial convexity of unions of spheres and polydiscs, see Kallin [2]. The example in 2.5 is a modification by Rudin [2] of the arc example of Wermer [9]. The discussion of circled sets is based on deLeeuw [1, 2] and on the exposition of Rossi [1]. The existence of the functional calculus for Banach algebras is due to Waelbroeck [1]. The account in sections 4 and 5 is based on techniques developed by Waelbroeck as expounded in Bingen, Tits, and Waelbroeck [1], with some improvements due to Bourbaki [1]. The Shilov idempotent theorem was proved for finitely generated Banach algebras by Shilov [1], using the Cauchy-Weil integral formula. The general case is due to Arens and Calderón [1] and Waelbroeck [1]. The implicit function theorem for Banach algebras, and its consequences, are in Arens and Calderón [1]. An elegant treatment of the Arens-Royden theorem is in Royden [3]. The local maximum modulus principle is due to Rossi [1]. The presentation here follows that of Gunning and Rossi [1]. The statements about the Shilov boundaries in 9.2 and 9.4 were proved by Glicksberg [1]. The statement about the maximal ideal space in 9.2 is due to Rickart [3], and this proof is an unpublished argument of Quigley. Corollary 9.5 is due to Stolzenberg [3].

EXERCISES

1. (a) The union of two disjoint compact convex subsets of C^n is polynomially convex.
(b) The union of two polydiscs in C^n need not be polynomially convex.

2. Fix $a > 2$. Let K be the set of $z \in C$ satisfying $a \geq |z| \geq 1/a$ and $|z - 1| \geq 1/a$. For $z \in K$, define $\Phi(z) = (z, 1/z, 1/(z-1)) \in C^3$.
(a) $R(K) \cong P(\Phi(K))$.
(b) The polynomial convex hull of $\Phi(bK)$ is $\Phi(K)$.
(c) The three square polycylinders in C^3 with radii a and centers at $(-a + 1/a, 0, a + 1 - 1/(a+1))$, $(a + 1 - 1/a, a + 1 - 1/(a+1), 0)$, and $(0, -a + 1/a, -a + 1 - 1/(a+1))$ are disjoint. Their union is not polynomially convex.

3. Under the assumptions of 3.2 and 3.3, the Shilov boundary of K is the closure of $L^{-1}(E)$, where E is the set of extreme points of $L(K)$.

4. If K is a compact rationally convex subset of C^n, then every function analytic in a neighborhood of K is in $R(K)$.

5. There is a uniform algebra A, and an $f \in A$, such that $\hat{f}$ has a continuous square root on M_A, while f has no square root in A.

6. If B is a commutative Banach algebra, and if M_B (cf. exercise 6, chapter I) is compact, then there is $e \in B$ such that $ef = f$ for all $f \in B$.

7. (a) Suppose $f \in C(b\Delta)$ has an absolutely convergent Fourier series, and f does not vanish on $b\Delta$. Then there is an integer n, and a function $h \in C(b\Delta)$, such that h has an absolutely convergent Fourier series, and $f(e^{i\theta}) = e^{in\theta}e^{h(e^{i\theta})}$.

(b) Suppose $f \in L^1(-\infty, \infty)$ and $\lambda \neq 0$ are such that $\hat{f}$ does not assume the value λ. Then there is an integer n and a function $h \in L^1(-\infty, \infty)$ such that

$$\lambda - \hat{f}(s) = \lambda\left(\frac{i-s}{i+s}\right)^n e^{\hat{h}(s)}.$$

Hint: Show that $[(i-s)/(i+s)]^n\widehat{L^1} \subseteq \widehat{L^1}$, and use the description of the maximal ideal space of L^1 stated in I.3.6.

8. (a) The only element of A^{-1}/e^A of finite order is the identity.

(b) If M_A is locally connected, then no element of A^{-1}/e^A is completely divisible.

(c) The maximal ideal space of $H^\infty(\Delta)$ is not locally connected.

9. Let A be a uniform algebra, and suppose that M_A is obtained from a compact space Y by identifying two points $y_0, y_1 \in Y$. Regard A as an algebra of functions on Y.

(a) There exists $h \in C(Y)$ such that $h^2 \in A^{-1}$, $h(y_0) = -1$, and $h(y_1) = 1$. *Hint:* Use the Arens-Royden theorem.

(b) $M_{[A,h]} = Y$.

(c) If $y_0, y_1 \notin \partial_A$, then $\partial_{[A,h]} = \partial_A$.

10. A subset E of M_A is an **A-variety** if for every $x \in E$, there is a neighborhood U of x and a subset S of A such that $U \cap E$ consists of the common zeros of $S|_U$. Show that every closed A-variety in M_A is A-convex.

11. If A and B are uniform algebras on X and Y respectively, then $A \otimes B$ is the closed subalgebra of $X \times Y$ generated by functions of the form $h(x, y) = f(x)g(y)$, $f \in A$, $g \in B$.

(a) $C(X) \otimes C(Y) = C(X \times Y)$.

(b) $M_{A\otimes B} = M_A \times M_B$.

(c) Let $f \in C(M_A)$ satisfy $\|f\| \leq 1$. Then $M_{[A,f]}$ is the $A \otimes P(\Delta)$-convex hull in $M_A \times \Delta$ of the graph of f.

(d) If the level sets of $f \in C(M_A)$ are not A-convex, then $M_{[A,f]} \neq M_A$.

(e) Identify $M_{[A,f]}$, for $A = P(\Delta)$ and $f(z) = |z|$.

12. Let A be a uniform algebra on X, and let $\{E_j\}_{j=1}^\infty$ be a closed cover of X.

(a) If $A|_{E_j}$ is closed in $C(E_j)$, $1 \leq j < \infty$, then $M_A = \bigcup_{j=1}^{\infty} \hat{E}_j$. *Hint:* Consider a Jensen measure for $\phi \in M_A$.

(b) If $A|_{E_j} = C(E_j)$, $1 \leq j < \infty$, then $A = C(X)$. *Hint:* Use the antisymmetric decomposition theorem and the Baire category theorem.

CHAPTER IV

HARDY SPACES

Throughout this chapter we will be considering a fixed uniform algebra A on a compact Hausdorff space X, and a fixed homomorphism $\phi \in M_A$.

By m we will denote a representing measure for ϕ. For $0 < p < \infty$, the Hardy space $\boldsymbol{H^p(m)}$ is the closure of A in $L^p(m)$. The weak-star closure of A in $L^\infty(m)$ is $\boldsymbol{H^\infty(m)}$. The Hardy spaces form a nested family of vector spaces, that is, $H^\infty \subseteq H^p \subseteq H^q$ if $\infty > p > q$.

Often it is possible to obtain important information about A by passing to the algebra H^∞, or by studying the other H^p-spaces. These methods are decisive in the problem, taken up in Chapter VI, of embedding an analytic structure in the maximal ideal space M_A of A.

In this chapter, we give a systematic discussion of the Hardy spaces. Sections 1 through 3 treat the general case, while sections 4 through 6 are aimed at the case of a finite dimensional set of representing measures. Some applications to rational approximation are developed in sections 7 and 8.

1. The Conjugation Operator

The kernel of ϕ will be denoted by A_0. That is, A_0 consists of the functions $f \in A$ such that $\int f dm = 0$. For $0 < p < \infty$, $\boldsymbol{H^p_0(m)}$ is the closure of A_0 in $L^p(m)$, while $\boldsymbol{H^\infty_0(m)}$ is the weak-star closure of A_0 in $L^\infty(m)$.

If $p \geq 1$, ϕ extends to be a continuous functional on H^p. This extension also will be denoted by ϕ. Evidently H^p_0 is the kernel of ϕ on H^p, $p \geq 1$.

The intersection $H^1 \cap L^\infty$ is a weak-star closed subalgebra of L^∞ (to prove this, see 2.1) which contains H^∞, but which does not coincide with H^∞ in general. If $f \in H^1$ and $g \in H^1 \cap L^\infty$, one obtains, by first approximating g, then f, that $\int fg dm = \int f dm \int g dm$. Hence ϕ is multiplicative on $H^1 \cap L^\infty$, and, in particular, on H^∞.

The functions in H^∞ can be regarded (cf. I.9) as continuous functions on the maximal ideal space Y of $L^\infty(m)$. Since H^∞ is closed under uniform convergence, H^∞ becomes a uniform algebra on the quotient space obtained by identifying the points of Y which are identified by all functions in H^∞. It will be convenient to consider H^∞ as a subalgebra of $C(Y)$, even though it may not separate the points of Y.

The measure m on X determines a unique measure $\tilde{m}$ on Y such that $m(E) = \int \chi_E \, d\tilde{m}$ for every Borel subset E of X. Here χ_E is the characteristic function of E, regarded as a function in $C(Y)$.

The measure $\tilde{m}$ is a representing measure for the homomorphism ϕ of $H^\infty(m)$, regarded as a subalgebra of $C(Y)$. What is the space $H^\infty(\tilde{m})$ associated with the algebra $H^\infty(m)$ and the representing measure $\tilde{m}$?

Every function in $L^\infty(\tilde{m})$ is actually continuous on Y, after adjustment on a set of $\tilde{m}$-measure zero. This allows us to identify $L^\infty(\tilde{m})$ and $C(Y) = L^\infty(m)$ in a natural way (cf. I.9.5). Similarly we can identify $L^p(\tilde{m})$ and $L^p(m)$ for $p > 0$.

Having identified $L^1(m)$ and $L^1(\tilde{m})$, we see that $H^\infty(m)$ is already weak-star closed in $L^\infty(\tilde{m})$. So $H^\infty(\tilde{m}) = H^\infty(m)$. And all the Hardy spaces of H^∞ coincide with the respective Hardy spaces of A.

Now $\int fg dm = \int f dm \int g dm$ also holds true if f and g belong to H^2. In particular, if $f \in H^2$ and $g \in H^2_0$, then $\int fg dm = 0$. So $H^2 \perp \overline{H^2_0}$ in L^2. One consequence of this relation is that the only real-valued functions in H^2 are the constants. In fact, if u is any such function, then $u - \int u dm$ is orthogonal to itself, so $u = \int u dm$ is constant.

The **conjugate function** of $u \in \text{Re}\,(H^2)$ is the function $^*u \in \text{Re}\,(H^2_0)$ such that $u + i^*u \in H^2$. In view of the preceding remark, $^*u$ is uniquely determined (a.e. dm) by u. If $u \in \text{Re}\,(A)$, then $^*u$ can be chosen so that $u + i^*u \in A$.

The conjugation operator $u \to {}^*u$ is real linear. If $u \in \text{Re}\,(H^2_0)$, then $u + i^*u \in H^2_0$, and

$$\begin{aligned} 0 &= \int (u + i^*u)^2 \, dm \\ &= \int u^2 dm - \int (^*u)^2 dm + 2i \int u^*u \, dm. \end{aligned}$$

Equating real and imaginary parts to zero, we make two deductions. First, $^*u \perp u$ for all $u \in \text{Re}\,(H^2)$. Secondly, the conjugation operator is an isometry of $\text{Re}\,(H^2_0)$. We state this result formally for emphasis.

1.1 Theorem. The conjugation operator $u \to {}^*u$ is an isometry of $\text{Re}\,(H^2_0)$ onto itself.

It follows, in particular, that $\text{Re}\,(H^2)$ is the closure of $\text{Re}\,(A)$ in L^2.

In some situations, the conjugation operator can be extended beyond $\text{Re}\,(H^2)$. One result, which will eventually allow us in special cases to extend the conjugation operator to the closure of $\text{Re}\,(A)$ in L^1, is the following extension of the classical theorem of Kolmogoroff on conjugate Fourier series.

1.2 Theorem. If $u + i{}^*u \in A$, and $u > 0$, then

$$\|{}^*u\|_p \leq \|u\|_1 \Big/ \left(\cos \frac{\pi p}{2}\right)^{1/p}, \qquad 0 < p < 1.$$

Proof. By composing $u + i{}^*u$ with that branch of z^p in the right half-plane which is positive for $z > 0$, we can define $(u + i{}^*u)^p \in A$. This function satisfies

$$\int (u + i{}^*u)^p \, dm = \left\{\int (u + i{}^*u) \, dm\right\}^p = \left\{\int u dm\right\}^p.$$

Taking real parts, and noting that the argument of $(u + i{}^*u)^p$ does not exceed $\pi p/2$, we have

$$\cos\left(\frac{\pi p}{2}\right) \int |u^2 + ({}^*u)^2|^{p/2} dm \leq \int (u + i{}^*u)^p dm.$$

This yields

$$\cos\left(\frac{\pi p}{2}\right) \int |{}^*u|^p dm \leq \left\{\int u dm\right\}^p = \|u\|_1^p.$$

That proves the theorem.

The classical proof of the theorem of M. Riesz on conjugate Fourier series also extends to this generality. It shows that for each p in the range $1 < p < \infty$ which is not an odd integer, there is a universal constant c_p such that $\|{}^*u\|_p \leq c_p \|u\|_p$ whenever $u + i{}^*u \in A$ satisfies $u > 0$. Instead of proving this assertion, we give a version of the theorem due to Bochner, which does not assume $u > 0$.

1.3 Theorem (M. Riesz–Bochner Theorem). For each even integer $k \geq 2$, there is a universal constant c_k such that

$$\|{}^*u\|_k \leq c_k \|u\|_k$$

for all $u \in \mathrm{Re}\,(A)$.

Proof. If $u \in \mathrm{Re}\,(A_0)$, then

$$0 = \int (u + i{}^*u)^k dm = \sum_{n=0}^{k} \binom{k}{n} i^n \int u^{k-n} ({}^*u)^n dm.$$

Taking real parts, we obtain

$$0 = \sum_{j=0}^{k/2} \binom{k}{2j} (-1)^j \int u^{k-2j} ({}^*u)^{2j} dm.$$

Hence

$$\int ({}^*u)^k dm \leq \sum_{j=0}^{k/2-1} \binom{k}{2j} \left| \int u^{k-2j} ({}^*u)^{2j} dm \right|.$$

From Hölder's inequality,

$$\left| \int u^{k-2j} ({}^*u)^{2j} dm \right| \leq \left\{\int u^k dm\right\}^{1-2j/k} \left\{\int ({}^*u)^k dm\right\}^{2j/k}$$

Let $\xi = \|{*u}\|_k/\|u\|_k$. The preceding two inequalities yield

$$\xi^k \leq \sum_{j=0}^{k/2-1} \binom{k}{2j}\xi^{2j} \leq \sum_{n=0}^{k-1}\binom{k}{n}\xi^n = (1+\xi)^k - \xi^k.$$

Hence $2\xi^k \leq (1+\xi)^k$, or $\xi \leq 1/(2^{1/k}-1)$. So

$$\|{*u}\|_k \leq (2^{1/k}-1)^{-1}\|u\|_k, \qquad u \in \operatorname{Re}(A_0).$$

If $v \in \operatorname{Re}(A)$, then $v - \int v\,dm \in \operatorname{Re}(A_0)$, and

$$\|{*v}\|_k \leq (2^{1/k}-1)^{-1}\left\|v - \int v\,dm\right\|_k \leq 2(2^{1/k}-1)^{-1}\|v\|_k.$$

That does it.

One additional piece of information is that the constants c_k can be chosen to be of the order of k. In fact, we can take

$$c_k = \frac{2}{(2^{1/k}-1)} \leq \frac{2k}{\log 2}.$$

2. Representing Measures for H^∞

The Krein-Schmulian theorem, applied to L^∞, takes the following form.

2.1 Lemma. For a convex subset S of L^∞, the following are equivalent:

(i) S is weak-star closed.

(ii) If $\{f_n\}_{n=1}^\infty$ is a sequence in S which converges pointwise boundedly to f, then $f \in S$.

Proof. Since a sequence that converges pointwise boundedly also converges in the weak-star topology, (i) implies (ii).

Conversely, suppose the set S has the property (ii). We wish to prove that S is weak-star closed.

By the Krein-Schmulian theorem, it suffices to show that the intersection of S with every multiple of the unit ball of L^∞ is weak-star closed. Hence we can assume that S is bounded in L^∞.

Let f be a weak-star adherent point of S. Then f belongs to the weak closure of S in L^2, and consequently f belongs to the closure of S in L^2. Let f_n be a sequence in S which converges to f in L^2. A subsequence of the f_n will converge almost everywhere to f. Since S is assumed bounded, the subsequence converges pointwise boundedly to f. And $f \in S$. This proves the lemma.

Note that 2.1 does not say anything about the weak-star closure of a convex subset of L^∞.

Another fact concerning the weak-star topology which we shall have occasion to use is the following: If f is a weak-star cluster point of a sequence $f_n \in L^\infty$, then

$$|f| \leq \limsup |f_n|.$$

This is deduced by noting that f is a weak cluster point of the f_n in L^2, and by considering convex combinations of the f_n which converge almost everywhere to f_1.

By $\boldsymbol{P}$ we will denote the cone of non-negative functions in L^∞_R.

2.2 Lemma. For $u \in L^\infty_R$, the following are equivalent:

(i) For each $t > 0$, there is an $h_t \in H^\infty$ such that $\phi(h_t) = 1$ and $|h_t| \leq e^{tu}$.

(ii) u is in the weak-star closure of $\operatorname{Re}(A_0) + P$.

(iii) u is in the uniform closure of $\operatorname{Re}(H^\infty_0) + P$.

Proof. Let Q be the set of $u \in L^\infty_R$ satisfying (i). Then Q is a convex cone. We will apply 2.1 to show that Q is weak-star closed.

Let $\{u_n\}$ be a bounded sequence in Q such that $u_n \to u$ a.e. For fixed $t > 0$, choose $f_n \in H^\infty$ such that $\phi(f_n) = 1$ and $|f_n| \leq e^{tu_n}$. The f_n are bounded, and so they have a weak-star adherent point $h_t \in H^\infty$. Now $\phi(h_t) = 1$, and $|h_t| \leq \overline{\lim}\, |f_n| \leq \lim e^{tu_n} = e^{tu}$. Hence $u \in Q$. By 2.1, Q is weak-star closed.

Since Q is weak-star closed, and since Q contains $\operatorname{Re}(A_0)$ and P, (ii) implies (i). Since H^∞_0 is the weak-star closure of A_0, (iii) implies (ii).

It remains to show that (i) implies (iii). By the separation theorem for convex sets, it will suffice to show that every $\psi \in (L_R)^*$ which satisfies $\psi(\operatorname{Re}(H_0) + P) \geq 0$ also satisfies $\psi(Q) \geq 0$.

Let $\psi \in (L^\infty_R)^*$ satisfy $\psi(\operatorname{Re}(H^\infty_0) + P) \geq 0$, and let $u \in Q$. Choose h_t as in (i). Since $\psi(P) \geq 0$, we have

$$\psi(\operatorname{Re}(h_t)) \leq \psi(|h|) \leq \psi(e^{tu}).$$

Since $h_t - 1 \in H^\infty_0$, we obtain

$$0 \leq \psi\left(\frac{\operatorname{Re}(h_t - 1)}{t}\right) \leq \psi\left(\frac{e^{tu} - 1}{t}\right).$$

Letting $t \to 0$, we obtain $0 \leq \psi(u)$. This completes the proof.

The significance of the cone Q described in 2.2 is that a functional $\psi \in (L^\infty)^*$ satisfies $\psi(1) = 1$ and $\psi(Q) \geq 0$ if and only if ψ is a positive extension of ϕ from H^∞ to L^∞, that is, if and only if ψ gives rise to a representing measure on M_{L^∞} for ϕ.

2.3 Theorem. The set of representing measures on the maximal ideal space M_{L^∞} of L^∞ for the homomorphism ϕ of H^∞ is the weak-star closure in

$(L^\infty)^*$ of the set of representing measures for ϕ which are absolutely continuous with respect to m.

Proof. Let Q be the cone described in 2.2. Let K be the cone of $h \in L^1_R$ satisfying $\int whdm \geq 0$ for all $w \in Q$. Since Q is weak-star closed, Q is precisely the set of all $w \in L^\infty_R$ such that $\int whdm \geq 0$ for all $h \in K$.

Let μ be a representing measure on M_{L^∞} for ϕ. If $v \in L^\infty_R$ satisfies $\int vhdm \geq 0$ for all $h \in K$, then $v \in Q$, so $\int vd\mu \geq 0$. Consequently no weak-star continuous linear functional separates μ and K. So μ is in the weak-star closure of K, regarded as a subset of $(L^\infty)^*$.

Let $\{h_\alpha\}$ be a net in K converging weak-star to μ. Then $\{h_\alpha dm/\int h_\alpha dm\}$ is a net of absolutely continuous representing measures converging weak-star to μ. This proves the theorem.

The weak topology of a subset of L^1 is identical with its weak-star topology, regarded as a subset of $(L^\infty)^*$. This remark yields the following information.

2.4 Corollary. Every representing measure on M_{L^∞} for ϕ on H^∞ is absolutely continuous with respect to m if and only if the set of representing measures absolutely continuous with respect to m is a weakly compact subset of L^1.

2.5 Corollary. If the set of representing measures in M_ϕ which are absolutely continuous with respect to m is finite dimensional, then the representing measures for ϕ on H^∞ are precisely the representing measures in M_ϕ which are absolutely continuous with respect to m.

3. The Uniqueness Subspace

The **uniqueness subspace** of H^∞, denoted by $\boldsymbol{D(m)} = D$, consists of all $u \in L^\infty_R$ such that $\int ud\mu = \int udm$ for all representing measures μ on M_{L^∞} for ϕ on H^∞. That is, D consists of all $u \in L^\infty_R$ such that all norm-preserving extensions of ϕ from H^∞ to L^∞ agree on u. The uniqueness subspace D is a closed subspace of L^∞_R which contains $\mathrm{Re}\,(H^\infty)$.

By II.2.1, D consists of all $u \in L^\infty_R$ such that

$$\sup\{\phi(v): v \in \mathrm{Re}\,(H^\infty), v \leq u\} = \inf\{\phi(w): w \in \mathrm{Re}\,(H^\infty), w \geq u\}.$$

This common value coincides with $\int ud\mu$ for all representing measures μ on M_{L^∞} for ϕ.

The uniqueness subspace D can also be described in terms of the cone Q introduced in section 2. In view of the remarks preceding 2.3, $Q \cap (-Q)$ consists of all $u \in L^\infty_R$ such that $\psi(u) = 0$ for all norm-preserving extensions

ψ of ϕ from H^∞ to L^∞, that is, $Q \cap (-Q)$ consists of those $u \in D$ such that $\int udm = 0$. So D is the direct sum of $Q \cap (-Q)$ and the constants.

From 2.3, we conclude the following.

3.1 Theorem. A function $u \in L_R^\infty$ belongs to the uniqueness subspace D if and only if $\int udm = \int uhdm$ for all representing measures hdm which are absolutely continuous with respect to m. In particular, D is weak-star closed.

The following characterization of D can be given independently of the results in section 2.

3.2 Theorem. For $u \in L_R^\infty$, the following are equivalent:

(i) u belongs to the uniqueness subspace D.

(ii) For each real t, there is $h_t \in H^\infty$ such that $\int h_t dm = e^{t\int udm}$ and $|h_t| \leq e^{tu}$.

(iii) $u \in \operatorname{Re}(H^2)$, and $e^{t(u+i^*u)} \in (H^\infty)^{-1}$ for all real t.

Proof. First we show that (i) implies (ii).

Suppose that $u \in D$. Choose $u_n \in \operatorname{Re}(H^\infty)$ such that $u_n \leq u$ and $\int u_n dm \to \int udm$. Then $u_n \to u$ in L^1. Passing to a subsequence, we can assume that $u_n \to u$ a.e.

For fixed $t > 0$, the sequence $f_n = e^{t(u_n + i^*u_n)}$ is bounded in H^∞. Let h_t be a weak-star adherent point of $\{f_n\}$. Then $\lim \int f_n dm = \lim e^{t\int u_n dm} = e^{t\int udm}$ must be equal to $\int h_t dm$. Also,

$$|h_t| \leq \limsup |f_n| \leq \limsup e^{tu_n} = e^{tu}.$$

This proves (ii), for $t > 0$. If $t < 0$, we can apply the same reasoning to $-t$ and $-u$ to obtain the desired function h_t. In any event, we obtain (ii).

Now assume that (ii) is true. From $|h_{-s}| \leq e^{-su}$, $|h_{-t}| \leq e^{-tu}$, $|h_{s+t}| \leq e^{(s+t)u}$, and $\int h_{-s}h_{-t}h_{s+t}\, dm = 1$, we conclude that each of these inequalities must in fact be an equality. That is, $|h_t| = e^{tu}$ for all real t. Also, we must have $h_{-s}h_{-t}h_{s+t} = 1$. In particular, $h_{-s}h_s = 1$, so $h_s \in (H^\infty)^{-1}$, and $h_{-s} = 1/h_s$. Hence $h_s h_t = h_{s+t}$ for all real s and t.

From the estimate

$$\begin{aligned}\int |1 - h_t|^2\, dm &= 1 + \int |h_t|^2\, dm - 2 \operatorname{Re} \int h_t dm \\ &= 1 + \int e^{2tu} dm - 2e^{t\int udm} \\ &= t^2 \left\{2 \int u^2 dm - \left(\int udm\right)^2\right\} + 0(t^3),\end{aligned}$$

we see that $(1 - h_t)/t$ is bounded in L^2 as $t \to 0$. So $(1 - h_t)/t$ has a weak adherent point $f \in L^2$ as $t \to 0$. Since

$$\frac{\text{Re}\,(h_t - 1)}{t} \leq \frac{(e^{tu} - 1)}{t} \to u,$$

we have $\text{Re}\,(f) \leq u$. Also,

$$\int f dm = \lim_{t\to 0} \frac{\int h_t dm - 1}{t} = \int u dm,$$

so $\text{Re}\,(f) = u$ a.e. Hence $u \in \text{Re}\,(H^2)$, and $f = u + i{*u}$.

Now the net $\{(h_t - 1)/t\}$ has a unique weak adherent point as $t \to 0$, namely, f. So $(h_t - 1)/t$ converges weakly to f as $t \to 0$. Also, $\int |f|^2\, dm = \int u^2 dm + \int ({*u})^2 dm = 2 \int u^2 dm - (\int u dm)^2$. So $\|(h_t - 1)/t\|_2$ tends to $\|f\|_2$ as $t \to 0$. Hence

$$\left\| f - \frac{(h_t - 1)}{t} \right\|_2^2 = \|f\|_2^2 + \left\| \frac{(h_t - 1)}{t} \right\|_2^2 - 2\,\text{Re} \int \frac{f(\bar{h}_t - 1)}{t} dm$$

tends to 0 as $t \to 0$. That is, $(h_t - 1)/t \to f$ in L^2 as $t \to 0$.

Since $t \to h_t$ is a continuous group in L^2 which satisfies the differential equation $(d/dt)\, h_t = f h_t$, it seems reasonable that $h_t = e^{tf}$. That this is in fact true can be seen as follows.

Let $\{U_t\}$ be the unitary group of operators on L^2 defined by

$$U_t g = h_t e^{-tf} g, \qquad g \in L^2, -\infty < t < \infty.$$

If $g \in L^\infty$, then $U_t g$ moves continuously in L^2. Since such g are dense in L^2, U_t is a strongly continuous unitary group on L^2. By Stone's theorem, there is a unique self-adjoint operator T on L^2 such that $U_t = e^{itT}$. The domain of T consists of all functions $g \in L^2$ such that

$$\lim_{t\to 0} \frac{U_t - I}{t} g$$

exists. This limit, when it exists, is defined to be iTg. However, if g and fg are both bounded, then

$$\frac{U_t - I}{t} g = \left(\frac{h_t - 1}{t} - f\right) e^{-tf} g + \left(\frac{e^{-tf} - 1}{t}\right) g + f e^{-tf} g$$

converges to 0 in L^2. Since such functions are dense in L^2, $T = 0$. By the uniqueness of T, the group U_t must coincide with the identity I. Hence $h_t = e^{tf}$, $-\infty < t < \infty$.

This shows that (ii) implies (iii). That (iii) implies (i) is a consequence of Hoffman's uniqueness theorem II.4.2. That completes the proof.

For some problems (cf. II.9.6), it is crucial to have information concerning the real annihilator of A in L^p_R. These spaces will be denoted by

$$N^p(m) = A \cap L^p_R(m), \qquad 1 \leq p \leq \infty.$$

That is, N^p consists of all $w \in L^p_R$ such that $\int w f dm = 0$ for all $f \in A$.

By definition, $N^p = N^1 \cap L^p_R$. So the N^p form a nested family of spaces which decrease as p increases.

We will also need the annihilator $N(m)$ of $D(m)$, considered as a subset of $L^1_R(m)$. In other words, N consists of all $w \in L^1_R$ such that $\int wudm = 0$ for all $u \in D$. Since $D \supseteq \mathrm{Re}\,(A)$, we obtain $N \subseteq N^1$.

It is not known in general whether $N = N^1$, that is, whether $\mathrm{Re}\,(A)$ is weak-star dense in D. However, since $D \subseteq \mathrm{Re}\,(H^2)$ by 3.2, we obtain $N^2 \subseteq N$.

The spaces in which we are really interested are N^1, N^2, N^∞ and N. For this, the inclusions we have obtained are as follows.

3.3 Lemma. $N^\infty \subseteq N^2 \subseteq N \subseteq N^1$.

The complexifications of N^p and N will be denoted respectively by N^p_c and N_c. From the definition of N^2, we obtain the decomposition

$$L^2 = H^2 \oplus \overline{H^2_0} \oplus N^2_c,$$

where the bar indicates complex conjugation.

4. Enveloped Measures

We now wish to turn to the case in which the set of representing measures M_ϕ for ϕ is finite dimensional. But we will not make this assumption explicitly until section 6.

Motivated by a theorem we shall prove in section 7, we make the following definition.

The representing measure m is **enveloped** if whenever $u_n \in C_R(X)$ satisfy $u_n \geq 0$ and $\int u_n dm \to 0$, there is a subsequence $\{u_{n_k}\}$ and $f_k \in A$ such that $|f_k| \leq e^{-u_{n_k}}$ and $\phi(f_k) \to 1$. In particular, $|f_k| \leq 1$, and $f_k \to 1$ in measure.

4.1 Theorem. If m is enveloped, then every representing measure on X for the homomorphism ϕ on A is absolutely continuous with respect to m.

Proof. Let $\mu = hdm + \mu_s$ be the Lebesgue decomposition of a representing measure μ for ϕ with respect to m. Choose $u_n \in C_R(X)$ such that $u_n > 0$, $\int u_n dm \to 0$, and $u_n \to +\infty$ almost everywhere with respect to μ_s. Passing to a subsequence, we can find $f_n \in A$ such that $|f_n| \leq e^{-u_n}$ while $\phi(f_n) \to 1$. In particular, $|f_n| \leq 1$, and $\int f_n d\mu \to 1$, so f_n converges to 1 in μ-measure. However, $f_n \to 0$ almost everywhere with respect to μ_s. It follows that $\mu_s = 0$.

4.2 Theorem. Suppose m is enveloped. If $u \in L^\infty_R$, and u is in the L^1-closure of $\mathrm{Re}\,(A)$, then u belongs to the uniqueness subspace D.

Proof. Choose $u_n \in \mathrm{Re}\,(A)$ such that $u_n \to u$ in L^1. Let c be the essential supremum of u, and let $v_n = \max\,(c,u_n)$. Then $v_n - c \geq 0$, and $\int (v_n - c)dm \to 0$. Passing to a subsequence, we can assume there are $f_n \in A$ with $|f_n| \leq e^{-v_n+c}$ and $\phi(f_n) \to 1$. Let $h_n = f_n e^{u_n + i^*u_n} \in A$. Since $|h_n| \leq e^c$, the sequence $\{h_n\}$ has a weak-star adherent point $h \in H^\infty$. The function h satisfies $\phi(h) =$

$e^{\int udm}$, and $|h| \leq \limsup |f_n| e^{u_n} = e^u$. Applying the same argument to tu, we find, for each real t, a function $h_t \in H^\infty$ such that $|h_t| \leq e^{tu}$ and $\phi(h_t) = e^{t\int udm}$. By 3.2, $u \in D$.

The dual of 4.2 is the following.

4.3 Corollary. If m in enveloped, then N^∞ is L^1-dense in N.

Proof. Recall that $N^\infty = A^\perp \cap L_R^\infty$, and that N is the annihilator in L_R^1 of the uniqueness subspace D. So $N^\infty \subseteq N$. If $u \in L_R^\infty$ is orthogonal to N^∞, then u belongs to the L^1-closure of Re (A). By 4.2, u belongs to D, so also u is orthogonal to N. By the Hahn-Banach theorem, N^∞ is L^1-dense in N.

The following lemma will be extended later to other H^p-spaces.

4.4 Lemma. If m is enveloped, then $H^2 \cap L^\infty = H^\infty$.

Proof. Suppose $u + i^*u \in H_0^2 \cap L^\infty$. Then $u \perp N^2 \supseteq N^\infty$. By 4.3, $u \perp N$. Hence $u \in D$, and $e^{t(u+i^*u)} \in H^\infty$ for all real t. Since $(e^{t(u+i^*u)} - 1)/t \to u + i^*u$ uniformly, $u + i^*u \in H^\infty$.

4.5 Theorem. Suppose m is enveloped. If $f \in L^1$ and $\int fgdm = 0$ for all $g \in A_0$, then f is in the L^1-closure of $A + N_c^\infty$.

Proof. Suppose $h \in L^\infty$ is orthogonal to $A + N_c^\infty$. We must show that h is orthogonal to f, that is, that $\int hfdm = 0$. It suffices to show that $h \in H_0^\infty$.

By 4.3 and the inclusion $N^2 \subseteq N$, we have $\int hgdm = 0$ for all $g \in N_c^2$. In view of the decomposition $L^2 = H^2 \oplus \overline{H_0^2} \oplus N_c^2$, we obtain $h \in H_0^2$. By 4.4, $h \in H_0^\infty$.

5. Core Measures

According to one definition, an element x of a convex subset K of a vector space V is a **core point** of K if whenever $z \in V$ is such that $x + z \in K$, then also $x - \epsilon z \in K$ for $\epsilon > 0$ sufficiently small. The set of core points of K is the **core** of K.

The core of K is a convex subset of K, which may be empty. If K is a finite dimensional convex set, that is, if K is contained in a finite dimensional subspace of V, then K is the closure of its set of core points. If V is a linear topological vector space, and K is a convex subset of V which has an interior point, then the core of K coincides with the interior of K.

A **core measure** for ϕ is a measure which is a core point of the set M_ϕ of representing measures for ϕ. Core measures need not exist in general, although they exist if M_ϕ is finite dimensional. It is not known whether the existence of a core measure, or of an enveloped measure, already forces M_ϕ to be finite dimensional.

5.1 Theorem. The following are equivalent, for a representing measure m for ϕ.

(i) m is a core point of M_ϕ.

(ii) There is a constant $c > 0$ such that $\mu \leq cm$ for all $\mu \in M_\phi$.

(iii) Whenever $u_n \in C(X)$ is a sequence such that $u_n \geq 0$ and $\int u_n dm \to 0$, there is a sequence $v_n \in \text{Re}(A)$ such that $v_n \geq u_n$ and $\int v_n\, dm \to 0$.

Proof. Suppose that (ii) fails. Then there are representing measures $\mu_n \in M_\phi$ such that $\mu_n \not\leq n2^n m$. The series $\sum_{n=1}^{\infty} \mu_n/2^n$ converges in the total variation norm to $\mu \in M_\phi$. The measure μ is not boundedly absolutely continuous with respect to m. So $\mu = m + \nu$, where ν is not boundedly absolutely continuous with respect to m. Consequently $m - \epsilon\nu$ cannot be a positive measure, for any $\epsilon > 0$. And m cannot be a core point of M_ϕ. This shows that (i) implies (ii).

Now suppose that (ii) is true. Suppose $u_n \in C(X)$ satisfy $u_n \geq 0$ and $\int u_n dm \to 0$. By II.2.1,

$$\inf\left\{\int vdm \colon v \in \text{Re}(A), v \geq u_n\right\} = \sup\left\{\int u_n d\mu \colon \mu \in M_\phi\right\} \leq c\int u_n dm.$$

Consequently there are $v_n \in \text{Re}(A)$ satisfying $v_n \geq u_n$ and $\int v_n dm \leq c \int u_n dm + 1/n \to 0$. Thus (ii) implies (iii).

It suffices now to show that (iii) implies (i).

Suppose that (i) fails. Then there is a measure ν such that $m + \nu \in M_\phi$, while $m - \epsilon\nu \notin M_\phi$ for all $\epsilon > 0$. Consequently $m - \epsilon\nu$ is not positive, for all $\epsilon > 0$. Hence there exist $u_n \in C(X)$ such that $u_n \geq 0$ and $\int u_n dm < \int u_n d\nu/n$. Multiplying by a positive constant, we can assume that $\int u_n d(m + \nu) = 1$. Then $\int u_n dm < \int u_n d\nu/n \leq 1/n \to 0$. Suppose $v_n \in \text{Re}(A)$ satisfy $v_n \geq u_n$. Then $\int v_n dm = \int v_n d(m + \nu) \geq \int u_n d(m + \nu) = 1$. Consequently (iii) fails. This completes the proof.

5.2 Corollary. Core measures are enveloped.

Proof. Suppose $u_n \in C(X)$ satisfy $u_n \geq 0$ and $\int u_n dm \to 0$. Choose $v_n \in \text{Re}(A)$ as in 5.1(iii), and set $f_n = e^{-(v_n + i{}^*v_n)} \in A$. Then $|f_n| \leq e^{-u_n}$ while $\phi(f_n) \to 1$. So m is enveloped.

5.3 Theorem (Abstract Kolmogoroff Theorem). Suppose m is a core measure for ϕ. There are constants a_p and b_p, defined for $0 < p < 1$, such that

$$||{}^*u||_p \leq a_p ||u||_1, \qquad u \in \text{Re}(A),$$

$$||f||_p \leq b_p ||f + g||_1, \qquad f \in A, g \in A_0.$$

Proof. To show the existence of the constants a_p, we must show that whenever $u_n \in \text{Re}(A)$ converges to 0 in L^1, then ${}^*u_n$ converges to 0 in L^p, $0 < p < 1$.

So we assume that $u_n \in \text{Re}(A)$ and $\int |u_n|\, dm \to 0$. Let $w_n = \max(u_n, 0) \geq 0$.

Since $\int w_n dm \to 0$, there are $v_n \in \text{Re}(A)$ such that $v_n \geq w_n$ and $\int v_n dm \to 0$. By 1.2, $*v_n \to 0$ in L^p, $0 < p < 1$. Also, $v_n - u_n \geq 0$, and $\int (v_n - u_n) dm \to 0$, so $*v_n - *u_n \geq 0$ in L^p, $0 < p < 1$. Consequently $*u_n \to 0$ in L^p, $0 < p < 1$. That establishes the first inequality.

To obtain the second inequality from the first, one must repeatedly use the relation

$$||h + k||_p \leq 2^{1/p}(||h||_p + ||k||_p)$$

valid for $h, k \in L^p$, $0 < p < 1$. From the inequality already proved, we deduce that

$$||f||_p \leq \text{const.}\, ||f \pm \bar{f}||_1, \qquad f \in A_0.$$

Hence if f and g both belong to A_0,

$$\begin{aligned} ||f||_p &\leq \text{const.}\, (||f + g||_p + ||f - g||_p) \\ &\leq \text{const.}\, (||f + g + \bar{f} + \bar{g}||_1 + ||f - g - \bar{f} + \bar{g}||_1) \\ &\leq \text{const.}\, ||f + \bar{g}||_1. \end{aligned}$$

Finally, if $f \in A$ and $g \in A_0$ are arbitrary, we can apply the above inequality to $f - \phi(f)$ and g, thereby proving the theorem.

If m is a core measure, the abstract Kolmogoroff theorem allows us to extend the domain of definition of the conjugation operator $u \to *u$ to all functions u in the L^1-closure of Re (A), so that $u + i*u \in H^p$ for $0 < p < 1$. If $f = u + iv \in H^1$, and $\int v dm = 0$, then $v = *u$. In particular, if $f \in H_0^1$ and Re $(f) = 0$, then $f = 0$. Rephrasing this conclusion, we obtain the following.

5.4 Corollary. If m is a core measure for ϕ, then every real-valued function in H^1 is constant.

6. The Finite Dimensional Case

Now we are ready to assault the case in which M_ϕ is finite dimensional.

6.1 Theorem. Suppose the convex set M_ϕ of representing measures for ϕ is finite dimensional. Then the measure $m \in M_\phi$ is a core point of M_ϕ if and only if m is enveloped.

Proof. We have already noted in 5.2 that core measures are enveloped. Suppose that m is enveloped, and let $m + \nu \in M_\phi$. By 4.1, $m + \nu = (1 + h)m$, where $h \in L^1$. Since h is orthogonal to the uniqueness subspace D, h belongs to N.

Now M_ϕ is finite dimensional. By 2.5, the set of representing measures on M_{L^∞} for ϕ, as a homomorphism of H^∞, is also finite dimensional. Consequently the uniqueness subspace D has finite codimension in L^∞_R. And $D^\perp = N$ is finite dimensional.

By 4.3, N^∞ is L^1-dense in N, so $N^\infty = N$. In particular, h is bounded. It follows that $m - \epsilon v = (1 - \epsilon h)m \geq 0$ for $\epsilon > 0$ small. So $m - \epsilon v \in M_\phi$ for $\epsilon > 0$ small, and m is a core point of M_ϕ.

6.2 Theorem. Suppose that the set M_ϕ of representing measures for ϕ is finite dimensional, and that m is a core point of M_ϕ.

(i) If $h \in L^1$ and $\int hf dm = 0$ for all $f \in A_0 + N^\infty$, then $h \in H^1$.

(ii) If $0 < p < 1$, there are constants a_p such that

$$\|{}^*u\|_p \leq a_p \|u\|_1, \qquad u \in \mathrm{Re}\,(A).$$

(iii) The only real-valued functions in H^1 are the constants.

(iv) $A + \bar{A} + N^\infty_c$ is weak-star dense in L^∞.

(v) $N^1 = N^\infty = N$.

(vi) If $1 < p < \infty$, there are constants c_p and d_p such that

$$\|{}^*u\|_p \leq c_p \|u\|_p, \qquad u \in \mathrm{Re}\,(A)$$

$$\|f\|_p \leq d_p \|f + \bar{g}\|_p, \qquad f \in A, g \in A_0.$$

(vii) If $1 < p < \infty$, then $L^p = H^p \oplus \overline{H^p_0} \oplus N^\infty_c$.

(viii) If $1 \leq p \leq q \leq \infty$, then $H^p \cap L^q = H^q$.

Proof. If $h \perp A_0$, then h is in the L^1-closure of $A + N^\infty_c$, by 4.5. This closure is $H^1 + N^\infty_c$. If also $h \perp N^\infty$, then $h \in H^1$. This proves (i).

Facts (ii) and (iii) are proved in section 5. Fact (iv) is a corollary of (i) and (iii), and (v) follows from (iv).

Now we turn to the inequalities in (vi). By 1.3, the conjugation operator is bounded in the L^p-norm when p is an even integer. Consider the operator defined on the dense subspace Re $(A) + N^\infty$ by $T(u + w) = {}^*u$, $u \in \mathrm{Re}\,(A)$, $w \in N^\infty$. Since N^∞ is finite dimensional and orthogonal to Re(A), T remains bounded from L^p to L^p, p an even integer. By the M. Riesz convexity theorem, T is bounded in L^p for $2 \leq p < \infty$. The adjoint T^* of T is then bounded on L^p for $1 < p \leq 2$. A short calculation shows that $T^* = -T$. This shows that the required constants c_p exist, for $1 < p < \infty$. An estimate of the form $\|f\|_p \leq d_p \|f + \bar{g}\|_p$, $f \in A$, $g \in A_0$, is a consequence of the estimate $\|{}^*u\|_p \leq c_p \|u\|_p$, as indicated in the proof of 5.3. This completes the proof of (vi).

Fact (vii) is a consequence of (vi).

For (viii), it suffices to consider the case $p = 1$.

Suppose first that $1 < q < \infty$. If $f \in H^1 \cap L^q$, we can write $f = g + \bar{h} + w$, where $g \in H^q$, $h \in H^q_0$ and $w \in N^\infty_c$. Since $0 = \int f w dm = \int |w|^2\, dm$, $w = 0$. Hence $h \in H^1_0$ and $\bar{h} = f - g \in H^1$. By (iii), $h = 0$. So $f \in H^q$.

Suppose now that $q = \infty$. Let $f = u + i^*u \in H_0^1 \cap L^\infty$. Since $u \perp N^\infty = N$, u belongs to D, and $u + i^*u \in H^2$. By 4.3, $u + i^*u \in H^\infty$. This completes the proof of the theorem.

Now we wish to study the additive group $\log|(H^\infty)^{-1}|$, under the assumptions that M_ϕ be finite dimensional and that $m \in M_\phi$ be a core measure. In this case, $L_R^\infty = D \oplus N$. By 3.2, $D \subseteq \log|(H^\infty)^{-1}|$. Hence

$$\log|(H^\infty)^{-1}| = D \oplus [N \cap \log|(H^\infty)^{-1}|].$$

6.3 Lemma. Suppose that M_ϕ is finite dimensional, and that m is a core point of M_ϕ. Then $N \cap \log|(H^\infty)^{-1}|$ is a discrete additive subgroup of N.

Proof. If $N \cap \log|(H^\infty)^{-1}|$ is not discrete, its closure will contain a nontrivial subspace $\{tw: -\infty < t < \infty\}$ of N. By Hoffman's uniqueness theorem II.4.2, all representing measures for ϕ on H^∞ agree on w, that is, $w \in D$. This contradiction establishes the lemma.

6.4 Theorem. Suppose that the set of representing measures M_ϕ for ϕ is finite dimensional, and that m is a core point of M_ϕ. Then the first Čech cohomology group of M_{H^∞}, with coefficients in the integers Z, is isomorphic to the direct sum

$$H^1(M_{H^\infty}, Z) \cong D' \oplus Z^r,$$

where the group D' is infinitely divisible, and the integer r is the dimension of the subspace of N spanned by $N \cap \log|(H^\infty)^{-1}|$.

Proof. The algebra H^∞ is antisymmetric. From the corollary to the Arens-Royden theorem, III.7.5, we have the isomorphism

$$H^1(M_{H^\infty}, Z) = \frac{\log|(H^\infty)^{-1}|}{\mathrm{Re}\,(H^\infty)}.$$

Since $\mathrm{Re}\,(H^\infty) \subseteq D$, we obtain

$$H^1(M_{H^\infty}, Z) = \left[\frac{D}{\mathrm{Re}(H^\infty)}\right] \oplus [N \cap \log|(H^\infty)^{-1}|].$$

This is the desired result, with $D' = D/\mathrm{Re}\,(H^\infty)$.

Note that the subgroup D' is characterized intrinsically as the largest infinitely divisible subgroup of $H^1(M_{H^\infty}, Z)$. The integer r is also characterized intrinsically as the number of generators of $H^1(M_{H^\infty}, Z)/D'$.

7. Logmodular Measures

A positive measure μ on X is a **logmodular measure**, or **Arens-Singer measure**, for ϕ if

$$\log|\phi(f)| = \int \log|f|\, d\mu$$

for all $f \in A^{-1}$. In particular, if $u \in \operatorname{Re}(A)$, then

$$\phi(u) = \log|\phi(e^{u+i^*u})| = \int u\,d\mu.$$

Hence logmodular measures for ϕ are representing measures for ϕ.

Every Jensen measure for ϕ is a logmodular measure for ϕ. This can be seen by applying Jensen's inequality to f and $1/f$, $f \in A^{-1}$.

7.1 Theorem. The functional L on $\log|A^{-1}|$, defined by

$$L(\log|f|) = \log|\phi(f)|, \qquad f \in A^{-1},$$

extends uniquely to be a linear functional (also denoted by L) on the linear span of $\log|A^{-1}|$. The functional L is positive on the linear span of $\log|A^{-1}|$. The logmodular measures for ϕ coincide with the positive (= norm-preserving) extensions of L from the linear span of $\log|A^{-1}|$ to $C_R(X)$.

Proof. If $u = \sum_{j=1}^{n} a_j \log|f_j|$ belongs to the linear span of $\log|A^{-1}|$, we set

$$L(u) = \sum_{j=1}^{n} a_j \log|\phi(f_j)|.$$

We will show that with this definition of L, we have

$$|L(u)| \leq \|u\|_X$$

for u in the linear span of $\log|A^{-1}|$. From this, it follows that L is well-defined and linear.

Suppose for some $u = \sum a_j \log|f_j|, f_j \in A^{-1}$, it is true that

$$|L(u)| = |\sum a_j \log|\phi(f_j)|| > \|u\|_X.$$

Wiggling the coefficients a_j slightly, we can assume that they are rational. Multiplying by an integer, we can assume that the a_j are integers. Then $u = \log|f|$, where $f = f_1^{a_1} \ldots f_1^{a_n} \in A^{-1}$. Also, $L(u) = \sum a_j \log|\phi(f_j)| = \log|\phi(f)|$. Consequently the inequality $|L(u)| > \|u\|_X$ leads to the inequality $|\phi(f)| > \|f\|_X$, which is a contradiction.

Since $L(1) = 1$, the norm of L on the linear span of $\log|A^{-1}|$ is 1. If μ is any norm-preserving extension of L to $C_R(X)$, then $\int d\mu = L(1) = 1 = \|L\| = \int d|\mu|$. Consequently $\mu \geq 0$, and μ is a logmodular measure for ϕ.

It follows that the functional L is positive on the linear span of $\log|A^{-1}|$, and the positive and the norm-preserving extensions of L to $C_R(X)$ coincide. And these must be the logmodular measures on X for ϕ. This proves the theorem.

The relation between logmodular measures and algebras of analytic functions in the plane is indicated by the following theorem.

7.2 Theorem. Let K be a compact subset of the plane, and let $z_0 \in K$.

A positive measure μ on K is a logmodular measure for z_0 on the algebra $R(K)$ if and only if

$$u(z_0) = \int u\,d\mu$$

for all functions u which are harmonic in a neighborhood of K.

Proof. In view of 7.1, it suffices to show that the closed linear span of $\log|R(K)^{-1}|$ coincides with the closure in $C_R(K)$ of the functions harmonic in a neighborhood of K.

Suppose first that $f \in R(K)^{-1}$. There is a sequence $f_n \in R(K)^{-1}$ such that f_n is analytic in a neighborhood of K, and $f_n \to f$ uniformly on K. Then $\log|f_n|$ is harmonic in a neighborhood of K, and $\log|f_n|$ tends uniformly to $\log|f|$ on K. Consequently every function in the closed linear span of $\log|R(K)^{-1}|$ is uniformly approximable on K by functions harmonic in a neighborhood of K.

Conversely suppose that u is harmonic in a neighborhood of K. We will show that u belongs to the linear span of $\log|R(K)^{-1}|$.

Let J be a compact set whose boundary consists of a finite number of disjoint simple closed Jordan curves $\Gamma_1, \ldots, \Gamma_n$, such that $K \subseteq J$, and u is harmonic in a neighborhood of J. Let a_j be the increment of the multivalued harmonic conjugate $*u$ of u along the curve Γ_j, $1 \leq j \leq n$. Let z_j be any complex number in the component of the complement of Γ_j which does not meet J. Then the increment of the harmonic conjugate of $u - 1/2\pi \sum_{j=1}^{n} a_j \log|z - z_j|$ along each curve Γ_j is 0. Hence the function $f = e^{u+i*u}(z - z_1)^{-a_1/2\pi} \cdots (z - z_n)^{-a_n/2\pi}$ is analytic in a neighborhood of K. The functions $f, z - z_1, \ldots, z - z_n$ all belong to $R(K)^{-1}$, and

$$u = \log|f| + \frac{1}{2\pi}\sum_{j=1}^{n} a_j \log|z - z_j|$$

belongs to the linear span of $\log|R(K)^{-1}|$. This proves the theorem.

Recall that a point $z \in bK$ satisfies Lebesgue's condition if $\int_S dr/r = +\infty$, where S consists of all r, $0 < r < 1$, such that the circle of radius r and center z meets the complement of K(cf. II.3). As a simple consequence of II.3.2, we obtain the following.

7.3 Theorem. Let K be a compact subset of the complex plane, such that every point on bK satisfies Lebesgue's condition. Then every real-valued continuous function on bK can be approximated uniformly on bK by functions harmonic in a neighborhood of K. In particular, every point $z_0 \in K$ has a unique logmodular measure μ on bK for the algebra $R(K)$.

Theorem 7.3 applies, for instance, to any compact set K whose complement has a finite number of components.

Now we return to the uniform algebra A on the compact space X, and to $\phi \in M_A$.

7.4 Theorem. Suppose that the set M_ϕ of representing measures for ϕ is finite dimensional. Then there is a core measure for ϕ which is a logmodular measure for ϕ.

Proof. Let m be a core point of M_ϕ. By 2.5, the set of representing measures for ϕ, as a homomorphism of $H^\infty(m)$, coincides with M_ϕ.

Let S be the convex subset of M_ϕ consisting of the measures which are logmodular on $H^\infty(m)$. Let μ be a core point of S. Since S is finite dimensional, there exists $\epsilon > 0$ such that $\mu - \epsilon\eta \in S$ whenever $\mu + \eta \in S$. In particular, if $\mu + \eta \in S$, then $\mu - \epsilon\eta \geq 0$, and $\mu + \eta \leq (1 + 1/\epsilon)\mu$. In other words, there exists $c > 0$ such that $\nu \leq c\mu$ for all $\nu \in S$.

We will show that μ is enveloped by $H^\infty(m)$. By 6.1, μ is then a core measure.

Suppose $u_n \in L_R^\infty$ satisfy $u_n \geq 0$ and $\int u_n d\mu \to 0$. Since S consists of the positive extensions to L_R^∞ of the linear functional L of 7.1, the infimum of the values $L(v)$, for v in the linear span of $\log|(H^\infty)^{-1}|$ satisfying $v \geq u_n$, must coincide with the supremum of $\int u_n d\nu$ over $\nu \in S$, just as in II.2.1. Since

$$\sup\left\{\int u_n d\nu : \nu \in S\right\} \leq c \int u_n d\mu \to 0,$$

there are functions v_n in the linear span of $\log|(H^\infty)^{-1}|$ such that $v_n \geq u_n$ and $\int v_n d\mu \to 0$.

Now choose $h_1, \cdots, h_r \in (H^\infty)^{-1}$ so that $\{\log|h_j|\}_{j=1}^r$ is a basis for the subspace of N spanned by $N \cap \log|(H^\infty)^{-1}|$. We can assume that $\phi(h_j) = 1$, $1 \leq j \leq r$. Also, set

$$b = \max_{1 \leq j \leq r} \|\log|h_j|\|_\infty.$$

Choose measures $\nu_1, \ldots, \nu_r$ orthogonal to D, such that

$$\int \log|h_j|\, d\nu_k = \delta_{jk}, \qquad 1 \leq j, k \leq r.$$

Then

$$v_n = w_n + \sum_{j=1}^{r} c_{nj} \log|h_j|,$$

where $w_n \in D$, and $c_{nj} = \int v_n d\nu_j$, $1 \leq j \leq r$. Let m_{nj} be the largest integer which does not exceed c_{nj}, $1 \leq j \leq r$. Passing to a subsequence, we can assume that $\boldsymbol{c}_n - \boldsymbol{m}_n = (c_{n1}, \ldots, c_{nr}) - (m_{n1}, \ldots, m_{nr})$ converges to $\boldsymbol{a} = (a_1, \ldots, a_r)$. In fact, by passing to a subsequence, we can suppose that

$$\|\boldsymbol{c}_n - \boldsymbol{m}_n - \boldsymbol{a}\| \leq n^{-r-1}, \qquad 1 \leq n < \infty,$$

$$L(v_n) \leq n^{-r-1}, \qquad 1 \leq n < \infty,$$

where $\|(b_1, \ldots, b_r)\| = \max(|b_1|, \ldots, |b_r|)$ is the square norm on R^r. By Dirichlet's theorem on Diaphantine approximation (putting $k+1$ objects in k boxes), there are integers $0 < q_n \leq n^r$ and vectors $\boldsymbol{p}_n$ with integral coordinates, such that

$$\|q_n\boldsymbol{a} - \boldsymbol{p}_n\| < \frac{1}{n}, \qquad 1 \leq n < \infty.$$

It follows that

$$\|q_n\boldsymbol{c}_n - q_n\boldsymbol{m}_n - \boldsymbol{p}_n\| \leq q_n\|\boldsymbol{c}_n - \boldsymbol{m}_n - \boldsymbol{a}\| + \|q_n\boldsymbol{a} - \boldsymbol{p}_n\| \leq \frac{2}{n}.$$

Consequently

$$\left|\sum_{j=1}^{r} (q_n c_{nj} - q_n m_{nj} - p_{nj}) \log|h_j|\right| \leq \frac{2rb}{n}.$$

Now set

$$f_n = e^{-2rb/n} e^{-q_n(w_n + i^*w_n)} \prod_{j=1}^{r} h_j^{-(p_{nj} + q_n m_{nj})}, \qquad 1 \leq n < \infty.$$

Then

$$\begin{aligned} u_n &\leq v_n \leq q_n v_n \\ &\leq q_n w_n + \sum_{j=1}^{r} (p_{nj} + q_n m_{nj}) \log|h_j| + \sum_{j=1}^{r} (q_n c_{nj} - q_n m_{nj} - p_{nj}) \log|h_j| \\ &\leq -\log|f_n|. \end{aligned}$$

Hence

$$|f_n| \leq e^{-u_n}, \qquad 1 \leq n < \infty.$$

Also,

$$\phi(f_n) = e^{-2rb/n} e^{-q_n \int w_n d\mu} = e^{-2rb/n} e^{-q_n \int v_n d\mu} \to 1.$$

It follows that μ is enveloped.

7.5 Corollary. If M_ϕ is finite dimensional, and ϕ has a unique logmodular measure m, then m is a core measure for ϕ.

7.6 Corollary. Suppose M_ϕ is finite dimensional, and m is a core measure for ϕ. Then ϕ has a unique logmodular measure on H^∞ if and only if the linear span of $\log|(H^\infty)^{-1}|$ is L^∞_R.

Proof. By 7.4, there is a core measure $\mu \in M_\phi$ which is logmodular on H^∞. Replacing m by μ does not alter H^∞. Hence we can assume that m is logmodular on H^∞.

Now L^∞_R is the direct sum of D and N, while $D \subseteq \log|(H^\infty)^{-1}|$. Consequently the linear span of $\log|(H^\infty)^{-1}|$ is L^∞_R if and only if the linear span of $N \cap \log|(H^\infty)^{-1}|$ is N. This occurs if and only if the only $u \in N$ which is orthogonal to $\log|(H^\infty)^{-1}|$ is $u = 0$. But the functions $u \in N$ which are orthogonal to $\log|(H^\infty)^{-1}|$ are precisely those $u \in N$ such that $(1 + \epsilon u)\, dm$ is a logmodular measure for ϕ, for ϵ sufficiently small. That does it.

We now prove a maximality theorem for H^∞. The proof here is based on the Arens-Royden theorem. An elementary proof can also be given, but the details are tedious.

7.7 Theorem. Suppose that the set M_ϕ of representing measures for ϕ is finite dimensional, and suppose that ϕ has a unique logmodular measure m. If B is a subalgebra of L^∞ such that $B \supseteq H^\infty$ and m is multiplicative on B, then $B = H^\infty$.

Proof. The Hardy spaces, etc., associated with the representing measure m as a homomorphism of B will be denoted by $H^p(B)$. Theorem 6.2 yields two decompositions of L^2:

$$\begin{aligned} L^2 &= H^2 \oplus \overline{H_0^2} \oplus N_c \\ &= H^2(B) \oplus \overline{H_0^2(B)} \oplus N_c(B). \end{aligned}$$

where $H^2 \subseteq H^2(B)$, and $\overline{H_0^2} \subseteq \overline{H_0^2(B)}$. Since N_c is finite dimensional, H^2 has finite codimension in $H^2(B)$.

Since $H^2 \cap L^\infty = H^\infty$, H^∞ has finite codimension in B. In particular, B is integral over H^∞, that is, every function in B satisfies a monic polynomial relation with coefficients in H^∞. By a theorem concerning integral extensions of rings (cf. Zariski and Samuel, p. 257) every maximal ideal in H^∞ can be extended to a maximal ideal in B.

By 7.6, the linear span of $\log|(H^\infty)^{-1}|$ is L_R^∞. By 7.1, every $\psi \in M_{H^\infty}$ has a unique logmodular measure m_ψ. If $\tilde{\psi}$ is any extension of ψ from H^∞ to B, then $\tilde{\psi}$ has a logmodular measure, which must coincide with m_ψ. Hence every $\psi \in M_{H^\infty}$ is uniquely extendable to B. It follows that $M_{H^\infty} = M_B$. In particular, M_{H^∞} and M_B have the same cohomology.

By 6.4 and 7.6, the first Čech cohomology groups of M_{H^∞} and M_B respectively are $D' \oplus Z^{\dim N}$ and $D'(B) \oplus Z^{\dim N(B)}$, where D' and $D'(B)$ are completely divisible. It follows that N and $N(B)$ have the same dimension. Since $N(B) \subseteq N$, we have $N(B) = N$. From the decomposition of L^2, we obtain $H^2 = H^2(B)$. Intersecting with L^∞, we obtain $H^\infty = B$. That proves the theorem.

The preceding proof simplifies considerably when m is a unique representing measure for ϕ. In this case, $N = \{0\}$, and the decompositions of L^2 yield immediately $H^2 = H^2(B)$.

8. Hypodirichlet Algebras

A uniform algebra A on a compact Hausdorff space X is **hypodirichlet on X** if the uniform closure of Re (A) has finite codimension in $C_R(X)$, and the linear span of $\log|A^{-1}|$ is dense in $C_R(X)$. The uniform algebra A is **hypodirichlet** if it is hypodirichlet on its Shilov boundary.

If A is hypodirichlet on X, every $\phi \in M_A$ has a finite dimensional set of representing measures M_ϕ. And every $\phi \in M_A$ has a unique logmodular measure $m_\phi \in M_\phi$. By 7.5, m_ϕ is a core measure for ϕ.

If $x \in X$, the logmodular measure for x must be the point mass at x. Consequently every $x \in X$ has a unique representing measure, and X coincides with the strong boundary of A(cf. II.12). In particular, if A is hypodirichlet on X, then X is the Shilov boundary of A.

The maximality theorem for H^∞ allows us to extend the abstract version of Mergelyan's theorem proved in II.9.2 to hypodirichlet algebras. Recall that a uniform algebra A is relatively maximal if whenever B is a subalgebra of $C(M_A)$ such that $B \supseteq A$ and $\partial_B = \partial_A$, then $B = A$.

8.1 Theorem. Let A be a hypodirichlet algebra on X. Suppose that the only measure in $A^\perp$ which is singular to all representing measures for homomorphisms $\phi \in M_A$ is the zero measure. Then A is relatively maximal.

Proof. Let B be a uniform algebra on $C(M_A)$ such that $B \supseteq A$ and $\partial_B = X$. Let μ be a measure on X which is orthogonal to A. We must show that μ is orthogonal to B. By the decomposition theorem for orthogonal measures II.7.11, it suffices to do this for measures of the form $\mu = hdm$, where m is a representing measure for some homomorphism $\phi \in M_B$. We can assume that m is the logmodular measure for ϕ.

Since ϕ extends to B, and since m is a unique logmodular measure for ϕ on A, m must also be multiplicative on B. By the maximality theorem 7.7, B is contained in H^∞, the weak-star closure of A in $L^\infty(m)$, and hdm is also orthogonal to B. That proves the theorem.

8.2 Corollary Let K be a compact subset of the plane such that $R(K)$ is hypodirichlet. Then $R(K) = A(K)$.

Proof. This follows from 8.1 and Wilken's theorem II.8.5.

8.3 Theorem. Let K be a compact subset of the complex plane whose complement has a finite number of components. Then the algebra $R(K)$ is a hypodirichlet algebra.

Proof: Choose one point z_j, $1 \leq j \leq r$, from each component of K^c. It suffices to show that the linear span of Re $R(K)$ and the functions $\log|z - z_j|$, $1 \leq j \leq r$, is dense in $C_R(bK)$.

Let μ be a real measure on bK which is orthogonal to Re $R(K)$ and to $\log|z - z_j|$, $1 \leq j \leq r$. We must show that μ is orthogonal to $C_R(bK)$. By 7.3, it suffices to show that μ is orthogonal to every real function u which is harmonic in a neighborhood of K. We can assume that u is infinitely differentiable on the complex plane, and that u has compact support. By Green's formulas,

$$u(\zeta) = -\frac{1}{2\pi}\iint (\nabla^2 u)(z) \log|\zeta - z|\, dxdy.$$

Since $\nabla^2 u$ vanishes in a neighborhood of K, this integral can be taken over a compact subset of K^c.

Suppose $z \in K^c$. Then a branch of $\log[(\zeta - z)/(\zeta - z_j)]$ is analytic in a neighborhood of K, providing z_j belongs to the same component of K^c as z. Consequently $\log|\zeta - z| - \log|\zeta - z_j|$ belongs to Re $R(K)$, and

$$\int \log|\zeta - z|\, d\mu(\zeta) = 0, \qquad \zeta \in K^c.$$

Applying Fubini's theorem to the representation formula for u, we obtain

$$\int u(\zeta) d\mu(\zeta) = -\frac{1}{2\pi}\iint_{K^c} (\nabla^2 u)(z)\left[\int \log|\zeta - z|\, d\mu(\zeta)\right] dxdy = 0.$$

This proves the theorem.

Combining 8.2 and 8.3, we find that $R(K) = A(K)$ when K^c has a finite number of components. This was also proved in II.10.4, using the localization theorem and Mergelyan's polynomial approximation theorem.

Theorem 8.1 also applies to uniform approximation theory on any open Riemann surfaces S. In fact, let K be a compact subset of S, and let A be the subalgebra of $C(K)$ of functions which can be approximated uniformly on K by functions analytic on S. The maximal ideal space $\hat{K}$ of K is the union of K and the relatively compact components of $S \backslash K$. The Shilov boundary of A is the topological boundary $b\hat{K}$ of $\hat{K}$. It can be shown that A is a hypo-dirichlet algebra on $b\hat{K}$ which has no singular orthogonal measures. From 8.1 we deduce that every continuous function on $\hat{K}$ which is analytic on the interior of $\hat{K}$ can be approximated uniformly on $\hat{K}$ by functions analytic on S.

NOTES

The uniqueness subspace first appears in Lumer [2], where the implication (i) $\Rightarrow$ (iii) of 3.2 is proved. Most of the results of sections 2 and 3 are due to Hoffman and Rossi [3], and we have followed their presentation rather closely. Some of the ideas stem from Ahern and Sarason [1]. Sections 4, 5, and 6 are based on Ahern and Sarason [1], Glicksberg [8], and Gamelin and Lumer [1]. Logmodular measures (= Arens-Singer measures) were introduced in Arens and Singer [1]. The proof of 7.4 stems from O'Neill [1]. A sharpened version of the proof was used by Ahern and Sarason [1] to obtain 7.5. The maximality theorem 7.7 was proved for logmodular algebras by Srinivasan and Wang (cf. the proof of VII. 8.2, which comes from Srinivasan [2]). This version is due to Gamelin [4]. An elementary proof, together with applications to algebras of analytic functions on Riemann surfaces, is in Gamelin and Lumer [1]. For more on the evolution of the abstract Hardy space theory, see the notes to chapter V.

EXERCISES

1. The representing measure m for ϕ is an extreme point of M_ϕ if and only if $A + \bar{A}$ is dense in $L^1(m)$.

2. If $u \in L_R^\infty(m)$, and u belongs to the closure of Re (A) in $L^p(m)$ for all finite p, then $e^{u+i^*u} \in H^p(m)$ for all finite p.

3. If $0 < p < 1$, there is a constant c such that for all $u \in \text{Re}(A)$,

$$\sup_{m \in M_\phi} \left\{ \int |{}^*u|^p dm \right\}^{1/p} \leq c \sup_{m \in M_\phi} \int |u|\, dm.$$

4. If $0 < p < 1$, there are real-valued functions in $H^p(d\theta/2\pi)$. *Hint:* Look at $i(z+1)/(z-1)$.

5. Let A be the algebra of bounded analytic functions in the half-plane $\{\text{Im}\,(z) > -1\}$. Let K_n be the solid rectangle consisting of z such that $|\text{Re}\,(z)| \leq n$ and $|\text{Im}\,(z)| \leq 1 - 1/n$. Let μ_n be harmonic measure for 0 on bK_n, $n \geq 2$. And let $\mu = \sum_{n=2}^{\infty} a_n \mu_n$, where $a_n \geq 0$ and $\sum_{n=2}^{\infty} a_n = 1$. Then A is weak-star closed in $L^\infty(\mu)$. If $\sum a_n$ converges rapidly, then $H^2(\mu) \cap L^\infty(\mu)$ coincides with the algebra of bounded analytic functions on the strip $\{|\text{Im}\,(z)| < 1\}$. *Hint:* Use the Paley-Wiener theorem.

6. There is a uniform algebra A and a dominant representing measure m for $\phi \in M_A$ such that m is not enveloped, and m is not dominant when regarded as a representing measure for $H^\infty(m)$.

7. (Gleason-Whitney Theorem): Suppose ϕ has a unique representing measure m. If $g \in L^1(m)$, the functional $f \to \int fg dm$ has a unique norm-preserving extension from A to $C(X)$. The norm-preserving extension is absolutely continuous with respect to m.

8. Let K be an annulus $\{r \leq |z| \leq 1\}$. Every real measure on bK orthogonal to $R(K)$ is a multiple of dz/iz. If m is an endpoint of the interval of representing measures for any point in int (K), and $p \leq 3$, then $A + \bar{A}$ is dense in $L^p(m)$. Identify $N^q(m)$, $1 \leq q \leq \infty$.

9. Let A be the subalgebra of $f \in P(b\Delta)$ such that $f(0) = f(\frac{1}{2})$.

(a) The set of representing measures M_0 for evaluation at 0 is two-dimensional.

(b) The logmodular measures for 0 are the intersection of M_0 with the line through $d\theta/2\pi$ and $P_{1/2}(\theta)d\theta/2\pi$, P the Poisson kernel.

(c) The Jensen measures for 0 are the convex combinations of $d\theta/2\pi$ and $P_{1/2}(\theta)d\theta/2\pi$.

(d) $N^\infty \cap \log|(H^\infty)^{-1}|$ is generated by $\log|f|$, where $f(z) = e^{2\pi i z}$.

10. Theorem 8.1 is valid under the assumptions that every M_ϕ be finite dimensional, that every ϕ have a unique logmodular measure, and that no extreme point of the unit ball of $A^\perp$ be singular to all the M_ϕ.

11. If M_ϕ is finite dimensional, and ϕ has a unique logmodular measure m, then all representing measures in M_ϕ are mutually absolutely continuous.

12. Suppose all representing measures for ϕ are mutually absolutely continuous. If $w \in L_R^\infty$, there is $f \in H^\infty$ such that $|f| = e^w$. *Hint:* Take a norm-preserving extension of the functional $\psi(e^{-w}f) = \phi(f)$ from $e^{-w}A$ to $C(X)$.

CHAPTER V

INVARIANT SUBSPACE THEORY

In this chapter we prove several of the theorems which are prominent in the historical development of the abstract Hardy space theory: Beurling's invariant subspace theorem, Wermer's embedding theorem, Szegö's theorem, and the factorization theorems. A natural setting for this circle of ideas is provided by the Hardy algebra, which we define and discuss in the first half of the chapter.

Throughout this chapter, A will be a fixed uniform algebra, and m will be a representing measure for a fixed homomorphism $\phi \in M_A$. The kernel of ϕ will be denoted by A_0. The (weak-star, if $p = \infty$) closure of A_0 in $L^p(m)$ will be denoted by $H_0^p(m)$.

1. Uniform Integrability

Let μ be a finite positive measure. A subset K of $L^1(\mu)$ is **uniformly integrable** if

$$\lim_{c \to +\infty} \sup_{f \in K} \int_{\{|f| \geq c\}} |f| \, d\mu = 0.$$

Here $\{|f| \geq c\}$ denotes the set of x satisfying $|f(x)| \geq c$.

A theorem of Dunford states that K is uniformly integrable if and only if the weak closure of K in $L^1(\mu)$ is weakly compact. We will not need this deeper result, but instead give the following characterizations of uniform integrability.

1.1 Theorem. The following conditions are equivalent, for a subset K of $L^1(\mu)$:

(i) K is uniformly integrable.

(ii) K is bounded, and $\lim_{\mu(E) \to 0} \int_E |f| \, d\mu = 0$, uniformly for $f \in K$.

(iii) There is a continuous real function $\Phi(t)$ such that $\lim_{t \to \infty} \Phi(t)/t = +\infty$, while

$$\sup_{f \in K} \int \Phi(|f|)d\mu < \infty.$$

Proof. That (i) implies (ii) follows from the inequality

$$\int_E |f|d\mu \leq c\mu(E) + \int_{\{|f| \geq c\}} |f|d\mu,$$

valid for any $c > 0$ and all measurable sets E. That (ii) implies (i) follows from the inequality

$$\mu(\{|f| \geq c\}) \leq \int |f| \frac{d\mu}{c}, \qquad c > 0.$$

That (iii) implies (i) follows from the inequality

$$\int_{\{|f| \geq c\}} |f|d\mu \leq \sup_{t \geq c} \left[\frac{t}{\Phi(t)}\right] \int_{\{|f| \geq c\}} \Phi(|f|)d\mu.$$

The details of these first three implications are left to the reader.

Now suppose that (i) is true. We wish to prove (iii).

Let $\{n_k\}_{k=1}^{\infty}$ be a strictly increasing sequence of integers such that

$$\int_{\{|f| \geq n_k - 1\}} |f|d\mu \leq 2^{-k}, \qquad f \in K.$$

For each $t > 0$, let $\Psi(t)$ be the number of integers k such that $n_k < t$. Let

$$\Phi(t) = \int_0^t \Psi(s)ds, \qquad t > 0.$$

Since Ψ is non-decreasing, and $\lim_{t \to \infty} \Psi(t) = +\infty$, we also have

$$\lim_{t \to \infty} \frac{\Phi(t)}{t} = +\infty.$$

If $f \in K$, and $a_m = \mu(\{m - 1 < |f| \leq m\})$, then

$$\begin{aligned} \int \Phi(|f|)d\mu &\leq \sum_{m=1}^{\infty} \Phi(m)a_m \leq \sum_{m=1}^{\infty} m\Psi(m)a_m \\ &= \sum_{m=1}^{\infty} \sum_{k=1}^{\Psi(m)} ma_m = \sum_{k=1}^{\infty} \sum_{m=n_k}^{\infty} ma_m \\ &\leq \sum_{k=1}^{\infty} \int_{\{|f| \geq n_k - 1\}} |f|\, d\mu \leq 1. \end{aligned}$$

This proves the theorem.

1.2 Corollary. If K is a bounded subset of $L^p(\mu)$ for some $p > 1$, then K is uniformly integrable. If K is a bounded subset of $L^p(\mu)$ for some $p > 0$, then $\log^+ |K|$ is uniformly integrable.

Proof. For the first statement, take $\Phi(t) = t^p$. For the second, take $\Phi(t) = e^{pt}$.

1.3 Theorem. Let $\{f_n\}_{n=1}^{\infty}$ be a sequence in $L^1(\mu)$. If f_n converges to f

in $L^1(\mu)$, then $\{f_n\}_{n=1}^{\infty}$ is uniformly integrable. Conversely, if $\{f_n\}_{n=1}^{\infty}$ is uniformly integrable, and f_n converges to a function f in measure, then $f \in L^1(\mu)$, and f_n converges to f in $L^1(\mu)$.

Proof. Suppose that f_n converges to f in $L^1(\mu)$. For every measurable set E,

$$\int_E |f_n|\,d\mu \leq \int_E |f|\,d\mu + \|f_n - f\|_1.$$

From this it follows that $\sup_n \int_E |f_n|\,d\mu$ tends to 0 as $\mu(E)$ tends to 0. By 1.1, $\{f_n\}$ is uniformly integrable.

Conversely, suppose that f_n converges to f in measure, and $\{f_n\}_{n=1}^{\infty}$ is uniformly integrable. By Fatou's lemma,

$$\int |f|\,d\mu \leq \liminf_{n\to\infty} \int |f_n|\,d\mu,$$

so $f \in L^1(\mu)$. For each $c > 0$,

$$\|f_n - f\|_1 \leq \int_{\{|f_n - f| \leq 2c\}} |f_n - f|\,d\mu + 2\int_{\{|f_n| \geq c\}} |f_n|\,d\mu + 2\int_{\{|f| \geq c\}} |f|\,d\mu.$$

The latter two integrals are small when c is large, independently of n. The first integral tends to zero for each fixed c, by the bounded convergence theorem of Lebesgue. Hence f_n converges to f in $L^1(\mu)$. This proves the theorem.

1.4 Lemma. Let $f \in L^1(\mu)$. Suppose $\{f_n\}_{n=1}^{\infty}$ is a sequence in $L^1(\mu)$ such that $f_n \geq 0$, f_n converges to f in measure, and $\|f_n\|_1$ tends to $\|f\|_1$. Then f_n converges to f in $L^1(\mu)$.

Proof. Note that $f \geq 0$. By the dominated convergence theorem, $\int \min(f_n, f)\,d\mu$ tends to $\int f\,d\mu$. Now if we integrate the identity $|f_n - f| = f_n + f - 2\min(f_n, f)$, we obtain $\|f_n - f\|_1 \to 0$.

2. The Hardy Algebra

Let μ be a finite positive measure. Let $\boldsymbol{L(\mu)}$ be the set of measurable functions f, modulo null functions, such that $\log^+|f|$ is integrable. We define a metric in $L(\mu)$ by setting

$$d(f, g) = \inf_{t>0}\,[t + \mu(\{|f - g| \geq t\})] + \int |\log^+|f| - \log^+|g||\,d\mu.$$

A sequence f_n converges to f in $L(\mu)$ if and only if f_n converges to f in measure, and $\log^+|f_n|$ converges to $\log^+|f|$ in $L^1(\mu)$. With the metric d, $L(\mu)$ becomes a complete metric space.

As an immediate consequence of 1.3 and 1.4, we obtain the following lemma.

2.1 Lemma. Suppose $\{f_n\}_{n=1}^{\infty}$ is a sequence in $L(\mu)$ which converges to f in measure. Then the following are equivalent:

(i) f_n converges to f in $L(\mu)$,
(ii) $\{\log^+ |f_n|\}_{n=1}^{\infty}$ is uniformly integrable,
(iii) $\lim_{n\to\infty} \int \log^+ |f_n|\, d\mu = \int \log^+ |f|\, d\mu < \infty$.

2.2 Corollary. If $f_n \to f$ in $L^p(\mu)$ for some $p > 0$, then $f_n \to f$ in $L(\mu)$.

Proof. This follows from 2.1 and the second statement of 1.2.

2.3 Theorem. $L(\mu)$ is a topological algebra.

Proof. We will use the inequalities

$$(*) \qquad \log^+ |ab| \le \log^+ |a| + \log^+ |b|$$

$$(**) \qquad \log^+ |a + b| \le \log^+ |a| + \log^+ |b| + \log 2,$$

where a and b are complex numbers. These inequalities show that if f and g belong to $L(\mu)$, then fg and $f + g$ belong to $L(\mu)$. So $L(\mu)$ is an algebra.

Suppose f_n and g_n converge respectively to f and g in $L(\mu)$. Then $f_n + g_n$ and $f_n g_n$ converge respectively to $f + g$ and fg in measure. From (*) and (**) it follows that $\{\log^+ |f_n + g_n|\}$ and $\{\log^+ |f_n g_n|\}$ are uniformly integrable. By 2.1, $f_n + g_n$ and $f_n g_n$ converge in $L(\mu)$. This proves the theorem.

Now let A be a uniform algebra, let $\phi \in M_A$, and let m be a representing measure for ϕ. The **Hardy algebra $H(m)$** is the closure of A in $L(m)$. Since $L(m)$ is a topological algebra, $H(m)$ is a closed subalgebra of $L(m)$. Since convergence in L^p implies convergence in L, the Hardy algebra $H(m)$ contains each of the Hardy spaces $H^p(m)$, $0 < p \le \infty$.

The set of invertible functions in the algebra $H(m)$ will be denoted by $\boldsymbol{H(m)^{-1}}$.

2.4 Lemma. If $u_n \in \mathrm{Re}\,(A)$, $u_n \to u$ in $L^1(m)$, and $^*u_n \to v$ in measure, then $e^{u+iv} \in H(m)^{-1}$.

Proof. For each real t, $e^{t(u_n + i^*u_n)}$ converges in $L(m)$ to $e^{t(u+iv)}$. So $e^{t(u+iv)} \in H(m)$ for all real t.

2.5 Corollary. Suppose the set of representing measures for ϕ is finite dimensional, and m is a core representing measure for ϕ. If $u \perp N^\infty(m)$, that is, if u is in the L^1-closure of $\mathrm{Re}\,(A)$, then $e^{u+i^*u} \in H(m)^{-1}$.

Proof. In this case the convergence of $u_n \in \mathrm{Re}\,(A)$ to u in the L^1-norm implies the convergence of $^*u_n$ to $^*u$ in the L^p-norm, $0 < p < 1$, by the abstract Kolmogoroff theorem IV.5.3. In particular, $^*u_n$ converges to $^*u$ in measure.

3. Jensen Measures

Recall that a representing measure m for ϕ is a Jensen measure if

$$\log|\phi(f)| \leq \int \log|f|\,dm, \qquad f \in A.$$

For $p > 0$, we define $\|\phi\|_p$ to be the norm of ϕ on $H^p(m)$. In other words,

$$\|\phi\|_p = \sup\{|\phi(f)| : f \in A, \|f\|_p \leq 1\}.$$

Then $\|\phi\|_p \geq 1$ for all $p > 0$, and $\|\phi\|_p = 1$ if $p \geq 1$. Also, ϕ extends continuously to H^p if and only if $\|\phi\|_p < \infty$.

3.1 Theorem. The representing measure m for ϕ is a Jensen measure if and only if $\|\phi\|_p = 1$ for all $p > 0$.

Proof. The theorem is a simple consequence of the classical fact that $\|f\|_p$ decreases to $\exp\{\int \log|f|\,dm\}$ as p decreases to 0, for all $f \in L^\infty$.

This latter fact can be proved as follows. That $\|f\|_p$ decreases as p decreases follows from Hölder's inequality. If $f \in L^\infty$ is a simple function (finite range), then the limit can be evaluated by applying L'Hospital's rule to the indeterminate form

$$\log\|f\|_p = \frac{\log \int |f|^p dm}{p}.$$

For an arbitrary $f \in L^\infty$, one takes simple functions $f_n \in L^\infty$ such that $|f_n| \geq |f|$ and $f_n \to f$ uniformly, and one justifies passing to the limit.

3.2 Theorem. If m is a Jensen measure for ϕ, then ϕ extends continuously to the Hardy algebra $H(m)$, and

$$\log|\phi(f)| \leq \int \log|f|\,dm, \qquad f \in H(m).$$

In particular, if f is an invertible element of $H(m)$, then

$$-\infty < \log|\phi(f)| = \int \log|f|\,dm.$$

Proof. Suppose $f_n \in A$ tends to 0 in H. Then $\int \log|f_n|\,dm \to -\infty$. By Jensen's inequality, $\log|\phi(f_n)| \to -\infty$, and $\phi(f_n) \to 0$. Hence ϕ extends continuously to $H(m)$.

Suppose now $f_n \in A$ converges to $f \in H$. For each $\epsilon > 0$, $\log(|f_n| + \epsilon)$ converges to $\log(|f| + \epsilon)$ in L^1. From

$$\log|\phi(f_n)| \leq \int \log|f_n|\,dm \leq \int \log(|f_n| + \epsilon)\,dm,$$

we obtain

$$\log|\phi(f)| \leq \int \log(|f| + \epsilon)\,dm.$$

Letting $\epsilon \to 0$, we obtain Jensen's inequality.

If $f \in H^{-1}$, we apply Jensen's inequality to both f and $1/f$ to obtain $-\infty < \log|\phi(f)| = \int \log|f|\, dm$. This completes the proof.

3.3 Corollary. Suppose the set of representing measures for ϕ is finite dimensional, and that m is a core measure for ϕ. Then ϕ extends continuously to the Hardy algebra $H(m)$. In particular, ϕ extends continuously to $H^p(m)$, $p > 0$.

Proof. By II.2.4, ϕ has a Jensen measure μ. Since m is a core measure, $\mu \leq cm$ for some constant c. Consequently convergence in $H(m)$ implies convergence in $H(\mu)$. Since ϕ extends continuously to $H(\mu)$, it extends continuously to $H(m)$.

The fact that $H(m)$ has any continuous linear functionals at all is rather interesting, in view of the fact that $L(m)$, or for that matter $L^p(m)$, $0 < p < 1$, has no non-zero continuous linear functionals in general.

4. Characterization of H

First we prove a lemma which is related to the modification technique used in the proof of the Hoffman-Wermer theorem II.7.2. The lemma allows us to replace convergence in L by as well-behaved convergence as the limit function will allow.

4.1 Lemma. Suppose M_ϕ is finite dimensional, and m is a core representing measure for ϕ. Suppose the sequence $f_n \in L(m)$ converges to f in $L(m)$. If $0 < p \leq \infty$, and $f \in L^p(m)$, then there are $g_n \in H^\infty(m)$ such that $f_n g_n \in L^\infty(m)$ and $f_n g_n$ converges (weak-star, if $p = \infty$) in $L^p(m)$ to f.

Proof. Let $w_1, \ldots, w_r$ be an orthonormal basis for N^∞.

First we note that if $u \in L^1_R$ is bounded above, and $u \perp N^\infty$, then $e^{u+i*u} \in H^\infty$. In fact, let $u_n = \max(u, -n)$, and set

$$a_{nj} = \int u_n w_j dm, \qquad 1 \leq j \leq r, \quad n \geq 1$$

$$v_n = u_n - \sum_{j=1}^{r} a_{nj} w_j, \qquad n \geq 1.$$

The v_n are bounded and orthogonal to N^∞, so $g_n = e^{v_n + i*v_n} \in H^\infty$. For each j, $a_{nj} \to 0$ as $n \to \infty$. Consequently the v_n are bounded above, independently of n, and the g_n are uniformly bounded in H^∞. Also, $v_n \to u$ in L^1, so $*v_n \to *u$ in measure, and $g_n \to e^{u+i*u}$ in measure. Hence $e^{u+i*u} \in H^\infty$, as was asserted.

Now suppose that $f \in L^p$ for some $0 < p \leq \infty$. Set $h_n(x) = |f(x)|$ if $1 \leq |f(x)| \leq n$, and set $h_n(x) = 1$ otherwise. Let $u_n = \max(0, \log^+|f_n| - \log h_n)$. It is easy to check that

$$e^{-u_n}|f_n| \leq h_n.$$

Now $\{u_n\}_{n=1}^{\infty}$ is uniformly integrable, and $u_n \to 0$ in measure. By 1.3, $u_n \to 0$ in L^1. Set

$$c_{nj} = \int u_n w_j dm, \qquad 1 \leq j \leq r, \quad n \geq 1,$$

$$v_n = u_n - \sum_{j=1}^{r} c_{nj} w_j, \qquad n \geq 1.$$

The first part of the proof shows that $g_n = e^{-(v_n + i^*v_n)} \in H^\infty$. For each fixed j, $c_{nj} \to 0$ as $n \to \infty$. Hence $v_n \to 0$ in L^1, and $^*v_n \to 0$ in measure. So $g_n \to 1$ in measure. Now $\{u_n - v_n\}$ is uniformly bounded, say $\|u_n - v_n\|_\infty \leq c$, so

$$|g_n f_n| = e^{u_n - v_n} e^{-u_n} |f_n| \leq e^c h_n.$$

This shows that $g_n f_n \in L^\infty$. Also, $|g_n f_n| \leq e^c h$, and $g_n f_n \to f$ in measure. If $0 < p < \infty$, then $g_n f_n \to f$ in L^p, by the dominated convergence theorem. If $p = \infty$, then $g_n f_n$ converges weak-star to f in L^∞. That completes the proof.

As an immediate consequence of 4.1, we have the following extension of IV.6.2(viii).

4.2 Theorem. Suppose M_ϕ is finite dimensional, and m is a core representing measure for ϕ. Then

$$H(m) \cap L^p(m) = H^p(m), \qquad 0 < p \leq \infty.$$

In particular,

$$H^p(m) \cap L^q(m) = H^q(m), \qquad 0 < p \leq q \leq \infty.$$

Proof. Clearly $H^p \subseteq H \cap L^p$.

Suppose $f_n \in A$ converges to f in H, and $f \in L^p$. By 4.1, there are $g_n \in H^\infty$ such that $g_n f_n$ converges (weak-star, if $p = \infty$) in L^p to f. Consequently $f \in H^p$, and $H \cap L^p \subseteq H^p$. That proves the theorem.

The first characterization of the Hardy algebra is the following.

4.3 Theorem. Suppose that M_ϕ is finite dimensional, and that m is a core measure for ϕ. If $f \in H(m)$, then $f = g/h$, where g and h belong to $H^\infty(m)$, and h is invertible in $H(m)$.

Proof. Let $f \in H$, and let $w_1, \ldots, w_r$ be an orthonormal basis for N^∞. Choose constants $c_1, \ldots, c_r$ such that $u = \log^+ |f| - \sum_{j=1}^{r} c_j w_j$ is orthogonal to N^∞. Then $h = e^{-(u + i^*u)}$ belongs to H^∞. Also, $g = fh \in H$ is bounded, so $g \in H^\infty$ also.

Another characterization of H is as follows.

4.4 Theorem. Suppose that M_ϕ is finite dimensional, and that m is a core measure for ϕ. Let $f \in L(m)$. Then $f \in H(m)$ if and only if there exists $w \in L^\infty$ such that $w \geq 0$, $\log w \in L^1$, and f belongs to the closure of A in $L^2(wdm)$.

Proof. Suppose first that $f \in H$, and express $f = g/h$ as above, where $g \in H^\infty$, and $h = e^{-(u+i^*u)} \in H^\infty$. Since $h \in H^{-1}$, there are $f_n \in A$ such that $f_n \to 1/h$ in H. Consequently $f_n h \to 1$ in H. By 4.1, there are $g_n \in H^\infty$ such that $g_n f_n h \in L^\infty$, and $g_n f_n h \to 1$ weak-star in L^∞. Let $h_n = g_n f_n g$, and let $w = e^{-2u} = |h|^2$. Then

$$\int |f - h_n|^2 \, wdm = \int |g|^2 \, |1 - f_n g_n h|^2 \, dm$$

tends to 0 as $n \to \infty$. Consequently f is in the closure of H^∞, and hence of A, in $L^2(wdm)$.

Conversely, suppose that w is as above, and that $f_n \in A$ converge to f in $L^2(wdm)$. Then $|f|^2 w \in L^1$, so $\log^+|f|$ is integrable, and $f \in L(m)$. Choose constants $c_1, \ldots, c_r$ such that

$$u = \log w - \sum c_j w_j$$

is in the closure of Re (A) in L^1. Set $h = e^{(u+i^*u)/2}$. Then

$$\int |f_k - f|^2 \, wdm = \int |f_k h - fh|^2 \, we^{-u} dm \to 0.$$

Since we^{-u} is bounded and bounded away from zero, $f_k h$ converges to fh in $L^2(m)$. Consequently $fh \in H$. Since h is invertible in H, f belongs to H. That proves the theorem.

Combining 4.4 and 4.2, we obtain the following abstract Phragmén-Lindelöf theorem.

4.5 Corollary. With m as above, suppose $0 < p < \infty$ and $f \in L^p(m)$. Suppose there is $w \in L^\infty(m)$ such that $w \geq 0$, $\log w \in L^1(m)$, and f is in the closure of A in $L^2(wdm)$. Then $f \in H^p(m)$.

5. Invertible Elements of H

Suppose that M_ϕ is finite dimensional, and that m is a core representing measure for ϕ. By IV.6.3, the additive group $\mathscr{L} = N^\infty \cap \log|(H^\infty)^{-1}|$ is a discrete subgroup of N^∞. Fix $h_1, \ldots, h_s \in (H^\infty)^{-1}$ such that $\{\log|h_j|\}_{j=1}^s$ forms a basis for $\mathscr{L}$, that is, such that every function in $\mathscr{L}$ can be expressed as a linear combination of the $\log|h_j|$ with integral coefficients.

5.1 Theorem. Suppose $m, h_1, \ldots, h_s$ are as above. Every $f \in H(m)^{-1}$ can be expressed uniquely in the form

$$f = \lambda e^{u+i^*u} \prod_{j=1}^{s} h_j^{q_j},$$

where λ is a constant of unit modulus, u belongs to the L^1-closure of Re (A), and $q_1, \ldots, q_s$ are integers.

Proof. If $f \in H^{-1}$, then $\log|f| \in L^1$. So $\log|f| = u + v$, where u belongs to the L^1-closure of Re (A), and $v \in N^\infty$. Then $g = fe^{-(u+i^*u)}$ is invertible in H, and both g and $1/g$ are bounded. By 4.2, $g \in (H^\infty)^{-1}$. Hence $\log|g| = v = \sum q_j \log|h_j|$, where the q_j are integers. Now $g\prod_{j=1}^{\infty} h_j^{-q_j}$ is an invertible function in H^∞ which has unit modulus, and hence must be a constant λ. This yields the desired factorization.

5.2 Theorem. Let m be as above, and let S be the set of logmodular measures for ϕ, regarded as a homomorphism of $H^\infty(m)$. Let $f \in H(m)$. Then

$$\log|\phi(f)| \leq \sup_{\mu \in S} \int \log|f|\, d\mu.$$

Moreover, $f \in H(m)^{-1}$ if and only if $\log|f| \in L^1(m)$ and

$$-\infty < \log|\phi(f)| = \sup_{\mu \in S} \int \log|f|\, d\mu.$$

Proof. For the first statement, note that there is always a Jensen measure in S, and apply 3.2.

From the description given in 5.1, any function in H^{-1} has the properties asserted by 5.2.

Conversely, suppose that $\log|f| \in L^1$ and $-\infty < \log|\phi(f)| = \sup_{\mu \in S} \int \log|f|\, d\mu$. Write $\log|f| = u + v$, where u belongs to the L^1-closure of Re (A), and $v \in N^\infty$. Set $g = fe^{-(u+i^*u)}$. Then $g \in H^\infty$, and

$$-\infty < \log|\phi(g)| = \sup_{\mu \in S} \int \log|g|\, d\mu.$$

It suffices to show that $g \in (H^\infty)^{-1}$.

By IV.7.4, we can assume that m itself is a logmodular measure. By the monotone extension principle (cf. II.2.1, or the proof of IV.7.4),

$$\sup_{\mu \in S} \int \log|g|\, d\mu = \inf_{v \geq \log|g|} \int v dm,$$

where the infimum is taken over all $v \in L_R^\infty$ such that all logmodular measures for ϕ agree on v. The set of such v coincides with the linear span of the $\log|h_j|$, $1 \leq j \leq s$, and the functions in L_R^∞ which are orthogonal to N^∞. Consequently there exist $v_n \perp N^\infty$ and real numbers $c_{j,n}$, $1 \leq j \leq s$, $1 \leq n \leq \infty$, such that

$$v_n + \sum_{j=1}^{s} c_{jn} \log|h_j| \geq \log|g|$$

$$\lim_{n\to\infty} \int \left(v_n + \sum_{j=1}^{s} c_{jn} \log|h_j| - \log|g|\right) dm = 0.$$

Since $\log|g| \in N^\infty$, and $v_n \perp N^\infty$, the functions $\sum_{j=1}^{s} c_{jn} \log|h_j|$ converge to $\log|g|$ in $L^1(m)$. Consequently $\lim_{n\to\infty} c_{jn} = c_j$ exists, and

$$\log|g| = \sum_{j=1}^{s} c_j \log|h_j|.$$

Choose integers $p_n \geq 1$ and integers q_{nj}, $1 \leq j \leq s$, $1 \leq n < \infty$, such that

$$\lim_{n\to\infty} \max_{1\leq j\leq s} |p_n c_j - q_{nj}| = 0.$$

Set

$$g_n = g^{p_n - 1} h_1^{-q_{n1}} \ldots h_s^{-q_{ns}}.$$

Then $g_n \in H^\infty$, and $\log|g_n g| = \sum_{j=1}^{s} (p_n c_j - q_{nj}) \log|h_j|$ converges uniformly to zero. Since $\log|\phi(g_n g)| = \sup_{\mu \in S} \int \log|g_n g|\, d\mu$, we obtain $|\phi(g_n g)| \to 1$.

Choose constants λ_n of unit modulus, such that $\phi(\lambda_n g_n g) \to 1$. Since $|\lambda_n g_n g| \to 1$ uniformly, the functions $\lambda_n g_n g$ converge to 1 in measure. Since $|g|$ is bounded away from zero, $\lambda_n g_n$ is a bounded sequence in H^∞, which converges to $1/g$ in measure. So $1/g \in H^\infty$, and $g \in (H^\infty)^{-1}$. That completes the proof.

Now we give a partial characterization of functions in H^{-1} which happen to belong to H^p.

5.3 Theorem. With m as above, suppose $0 < p < \infty$, $h \in H^p(m)$, and $\log|h| \in L^1(m)$. Then h is invertible in the Hardy algebra $H(m)$ if and only if Ah is dense in $H^p(m)$.

Proof. Suppose first that $h \in H^{-1}$. Choose $f_n \in A$ such that $f_n \to 1/h$ in $L(m)$. By 4.1, there are $g_n \in H^\infty$ such that $f_n g_n h \to 1$ in $L^p(m)$. Hence 1 is in the L^p-closure of $H^\infty h$, and the L^p-closure of Ah must be H^p.

For the converse, suppose that Ah is dense in H^p. Choose $f_k \in A$ such that $f_k h \to 1$ in L^p. In particular, $f_k \to 1/h$ in measure. From the inequality

$$\log^+|f_k| \leq \log^+|f_k h| + \log^+\left|\frac{1}{h}\right| + \log 2,$$

we see that $\{\log^+|f_k|\}$ is uniformly integrable. By 1.3, $f_k \to 1/h$ in H. So $h \in H^{-1}$. This completes the proof.

It is not known whether the density of Ah in $H^p(m)$ already implies that $\log|h| \in L^1(m)$. By 3.2, this is true when m is a Jensen measure.

We will now specialize these results to the case in which m is a unique representing measure for ϕ. In this case, m is a Jensen measure. The conjugation operator $u \to {}^*u$ is defined on all of L^1_R.

5.4 Theorem. Suppose ϕ has a unique representing measure m. If

$h \in H(m)$ and $\phi(h) \neq 0$, then $h = Fe^{u+i^*u}$, where $F \in H^\infty(m)$, $|F| = 1$ almost everywhere, and $u = \log|h| \in L^1(m)$. For $h \in H(m)$, the following are equivalent:

(i) $h \in H(m)^{-1}$;
(ii) $-\infty < \log|\phi(h)| = \int \log|h|\, dm$;
(iii) $h = \lambda e^{u+i^*u}$, where $u \in L^1_R$ and λ is a constant of modulus one.

If, in addition, $h \in H^p(m)$ for some $0 < p < \infty$, then each of these is equivalent to:

(iv) Ah is dense in $H^p(m)$.

Proof. For the first statement, we define $u = \log|h|$. Since $\log|\phi(h)| \leq \int \log|h|\, dm$, the function $u = \log|h|$ is integrable, and $e^{u+i^*u} \in H^{-1}$. Setting $F = he^{-(u+i^*u)}$, we obtain the desired factorization.

That (i) implies (ii) follows from 3.2, and the fact that m must be a Jensen measure.

Suppose that (ii) is true. Let $F = he^{-(u+i^*u)}$, where $u = \log|h|$. Then $|F| = 1$ almost everywhere, and

$$\log|\phi(F)| = \log|\phi(h)| + \log|\phi(e^{-(u+i^*u)})| = 0.$$

Consequently $|\phi(F)| = 1$, and F must be a constant of modulus one. This proves (iii).

That (iii) implies (i) is clear.

Theorem 5.3 shows that (iv) is equivalent to the other criteria if $f \in H^p$. This completes the proof.

Now we turn to the algebra $P(\Delta)$ of functions on the unit disc $\Delta = \{|z| \leq 1\}$ which are uniformly approximable by polynomials. The homomorphism "evaluation at 0" has a unique representing measure on $b\Delta$, namely, $d\theta/2\pi$.

It can be shown that any function f in the Hardy algebra $H(d\theta/2\pi)$ has an extension, also denoted by f, which is analytic on the interior of the unit disc, such that

$$\lim_{r \uparrow 1} f(re^{i\theta}) = f(e^{i\theta}) \text{ a.e.}$$

In fact, the Hardy algebra consists of precisely those functions f analytic on int Δ such that the integrals

$$F_r(e^{i\theta}) = \int_0^\theta \log^+|f(re^{i\psi})|\, d\psi, \qquad 0 \leq \theta \leq 2\pi$$

are uniformly absolutely continuous for $0 < r < 1$.

Suppose $f \in H(d\theta/2\pi)$ does not vanish identically. Then $f = z^m g$ for some $g \in H(d\theta/2\pi)$ satisfying $g(0) \neq 0$. The factorization theorem 5.4 applies to g. By incorporating z^m into the factor of g of modulus one, we see that every non-zero function $f \in H(d\theta/2\pi)$ can be factored in the form $f = FG$, where

$F \in H^\infty(d\theta/2\pi)$ satisfies $|F| = 1$ almost everywhere, and $G \in H(d\theta/2\pi)$ satisfies $-\infty < \log|G(0)| = \int \log|G|\, d\theta/2\pi$.

Functions in $H^\infty(d\theta/2\pi)$ of unit modulus are called **inner functions.** Functions in $H(d\theta/2\pi)$ satisfying $-\infty < \log|G(0)| = \int \log|G|\, d\theta/2\pi$, or any of the other equivalent properties of 5.4, are called **outer functions.** Our result can be stated as follows.

5.5 Theorem. Every non-zero function f in the Hardy algebra $H(d\theta/2\pi)$ can be factored in the form $f = FG$, where $F \in H^\infty(d\theta/2\pi)$ is inner, and $G \in H(d\theta/2\pi)$ is outer. The factors are unique, up to a multiplicative constant of modulus one.

6. Invariant Subspaces

Suppose that the set of representing measures for ϕ is finite dimensional, and that m is a core measure for ϕ.

A closed subspace M of $L^p(m)$ is **invariant** if $AM \subseteq M$. Since every function in H^∞ is a bounded pointwise limit of functions in A, also $H^\infty M \subseteq M$ whenever M is an invariant subspace of L^p, $0 < p < \infty$, or whenever M is a weak-star closed invariant subspace of L^∞.

The following theorem shows that, in considering invariant subspaces, it suffices to consider invariant subspaces of L^2, or alternatively, weak-star closed invariant subspaces of L^∞. The expression "weak-star closed" in parentheses will refer to the case $q = \infty$.

6.1 Theorem. Suppose the set of representing measures for ϕ is finite dimensional, and m is a core measure for ϕ. Suppose $0 < p < q \leq \infty$. There is a one-to-one correspondence between invariant subspaces M_p of $L^p(m)$ and (weak-star closed) invariant subspaces M_q of $L^q(m)$, such that $M_q = M_p \cap L^q$, and M_p is the closure in L^p of M_q.

Proof: It suffices to consider the case $q = \infty$.

Let M_p be an invariant subspace of L^p, and let $M = M_p \cap L^\infty$. If f_n is any sequence in M which converges pointwise boundedly to f, then certainly $f \in M$. By the Krein-Schmulian criterion IV.2.1, M is weak-star closed.

Suppose $f \in M_p$. If we apply 4.1 to the constant sequence $f_k = f$, we find $g_k \in H^\infty$ such that $g_k f \in L^\infty$ and $g_k f \to f$ in L^p. Since $g_k f \in M$, the closure of M in L^p is M_p.

To complete the proof, we must show that if M is a weak-star closed invariant subspace of L^∞, and M_p is the closure of M in L^p, then $M^p \cap L^\infty = M$. For this, suppose $f \in M_p \cap L^\infty$. Choose $f_n \in M$ such that $f_n \to f$ in L^p. By 4.1, there are $g_n \in H^\infty$ such that $g_n f_n \to f$ weak-star in L^∞. Since $g_n f_n \in M$, also $f \in M$. That completes the proof.

An invariant subspace M of L^p is **simply invariant** if A_0M is not (weak-star) dense in M.

6.2 Theorem. Let m be a unique representing measure for ϕ. Let M be a simply invariant subspace of L^p. Then there is $F \in M$ such that $|F| = 1$ almost everywhere and $M = FH^p(m)$.

Proof. Either all of the subspaces M_p of 6.2 are simply invariant, or none are. By that theorem, it suffices to consider the case $p = 2$.

Choose $F \in M$ such that $\int |F|^2\, dm = 1$ while $F \perp A_0M$. Since $F \perp A_0F$, $\int g|F|^2\, dm = 0$ for all $g \in A_0$. Hence $|F|^2\, dm$ is a representing measure for ϕ, and $|F| = 1$ almost everywhere.

Now $FH^2 \subseteq M$. If $f \in M$ and $f \perp FH^2$, then $\int f\bar{F}\bar{g}dm = 0$ for all $g \in A$. Also, $F \perp A_0f$, so $\int f\bar{F}gdm = 0$ for all $g \in A_0$. Hence $f\bar{F} \perp (\bar{A} + A_0)$. Since $L^2 = H^2 \oplus \overline{H_0^2}$, we obtain $f\bar{F} = 0$, and $f = 0$. Hence $M = FH^2$. That proves the theorem.

Now we consider the disc algebra $P(\Delta)$, the homomorphism $\phi(f) = f(0)$, and its unique representing measure $d\theta/2\pi$ on $b\Delta$. Since $P(\Delta)$ is generated by the coordinate function z, a subspace M of $L^p(d\theta/2\pi)$ is invariant if and only if $zM \subseteq M$. The invariant subspace M is simply invariant if and only if zM is not dense in M. Since zM is already closed, this occurs if and only if $zM \neq M$.

If M is any non-zero invariant subspace of $H^p(d\theta/2\pi)$, then $zM \neq M$, as can be seen by comparing the orders of the zeros of the functions in M and in zM at $z = 0$. From 6.2, we now obtain the L^p-version of Beurling's seminal theorem classifying the invariant subspaces of $H^2(d\theta/2\pi)$.

6.3 Theorem (Beurling's Theorem). Let $0 < p \leq \infty$, and let M be a (weak-star closed) invariant subspace of $H^p(d\theta/2\pi)$. Then there is a function $F \in H^\infty(d\theta/2\pi)$ such that $|F| = 1$ almost everywhere, and such that $M = FH^p(d\theta/2\pi)$.

We remark in passing that the invariant subspaces of $H^\infty(d\theta/2\pi)$ include the closed ideals of H^∞. Although much progress has been made on the maximal ideals of H^∞ by Carleson [2] and Hoffman [6], the closed ideals of H^∞ remain a mystery.

To complete the classification of the (weak-star closed) invariant subspaces of $L^p(d\theta/2\pi)$, we prove the following theorem.

6.4 Theorem. If M is a (weak-star closed) invariant subspace of $L^p(d\theta/2\pi)$ which is not simply invariant, then there is a measurable subset E of the circle $b\Delta$ such that $M = \chi_E L^p(d\theta/2\pi)$, where χ_E is the characteristic function of E.

Proof. Again we consider only the case $p = 2$.

In this case, $P_0(\Delta)M = zM$ is already closed. Since M is not simply invariant, $zM = M$. Since $|z| = 1$ on $b\Delta$, we obtain $\bar{z}M = M$. So $pM \subseteq M$ whenever p is a polynomial in z and $\bar{z}$. It follows that $L^\infty M \subseteq M$.

Let F be the orthogonal projection of 1 onto M. Then FL^∞ is orthogonal to $1 - F$. So $F - |F|^2$ is orthogonal to L^∞, and $F - |F|^2 = 0$. From this we deduce that F is real, and that $F^2 = F$. So $F = \chi_E$ for some measurable set E.

Since $\chi_E L^\infty \subseteq M$, we have $\chi_E L^2 \subseteq M$.

Suppose $g \in M$ is orthogonal to $\chi_E L^2$. Then $g = 0$ on E. Since $gL^\infty \subseteq M$, $1 - \chi_E$ is orthogonal to gL^∞. Since g vanishes on E, 1 is orthogonal to gL^∞. It follows that $g = 0$, and $M = \chi_E L^2$.

7. Embedding of Analytic Discs

One of the most striking results associated with invariant subspaces is the embedding theorem of Wermer. Here we give a short account of one formulation of the theorem. A more penetrating study of the embedding of analytic structure in maximal ideal spaces will be undertaken in the next chapter.

7.1 Theorem. Suppose that m is a unique representing measure for ϕ. Suppose that there exists $F \in H^\infty$ such that $|F| = 1$ almost everywhere, and $H_0^\infty = FH^\infty$. For $|\lambda| < 1$, set $m_\lambda = (1 - |\lambda|^2)/|1 - \bar{\lambda}F|^2\, dm$. Then

(i) m_λ is multiplicative on H^∞, $|\lambda| < 1$.

(ii) If $g \in H^\infty$, then $\tilde{g}(\lambda) = \int g\,dm_\lambda$ is a bounded analytic function on $\{|\lambda| < 1\}$.

(iii) $\lambda \to m_\lambda$ maps the open unit disc continuously into M_A. It maps the open unit disc homeomorphically into M_{H^∞}.

(iv) If $\psi \in M_A$ is weak-star continuous, then ψ has a unique representing measure, which coincides with m_λ for some λ, $|\lambda| < 1$.

Proof. If $|\lambda| < 1$, then

$$\frac{1 - |\lambda|^2}{|1 - \bar{\lambda}F|^2} = \sum_{n=1}^{\infty} \bar{\lambda}^n F^n + \sum_{n=0}^{\infty} \lambda^n \bar{F}^n.$$

If $g \in H^\infty$, and $a_n = \int g\bar{F}^n dm$, then

$$\tilde{g}(\lambda) = \int g\,dm_\lambda = \sum_{n=0}^{\infty} a_n \lambda^n.$$

Consequently $\tilde{g}$ is analytic on $\{|\lambda| < 1\}$.

If $|\lambda| < 1$, the multiplication operator $h \to (\lambda - F)h$, $h \in H^\infty$, has closed range, and it has null space consisting only of $\{0\}$. By the stability theory of the index of semi-Fredholm operators, or by an elementary direct argument, the range of the operators must have constant codimension (finite or infinite) in H^∞. Since $FH^\infty = H_0^\infty$, that codimension must be one. In other words, $H_\lambda^\infty = (\lambda - F)H^\infty$ has codimension one in H^∞ for all $|\lambda| < 1$.

Now H_λ^∞ is an ideal in H^∞, so it must be a maximal ideal. If $g \in H_\lambda^\infty$, then $h = g/(\lambda - F) \in H^\infty$, and

$$\int g\,dm_\lambda = \sum_{n=0}^{\infty} \lambda^n \int (\lambda - F)h\bar{F}^n dm = 0.$$

Hence m_λ is a representing measure corresponding to the maximal ideal H_λ^∞. This proves (i) and (ii).

The map $\lambda \to m_\lambda$ is clearly continuous. Since $\int F\,dm_\lambda = \lambda$, the inverse of the map is given by F, which is continuous on M_{H^∞}. This proves (iii).

To prove (iv), we first note that F cannot be constant on a set of positive measure. In fact, if $F = \lambda_0$ on a set of positive measure, then Jensen's inequality would yield $\log|\lambda_0| \le \int \log|F - \lambda_0|\,dm = -\infty$, or $\lambda_0 = 0$. However, $|F| = 1$ almost everywhere.

If ψ is weak-star continuous, then ψ has a representing measure of the form hdm. If $|\psi(F)| = |\int Fhdm| = 1$, then F would be constant almost everywhere with respect to hdm, that is, on a subset of positive measure. Consequently $|\psi(F)| < 1$.

If $\lambda = \psi(F)$, then ψ must vanish on $(\lambda - F)H^\infty = H_\lambda^\infty$. Hence $\psi(g) = \tilde{g}(\lambda)$ for all $g \in H^\infty$. In particular, m_λ represents ψ.

Suppose $\mu = kdm + \mu_s$ is the Lebesgue decomposition of another representing measure μ for ψ, regarded as a homomorphism of A. Then $\mu - m_\lambda \perp A$. By the abstract F. and M. Riesz theorem II.7.6, μ_s and $kdm - m_\lambda$ are orthogonal to A. Since $\mu_s \ge 0$, and $\int d\mu_s = 0$, we obtain $\mu_s = 0$. By IV.6.2, $A + \bar{A}$ is weak-star dense in $L^\infty(m)$. Consequently $kdm = m_\lambda$. This completes the proof.

We now give some information regarding the applicability of 7.1.

7.2 Theorem. Suppose m is a unique representing measure for ϕ. Suppose there is $\psi \in M_A$ such that $\psi \ne \phi$ and ψ is weak-star continuous. Then there is $F \in H^\infty$ such that $FH^\infty = H_0^\infty$ and $|F| = 1$ almost everywhere.

Proof. The homomorphism ψ extends to a weak-star continuous homomorphism of H^∞. The kernel H_ψ^∞ of ψ is a weak-star closed invariant subspace of L^∞. It is simply invariant, since $A_0 H_\psi^\infty \subseteq H_\psi^\infty \cap H_0^\infty$ cannot be weak-star dense in H_0^∞. Choose $G \in H^\infty$ such that $GH^\infty = H_\psi^\infty$ and $|G| = 1$ almost everywhere. Then $|\int Gdm| \ne 1$, or else G would be a constant. So $|\phi(G)| = |\int Gdm| < 1$. As in the proof of 7.1, $(\lambda - G)H^\infty$ has codimension one in H^∞ for $|\lambda| < 1$. In particular, we obtain $(G - \phi(G))H^\infty = H_0^\infty$. The desired function is then $F = (G - \phi(G))/(1 - \overline{\phi(G)}G)$.

8. Szegö's Theorem

In this section, we consider the problem of finding the distance from the constant function 1 to A_0 in $L^p(\mu)$, where μ is a positive measure on X. The classical theorem of Szegö states that if $h \in L^1(d\theta/2\pi)$, $h \geq 0$, then

$$\inf \int |1 - f(e^{i\theta})|^2 \frac{h(e^{i\theta})d\theta}{2\pi} = \exp\left(\int \log h(e^{i\theta}) \frac{d\theta}{2\pi}\right)$$

where the infimum is taken over all polynomials $f(z)$ satisfying $f(0) = 0$. This theorem extends to the case of a unique representing measure for a homomorphism (cf. 8.2). In fact, it can be taken as the cornerstone of the development of the invariant subspace theory in that case.

Kolmogoroff and Krein extended the Szegö theorem to positive measures μ. They showed that $\inf \int |1 - f(e^{i\theta})|^2 \, d\mu(\theta)$ depends only on the absolutely continuous part of μ with respect to $d\theta$, and so can be calculated via Szegö's theorem. We begin with this latter result.

8.1 Theorem (Kolmogoroff-Krein Theorem). Let $\mu = \mu_a + \mu_s$ be the Lebesgue decomposition of the positive finite measure μ with respect to the set M_ϕ of representing measures for ϕ. Then

$$\inf_{j \in A_0} \int |1 - f|^p \, d\mu = \inf_{f \in A_0} \int |1 - f|^p \, d\mu_a, \qquad 0 < p < \infty.$$

Proof. Let E be an F_σ-set such that $M_\phi(E) = 0$ while $\mu_s(X \backslash E) = 0$. By the Forelli lemma II.7.3, there is a sequence $g_k \in A$ such that $\|g_k\|_X \leq 1$, $g_k \to 1$ M_ϕ-almost everywhere, and $g_k(x) \to 0$ for $x \in E$. In particular, $\phi(g_k) \to 1$.

If $g \in A_0$, then $\{gg_k + \phi(g_k) - g_k\}_{k=1}^{\infty}$ is a bounded sequence in A_0 which converges to 1 μ_s-almost everywhere, and which converges to g μ_a-almost everywhere. Hence

$$\begin{aligned}\inf_{f \in A_0} \int |1 - f|^p \, d\mu &\leq \lim_{k \to \infty} \int |1 - gg_k - \phi(g_k) + g_k|^p \, d\mu \\ &= \int |1 - g|^p \, d\mu_a.\end{aligned}$$

Taking the infimum over $g \in A_0$, we obtain

$$\inf_{f \in A_0} \int |1 - f|^p \, d\mu \leq \inf_{g \in A_0} \int |1 - g|^p \, d\mu_a.$$

The reverse inequality is trivial, so the proof is complete.

For the next theorem, we need the inequality relating arithmetic and geometric means. The inequality is

$$e^{\int u \, dm} \leq \int e^u dm, \qquad u \in L^1_R,$$

with equality holding if and only if u is constant. To prove this, one notes that

$$1 - u(x) + \int udm \leq e^{u(x) - \int udm}, \qquad x \in X,$$

with equality only when $u(x) = \int udm$. Integrating this inequality, we obtain the desired result.

8.2 Theorem (Szegö's Theorem). Suppose ϕ has a unique representing measure m. Suppose $h \in L^1(m)$, $h \geq 0$, and $0 < p < \infty$. Then

$$\inf_{f \in A_0} \int |1 - f|^p \, hdm = \exp\left(\int \log hdm\right)$$

There is a unique function g belonging to the closure of A_0 in $L^p(hdm)$, such that

$$\inf_{f \in A_0} \int |1 - f|^p \, hdm = \int |1 - g|^p \, hdm.$$

If the infimum is positive, this function is

$$g = 1 - \exp\left(\frac{\int \log hdm - \log h - i^* \log h}{p}\right).$$

Proof. We agree that $e^{\int \log hdm} = 0$ if $\log h \notin L^1$.

Suppose first that $\log h \in L^1$. Let g be the function in $H_0(m)$ defined in 8.2. Then $g \in L^p(hdm)$. Define $u_k \in L_R^\infty$ by setting $u_k = \log h$ wherever $|\log h| < k$, and $u_k = 0$ elsewhere. Then $1 - \exp\left(\int u_k dm - u_k - i^* u_k\right)$ is a sequence in H_0^∞ converging to g in $L^p(hdm)$. Consequently g belongs to the closure of A_0 in $L^p(hdm)$. So

$$\inf_{f \in A_0} \int |1 - f|^p \, hdm \leq \int |1 - g|^p \, hdm = e^{\int \log hdm}.$$

If $f \in A_0$, then

$$\begin{aligned} \int |1 - f|^p \, hdm &= \int \exp(p \log |1 - f| + \log h) dm \\ &\geq \exp\left(p \int \log |1 - f| \, dm + \int \log hdm\right) \\ &\geq \exp\left(p \log |\phi(1 - f)| + \int \log hdm\right) \\ &= e^{\int \log hdm}, \end{aligned}$$

where we have used the inequality of arithmetic and geometric means, and Jensen's inequality. Consequently,

$$\inf_{f \in A_0} \int |1 - f|^p \, hdm = e^{\int \log hdm}.$$

If f belongs to the closure of A_0 in $L^p(hdm)$, then $f \in H_0(m)$, and this string of inequalities remains valid. If, in addition, $\int |1 - f|^p\, hdm = e^{\int \log h dm}$, then the inequalities become equalities. From this we make two deductions: $p \log |1 - f| + \log h$ is constant, and $\log |\phi(1 - f)| = \int \log |1 - f|\, dm$. By 5.4, $1 - f$ is a constant multiple of $e^{-(\log h + i^* \log h)/p}$. Since $\phi(f) = 0$, we find that f must coincide with g. That proves the uniqueness of g.

Now suppose $\log h \notin L^1$. If $\epsilon > 0$, then

$$\inf_{f \in A_0} \int |1 - f|^p\, hdm \leq \inf_{f \in A_0} \int |1 - f|^p (h + \epsilon) dm$$
$$= e^{\int \log (h+\epsilon) dm}.$$

Since $\int \log (h + \epsilon) dm \to -\infty$ as $\epsilon \to 0$, the infimum must be zero. That completes the proof.

9. Extremal Functions in H^1

A function $f \in H^1$ is **extremal** if $f \not\equiv 0$ and f satisfies the following condition: Whenever $k \in L^\infty_R$ is such that $kf \in H^1$, then k is constant on the set where $f \neq 0$. The following characterization extends to any subspace of an L^1-space.

9.1 Lemma. Suppose $f \in H^1$ is not identically zero. Then f is extremal if and only if $f/\|f\|_1$ is an extreme point of the unit ball of H^1.

Proof. Suppose, for simplicity, $\|f\|_1 = 1$.

If f is not extremal, there is $k \in L^\infty_R$ such that $kf \in H^1$ and kf is not a constant multiple of f. Replacing k by $\epsilon(k - \int k |f|\, dm)$, we can assume that $\int k |f|\, dm = 0$, $|k| \leq 1$, and $kf \not\equiv 0$. Then

$$\|f \pm kf\|_1 = \int |1 \pm k|\, fdm = \int (1 \pm k) |f|\, dm = 1.$$

Since f lies on the line segment joining $f + kf$ and $f - kf$, f is not an extreme point of the unit ball of H^1.

Conversely, suppose that f is not an extreme point. Then there is $g \in H^1$ such that $g \not\equiv 0$, and such that the line segment from $f - g$ to $f + g$ lies in the unit ball of H^1, that is, $\|f \pm g\|_1 \leq 1$. Write $g = kf + h$, where h vanishes where f does not. Then

$$0 \geq \|f + g\|_1 + \|f - g\|_1 - 2\|f\|_1$$
$$= \int (|1 + k| + |1 - k| - 2) |f|\, dm + 2 \int |h|\, dm.$$

Now $|1 + z| + |1 - z| - 2 \geq 0$, with equality only when z lies on the interval $[-1, 1]$. Hence $h = 0$, and $-1 \leq k \leq 1$. Since $\int k |f|\, dm = 0$, k cannot be constant $|f|\, dm$-almost everywhere. So f is not extremal. This proves the lemma.

9.2 Lemma. Suppose m is a unique representing measure for ϕ. Let E be a Borel subset of X such that $0 < m(E) < 1$. Then there exists $k \in H^\infty$ such that k is real on E while k is not constant on E.

Proof. Let u be the characteristic function of $X\backslash E$, and set $g = e^{u+i^*u}$. Then $g \in (H^\infty)^{-1}$. Since $u = 0$ on E, $|g| = 1$ on E. Also, $|\phi(g)| = e^{\int u dm} > 1$.

Let $k = g + 1/g$. Then $k \in H^\infty$, and k is real on E.

Suppose k is constant on E. Then g assumes at most two values on E, and g is a constant λ_0 on a subset of E of positive measure. Applying Jensen's inequality to $g - \lambda_0$, we obtain $\phi(g) = \lambda_0$. Since $|g| = 1$ on E, we obtain $|\phi(g)| = |\lambda_0| = 1$. This contradiction establishes the lemma.

9.3 Corollary. If m is unique, and $f \in H^1(m)$ is extremal, then f cannot vanish on a set of positive measure.

9.4 Lemma. Suppose m is a unique representing measure for ϕ. A function $f \in H^1$ is extremal if and only if $A_0 f + \overline{A_0 f} + C$ is dense in L^1.

Proof. Here C is the set of complex constants.

If f is not extremal, there exists $k \in L_R^\infty$ such that $k \not\equiv 0$, $kf \in H^1$, and $\int k dm = 0$. Then $k \perp A_0 f + \overline{A_0 f} + C$, so the set cannot be dense in L^1.

Conversely, suppose that $A_0 f + \overline{A_0 f} + C$ is not dense in L^1. If f vanishes on a set of positive measure, then f is not extremal, by 9.3. So we can assume that f does not vanish anywhere. Choose $k \in L_R^\infty$ such that $k \not\equiv 0$ and $k \perp A_0 f + \overline{A_0 f} + C$. In particular, $\int kfg dm = 0$ for all $g \in A_0$. By IV.6.2, $kf \in H^1$. Since $\int k dm = 0$, k cannot be a constant. Hence f is not extremal. This proves the lemma.

9.5 Theorem. Suppose that m is a unique representing measure for ϕ. Let $f \in H^1$. Then f is extremal if and only if $A_0 f$ is dense in H_0^1.

Proof. Suppose first that f is extremal. Let $g \in H_0^\infty$. By 6.4, there are $p_n, q_n \in A_0$ and complex constants a_n such that $p_n f + \overline{q_n f} + a_n \to g$ in L^1. Integrating the relation, we obtain $a_n \to 0$. Consequently $p_n f + \overline{q_n f} \to g$ in L^1. By the Kolmogoroff theorem IV.5.3, $p_n f \to g$ in L^p, $0 < p < 1$. By 4.1, there are $h_n \in H^\infty$ such that $h_n p_n f$ converges weak-star to g. So g belongs to the L^1-closure of $H_0^\infty f$. And $A_0 f$ is dense in H_0^1.

Conversely, if $A_0 f$ is dense in H_0^1, then $A_0 f + \overline{A_0 f} + C$ is dense in $H^1 + \overline{H^1}$. This, in turn, is dense in L^1, by IV.6.2. By 9.4, f is extremal. This proves the theorem.

It is not known, in general, whether the density of $A_0 f$ in H_0^1 implies the density of Af in H^1. It does, if $\phi(f) \neq 0$, or if $\log |f| \in L^1$. It also does if

$H_0^1 = FH^1$ for some unitary function F. In the case of the disc algebra $P(\Delta)$, we obtain the following.

9.6 Theorem (deLeeuw–Rudin Theorem). The extremal functions in $H^1(d\theta/2\pi)$ are the outer functions, that is, the functions in H^1 which satisfy one of the equivalent conditions of 5.4.

NOTES

The function theory which plays an integral role in this chapter stems from the theory of analytic functions on the unit disc, culminating in Nevanlinna's theory of functions of bounded characteristic. This theory was used by Beurling [1] to prove his invariant subspace theorem. The next major step was taken by Helson and Lowdenslager [1, 2], who passed beyond the framework of the unit disc, and developed a function theory in a group theoretic setting. Much of their function theory could be immediately carried over to dirichlet algebras. It was this observation which led Wermer [12] to his embedding theorem. That part of the function theory of Helson and Lowdenslager which is peculiar to the group theoretic setting will be dealt with in Chapter VII.

The next step was taken by Hoffman [3], who systematically developed a function theory in the context of logmodular algebras which allowed him to handle the algebra $H^\infty(\Delta)$. His paper contains a lucid treatment of many of the results touched on in sections 5–8. Hoffman takes Szegö's theorem as the cornerstone of the theory, and he relies heavily on the orthogonal projection techniques of Helson and Lowdenslager.

The circle of results from Hoffman [3] was extended to the context of a unique representing measure by Lumer [1], who introduced envelope considerations. In Hoffman and Rossi [2], and in König [1], the standard circle of theorems is proved in its ultimate generality, the situation described in exercise 4. Exercise 4 itself is from the article of Srinivasan and Wang in Birtel [1].

A function theory was developed for the case in which the space of representing measures is finite dimensional by Ahern and Sarason [1], who also assumed a unique logmodular measure. They proved a version of Szegö's theorem, and obtained 5.2, for instance, in that context. It was Glicksberg [8] who pointed out how the envelope techniques could be applied to core measures (= strongly dominant measures).

Meanwhile, Lumer [2] pointed out how the Hardy space theory could be based on the abstract conjugation operator. In Lumer [3], the Hardy algebra (= universal Hardy class) is introduced, and the notion of uniform integrability is employed (cf. also Gamelin and Lumer [1]). A general systematic development of the theory from this point of view is in Lumer [4], where the Hardy spaces are defined with respect to families of representing measures. In this setup, the abstract conjugation operator becomes the basic tool, and uniform integrability and envelope considerations replace projection methods and Szegö's theorem. The approach which we have followed is a simplified version of this program. For an alternative approach, see König [1], [2], [3].

The standard reference for function theory on the unit disc is Hoffman [4]. An account of various aspects of the theory is given in Helson [1], where it is shown how concrete Phragmén-Lindelöf theorems can be obtained from the abstract version 4.5. The deLeeuw-Rudin theorem is in deLeeuw and Rudin [1], and the abstract version 9.5 is in Gamelin [2].

EXERCISES

1. (Arens' Lemma). If μ is a probability measure, $a > 0$, and $h \in L^1_R(\mu)$ satisfies $\int \log|1 + th|\, d\mu \geq 0$ for $-a < t < a$, then $h = 0$. *Hint:* Observe that $u(z) = \int \log|1 + zh|\, d\mu$ is a positive harmonic function for $|\mathrm{Re}\,(z)| < a$ and $\mathrm{Im}(z) > 0$, satisfying $\lim_{y\to 0} u(iy)/y = 0$. Compare this with the Poisson integral representation of u on the disc $\{|z - ia| < a\}$.

2. If m is a Jensen measure, the only real-valued functions in $H^1(m)$ are the constants.

3. (a) If $A + \bar{A}$ is dense in $L^1(m)$, then $H(m) \cap L^\infty(m)$ is a logmodular algebra.
(b) If m is as in exercise 8, chapter IV, then $H(m) = L(m)$.

4. Show that the following are equivalent, for fixed $1 \leq p < \infty$, and $m \in M_\phi$:
(i) If $\mu \in M_\phi$ is absolutely continuous with respect to m, then $\mu = m$.
(ii) $A + \bar{A}$ is weak-star dense in $L^\infty(m)$.
(iii) $A + \bar{A}$ is dense in $L^q(m)$ for all $q < \infty$.
(iv) $H^\infty(m)$ is logmodular.
(v) Every simply invariant subspace of $L^p(m)$ is of the form $FH^p(m)$ for some unitary function F.
(vi) For all $w \geq 0$, $w \in L^1(m)$, it is true that

$$\inf_{f \in A_0} \int |1 - f|^p\, w\,dm = e^{\int \log w\,dm}.$$

Hint: For (iii) $\Rightarrow$ (iv), use exercise 2 of chapter IV. For (v) $\Rightarrow$ (vi), show first that $H^p \cap L^\infty$ is logmodular, so that m is a Jensen measure, then consider the closure of $w^{1/p}A$ in $L^p(m)$.

5. Suppose that M_ϕ is finite dimensional, and that ϕ has a unique logmodular measure m.
(a) There exists $c \geq 1$ such that

$$e^{\int \log h\,dm} \leq \inf_{f \in A_0} \int |1 - f|^p\, h\,dm \leq c e^{\int \log h\,dm}$$

for all $h \in L^1$, $h \geq 0$.
(b) Suppose $1 < p < \infty$, $h \geq 0$, $h \in L^1$, and $\log h \in L^1$. Then $h = |g|^p$ for some $g \in H^{-1}$ if and only if

$$\inf_{f \in A_0} \int |1 - f|^p\, h\,dm = e^{\int \log h\,dm}.$$

6. Suppose F is as in 7.1.
(a) Prove directly that $(\lambda - F)H^\infty$ has codimension one in H^∞ if $|\lambda| < 1$.

(b) The map $\sum_{n=0}^{\infty} a_n z^n \to \sum_{n=0}^{\infty} a_n F^n$ is an isometric isomorphism of $H^\infty(\Delta)$ and a subalgebra B of H^∞.

(c) If J is the ideal of $f \in H^\infty$ such that $\tilde{f}(\lambda) = 0$ for $|\lambda| < 1$, then H^∞ is the direct sum of B and J.

(d) J need not be empty (cf. exercise 4, chapter VII).

CHAPTER VI

PARTS

Let A be a uniform algebra. A point $\theta \in M_A$ belongs to the same part as $\phi \in M_A$ if there is a $c > 0$ such that Harnack's inequality is valid:

$$\frac{1}{c} < \frac{u(\theta)}{u(\phi)} < c, \qquad u \in \operatorname{Re}(A), u > 0.$$

Being in the same part of M_A is evidently an equivalence relation. The equivalence classes are the **parts** of M_A, or of A.

If a connected subset U of M_A is to be endowed with an analytic structure so that the functions in A become analytic on U, then U must be contained in the same part of M_A, by Harnack's inequality. One of our objects is to see to what extent one can introduce an analytic structure into the parts of M_A.

The notion of part has been especially successful in dealing with hypodirichlet algebras. According to a theorem of Wermer (7.2), every part of a dirichlet algebra is either an analytic disc, or it consists of one point. This result has been extended to hypodirichlet algebras, in which case every non-point part is a finite open (connected) Riemann surface (cf. 7.5).

The notion of part will also be useful in understanding the algebras of analytic functions associated with compact planar sets.

1. Representing Measures for a Part

Suppose θ and ϕ belong to the same part of M_A. Let $b(\theta,\phi)$ be the infimum of all c for which Harnack's inequality is valid: $1/c < u(\theta)/u(\phi) < c$, $u \in \operatorname{Re}(A)$, $u > 0$. Then b has the following properties:

$$\frac{1}{b(\theta,\phi)} \leq \frac{u(\theta)}{u(\phi)} \leq b(\theta,\phi), \qquad u \in \operatorname{Re}(A), u > 0$$

$$b(\theta,\phi) \geq 1$$

$$b(\theta,\phi) = 1 \quad \text{if and only if} \quad \theta = \phi$$

$$b(\theta,\phi) = b(\phi,\theta)$$

$$b(\theta,\phi)b(\phi,\psi) \leq b(\theta,\psi) \quad \text{if } \psi \text{ also belongs to the same part as } \theta.$$

In particular, $\log b(\theta,\phi)$ is a metric on each part of M_A.

1.1 Theorem. Suppose θ and ϕ belong to the same part of M_A. There are mutually absolutely continuous representing measures μ for θ and ν for ϕ such that

$$\frac{1}{b(\theta,\phi)} \leq \frac{d\mu}{d\nu} \leq b(\theta,\phi).$$

Proof. Write b for $b(\theta,\phi)$. Since $bu(\phi) - u(\theta) \geq 0$ for all positive $u \in \text{Re}(A)$, there is a positive measure α on X such that

$$bu(\phi) - u(\theta) = \int u\,d\alpha, \qquad u \in \text{Re}(A).$$

Similarly, there is a positive measure β on X such that

$$bu(\theta) - u(\phi) = \int u\,d\beta, \qquad u \in \text{Re}(A).$$

Solving these equations for $u(\theta)$ and $u(\phi)$, we find that $\mu = (b\beta + \alpha)/(b^2 - 1)$ and $\nu = (b\alpha + \beta)/(b^2 - 1)$ are representing measures for θ and ϕ respectively. These measures have the desired properties.

1.2 Corollary. If θ and ϕ lie in the same part of M_A, and η is a representing measure for θ, then there is a representing measure λ for ϕ such that η is absolutely continuous with respect to λ, and $d\eta/d\lambda \leq b(\theta,\phi)$.

Proof. The measure $\lambda = \eta/b + \nu - \mu/b$ does the job.

The map $\eta \rightarrow \eta/b + \nu - \mu/b$ is an affine homeomorphism of M_θ and a closed subset of M_ϕ. It shows that whatever M_θ might look like, M_ϕ cannot be much different. For instance, if M_ϕ is norm compact, then the image of M_θ in M_ϕ is norm compact, so M_θ is norm compact.

1.3 Corollary. Suppose θ and ϕ lie in the same part of M_A. Then M_θ is norm compact if and only if M_ϕ is norm compact, and M_θ is finite dimensional if and only if M_ϕ is finite dimensional. When finite, the dimensions of M_θ and M_ϕ are equal. In particular, θ has a unique representing measure if and only if ϕ has a unique representing measure.

2. Characterization of Parts

2.1 Theorem. The following are equivalent, for $\theta, \phi \in M_A$:

(i) θ and ϕ are in the same part of M_A.
(ii) $\|\theta - \phi\| < 2$, the norm being that of A^*.
(iii) The norm of the restriction of θ to A_ϕ is less than one.

(iv) Whenever $\{f_n\}$ is a sequence in A such that $||f_n|| \le 1$ and $|f_n(\theta)| \to 1$, then $|f_n(\phi)| \to 1$.

Proof. Suppose $||\theta - \phi|| = 2$. Choose $f_n \in A$ such that $||f_n|| < 1$ while $|f_n(\theta) - f_n(\phi)| \to 2$. We can assume that $f_n(\theta) \to 1$, so that $f_n(\phi) \to -1$. If $u_n = \text{Re}\,(1 - f_n)$, then $u_n > 0$, $u_n(\theta) \to 0$, and $u_n(\phi) \to 2$. Consequently θ and ϕ do not lie in the same part. This shows that (i) implies (ii).

Suppose the norm of θ on A_ϕ is 1. There are then $f_n \in A$ such that $||f_n|| < 1$, $f_n(\phi) = 0$, $f_n(\theta) > 0$, and $f_n(\theta) \to 1$. Let L_n be the conformal map of the disc onto itself such that $L_n(s)$ is real if s is real, and $0 < L_n(f_n(\theta)) = -L_n(0)$. The linear transformation L_n is given explicitly by

$$L_n(z) = \frac{z - t_n}{1 - t_n z}, \qquad t_n = (1 - \sqrt{1 - f_n(\theta)^2})/f_n(\theta).$$

If $g_n = L_n \circ f_n$, then $g_n \in A$, $||g_n|| < 1$, and $|g_n(\theta) - g_n(\phi)| = 2|L_n(0)| \to 2$. Consequently $||\theta - \phi|| = 2$. This shows that (ii) implies (iii).

Suppose there is a sequence $f_n \in A$ such that $||f_n|| < 1$, $f_n(\theta) \to 1$, and $|f_n(\phi)| \le c < 1$. Let $Q_n(z) = (z - f_n(\phi))/(1 - \overline{f_n(\phi)}z)$. Then $g_n = Q_n \circ f_n \in A$, $||g_n|| < 1$, $g_n(\phi) = 0$, and $|g_n(\theta)| \to 1$. So the norm of θ on A_ϕ is one. Hence (iii) implies (iv).

Suppose θ and ϕ are not in the same part. There is a sequence $f_n \in A$ such that $\text{Re}\,(f_n) > 0$, $\text{Re}\,f_n(\theta) \to 0$, and $\text{Re}\,f_n(\phi) \to \infty$. If $g_n = e^{-f_n}$, then $g_n \in A$, $||g_n|| < 1$, $|g_n(\theta)| \to 1$, and $g_n(\phi) \to 0$. This shows that (iv) implies (i), completing the proof of the theorem.

In contrast to 1.1, we now have the following.

2.2 Theorem. Suppose θ and ϕ are not in the same part of M_A. Then there are disjoint Borel sets E_1 and E_2 such that every representing measure for θ is supported on E_1, and every representing measure for ϕ is supported on E_2. In particular, representing measures for θ are singular to representing measures for ϕ.

Proof. By 2.1, there are $f_n \in A$ such that $||f_n|| < 1$, $|1 - f_n(\theta)| < 1/n^4$, and $|1 + f_n(\phi)| < 1/n^4$. Let E_1 be the set of x such that $f_n(x) \to 1$, and E_2 the set of x such that $f_n(x) \to -1$. Then E_1 and E_2 are disjoint Borel sets.

Let μ be a representing measure for θ. Then

$$\begin{aligned}\int |f_n - 1|^2\, d\mu &= \int |f_n|^2\, d\mu + 1 - 2\,\text{Re} \int f_n d\mu \\ &\le 2 - 2\,\text{Re}\,f_n(\theta) \\ &\le \frac{2}{n^4}.\end{aligned}$$

Hence

$$\int \sum_{n=1}^{\infty} |f_n - 1|^2\, d\mu < \infty.$$

It follows that $f_n \to 1$ almost everywhere with respect to μ, so that μ is supported on E_1. Similarly, every representing measure for ϕ is supported on E_2. That proves the theorem.

Let η be a finite measure on X. Theorem 2.2 shows that if θ and ϕ belong to different parts of M_A, and if η is absolutely continuous with respect to M_θ, then η is singular to M_ϕ. Corollary 1.2 shows that if θ and ϕ belong to the same part of M_A, then η is absolutely continuous (respectively singular) with respect to M_θ if and only if η is absolutely continuous (respectively singular) with repsect to M_ϕ. In particular, if θ and ϕ belong to the same part of M_A, then the Lebesgue decomposition of η with respect to M_θ coincides with the Lebesgue decomposition of η with respect to M_ϕ. In view of these remarks, the decomposition theorem for orthogonal measures II.7.11 can now be stated as follows.

2.3 Theorem. Let $\{\phi_\alpha\}$ be a subset of M_A containing exactly one homomorphism from each part of M_A. Let $\mu \in A^\perp$, and let μ_α be the absolutely continuous component of μ with respect to $M_{\phi\alpha}$. Then $\mu_\alpha \in A^\perp$, and the μ_α are mutually singular. Furthermore,

$$\mu = \mu_s + \sum_\alpha \mu_\alpha,$$

$$||\mu|| = ||\mu_s|| + \sum_\alpha ||\mu_\alpha||,$$

where $\mu_s \in A^\perp$, and μ_s is singular to all representing measures for A.

3. Parts of $R(K)$

Let K be a compact subset of the plane, and let $R(K)$ be the algebra of continuous functions on K which are uniform limits on K of rational functions. Harnack's inequality shows that if p and q belong to the interior of the same component of the interior of K, then p and q are in the same part of $R(K)$.

An example which illustrates the sort of adverse behavior we may expect is the "string of beads" (cf. fig. 3). In this example, K is the compact set obtained by deleting from the disc $\{|z| \leq 1\}$ a sequence of smaller disjoint open discs with centers on the real axis, so that the intersection of K and the real axis is a set of positive linear measure with no interior points. The boundary of each component of int (K) is a simple closed rectifiable Jordan arc. So harmonic measure for interior points of a component is comparable to arc length on the boundary of that component. In particular, harmonic measure for points in the top component is not singular to harmonic measure for points in the bottom component. Hence the two components of int (K) lie in the same part of K.

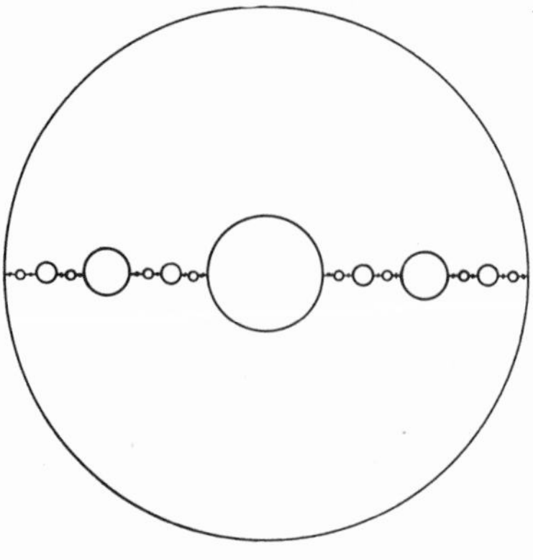

Figure 3. The string of beads.

Most of our knowledge of parts for $R(K)$ stems from the Cauchy transform $\hat{\mu}(z) = \int d\mu(\zeta)/(\zeta - z)$, defined for finite measures μ with compact support. Recall (cf. II.8) that the Cauchy transform converges absolutely $dxdy$-almost everywhere. If the Cauchy transform vanishes $dxdy$-almost everywhere on an open set U, then $|\mu|(U) = 0$.

3.1 Theorem. If $p \in K$ is not a peak point of $R(K)$, then the part of p has positive planar measure.

Proof. Let m be a representing measure on K for p which has no mass at p. Then $\mu = (z - p)m$ is orthogonal to $R(K)$. If $q \in K$, then $\mu/(z - q)$ is orthogonal to all rational functions in $R(K)$ which vanish at q. As in the proof of Wilken's theorem II.8.5, it follows that the measure $\mu(z)/[\hat{\mu}(q)(z - q)]$ is a complex representing measure for q whenever $\int |z - q|^{-1} d|\mu|(z) < \infty$ and $\hat{\mu}(q) \neq 0$. Since $\mu \not\equiv 0$, this occurs for q belonging to a set Q of positive planar measure. By theorem II.2.2, for any $q \in Q$ there is a representing measure for q which is absolutely continuous with respect to m. So Q is contained in the part of p. This proves the theorem.

3.2 Corollary. The number of non-trivial parts of $R(K)$ is at most countable.

A result stronger than 3.1 is true (cf. Browder [2]), and we mention it in passing. Suppose $p \in K$ is not a peak point for $R(K)$. For $\epsilon > 0$, let P_ϵ be the set of $z \in K$ such that $\|p - z\| < \epsilon$, the norm being taken in $R(K)^*$. Then each P_ϵ has planar density one at p.

3.3 Theorem. Let P be a part of $R(K)$, and let m be a representing measure for $p \in P$. Then m is supported on $\bar{P}$, and m represents p on $R(\bar{P})$.

Proof. We can assume that P is not a one-point part, and that $m(\{p\}) = 0$.

Let $\mu = (z - p)m$. By the proof of 3.1, $\hat{\mu}$ must vanish almost everywhere off P. By II.8.2, $|\mu|(C\backslash\bar{P}) = 0$. Hence μ and m are supported on $\bar{P}$.

Now $\hat{\mu}$ must vanish identically off $\bar{P}$. By II.8.1, $\mu \perp R(\bar{P})$. Hence m represents p on $R(\bar{P})$.

3.4 Theorem. Let $P_1, P_2, \ldots$ be the non-trivial parts of $R(K)$. There are disjoint Borel sets $E_1, E_2, \ldots$ such that

(i) $P_j \subset E_j \subset \bar{P}_j$.

(ii) Every representing measure for points in P_j is supported on E_j.

(iii) If $\mu \perp R(K)$, then $\mu_{E_j} \perp R(K)$, and $\mu = \sum \mu_{E_j}$, the series converging in norm.

Proof. By repeatedly applying 2.2, and taking an at most countable intersection of sets, we obtain disjoint Borel sets E_j for which (ii) is valid. By 3.3, we can replace E_j by $E_j \cap \bar{P}_j$, so that both (i) and (ii) are valid. Property (iii) follows from properties (i) and (ii), the decomposition theorem for orthogonal measures 2.3, and Wilken's theorem II.8.5.

Let's take another look at the string of beads K of fig. 3. Every point on bK which does not lie on the real axis is easily seen to be a peak point for $R(K)$, and so a one-point part. Theorem 3.1 shows that there is no room for any other parts, except peak points and the part containing int (K); any other part would lie on the real axis, which has zero area. It can be shown that the part of int(K) seeps across the real axis whenever the beads are very small. However, by choosing the beads carefully, it can be arranged so that every point of K which lies on the real axis is a peak point of $R(K)$. In this case, int(K) comprises a single disconnected part.

4. The Finitely Connected Case

In this section, we let K be a fixed compact subset of the complex plane whose complement has a finite number of components. In this case, $R(K)$ is a hypodirichlet algebra (IV.8.3). For $z \in \text{int}\, K$, m_z will denote the logmodular measure for z on bK, that is, m_z is the harmonic measure for z on bK.

Every point on bK is a peak point for $R(K)$. We wish to show that the nontrivial parts of K are precisely the components of int K. This amounts to showing that m_z is singular to m_w if z and w belong to distinct components of int K. That this is not completely obvious is indicated by the existence of compact sets J whose interior consists of two components U_0, U_1 such that $bJ = bU_0 = bU_1$.

First, we attend to some topological preliminaries.

Let U be a component of int K. A point $z \in bU$ is **accessible from U** if z is the endpoint of a continuous curve which has its other points in U.

4.1 Lemma. Two distinct components of int K have only a finite number of accessible points in common.

Proof. Fix $z_0 \in U$ and $z_1 \in V$ where U and V are different components of int K. Let $w_0, \ldots, w_k$ be distinct points which are accessible both from U and from V. There are then curves $\Gamma_0, \ldots, \Gamma_k$ such that Γ_j begins at z_0, passes through U to w_j, continues through V to z_1, and does not meet any other Γ_i en route. The Γ_j divide the complex plane into k components, each of which must contain points of K^c. So k cannot exceed the number of components of K^c. This proves the lemma.

Since bU has a finite number of components, there is a circle domain W and a conformal map f of W onto U. Let S be the subset of bW at which f has non-tangential boundary values. Then S is a Borel set, and f extends to be a Borel function on S. The extension of f to S will also be denoted by f. Note that every point of $f(S)$ is an accessible point of bU.

If C_n is a sequence of circles in W converging to a boundary circle, then $f(C_n)$ is a sequence of analytic curves in U converging to a boundary component of U. This establishes a one-to-one correspondence between boundary circles of W and boundary components of U.

A point $z_0 \in S$ is a **multiple point** of f if there is $z_1 \in S$ such that $f(z_0) = f(z_1)$.

4.2 Lemma. There are only a finite number of multiple points of f on bW.

Proof. Suppose $w_1, \ldots, w_p$ are multiple points on bW, such that $f(w_1) = \ldots = f(w_p) = z_0$. The w_j belong to the same circle of bW. We can assume that they are ordered consecutively on that circle. For convenience, set $w_0 = w_p$. For $1 \leq j \leq p$, let Γ_j be a curve connecting w_{j-1} and w_j in W, so that Γ_j tends to its endpoints non-tangentially, Γ_j is deformable in $\bar{W}$ to the arc from w_{j-1} to w_j, and Γ_j does not intersect the other curves Γ_k, except at its endpoints. Each $f(\Gamma_j)$ is a simple closed Jordan curve. The outside of $f(\Gamma_j)$ contains the curves $f(\Gamma_k)$, $k \neq j$, and the components of bU corresponding to the other circles. The inside of $f(\Gamma_j)$ contains at least one component of K^c—otherwise f would tend to z_0 almost everywhere on the arc from w_{j-1} to w_j, so f would be a constant. Consequently the integer p does not exceed the number of bounded components of K^c. And the number of multiple points corresponding to z_0 is finite.

We can now assume that $w_1, \ldots, w_p$ exhaust the multiple points corresponding to z_0.

Now let w_1^* be a different multiple point on the same boundary circle. Let $w_1^*, \ldots, w_q^*$ be the multiple points satisfying $f(w_j^*) = f(w_1^*) = z_0^*$. There is a k such that z_0^* is inside $f(\Gamma_k)$. Then $w_1^*, \ldots, w_q^*$ all belong to the arc from w_{k-1} to w_k. Proceeding as earlier, we find that there are at least q components of K^c contained inside $f(\Gamma_k)$.

It is clear now that every multiple point necessitates the existence of more components of K^c. So the number of multiple points must be finite. This proves the lemma.

Let μ_w be the harmonic measure on the boundary of the circle domain W for $w \in W$. That is, μ_w is the logmodular measure on bW for w, as a homomorphism of the algebra $R(\bar{W})$.

4.3 Lemma. There is a Borel set $T \subseteq bW$ such that

(i) f has non-tangential boundary values at every point of T, i.e., $T \subset S$.

(ii) f is one-to-one on T, $f(T) \subset bU$ is a Borel set, and f^{-1} is a Borel function on $f(T)$.

(iii) μ_w is supported on T for all $w \in W$.

(iv) m_z is supported on $f(T)$ for all $z \in U$. In particular, m_z is supported on the set of points of bU accessible from U.

(v) For all bounded Borel functions g on bU,

$$\int g dm_z = \int g \circ f d\mu_w, \qquad w \in W, z = f(w).$$

Proof. The set obtained by deleting the multiple points of f from S has full harmonic measure. By Lusin's theorem, there is an increasing sequence of closed subsets T_n of S such that T_n contains no multiple points of f, $f|_{T_n}$ is continuous, and $\mu_w(T_n) \to 1$. In particular, the restriction of f to T_n is a homeomorphism of T_n and $f(T_n)$.

Let $T = \bigcup_{n=1}^{\infty} T_n$. Then T evidently has properties (i), (ii), and (iii). Property (iv) is a consequence of (v). So it suffices to prove (v).

For $w \in W$ and $z = f(w)$, we define a measure ν_z on $f(T)$ by setting $\nu_z(E) = \mu_w(f^{-1}(E))$, $E \subseteq f(T)$. Then

$$\int g d\nu_z = \int g \circ f d\mu_w$$

for all bounded Borel functions g on $f(T)$.

Suppose $g \in R(K)^{-1}$. Then $g \circ f$ is a bounded analytic function on W. By Fatou's theorem, $g \circ f$ has boundary values almost everywhere on bW, which correspond to the boundary function $g \circ f$. Hence

$$\begin{aligned} \int \log |g| \, d\nu_z &= \int \log |g \circ f| \, d\mu_w \\ &= \log |g(f(w))| \\ &= \log |g(z)|. \end{aligned}$$

Hence ν_z is a logmodular measure for z, and $\nu_z = m_z$. This completes the proof.

4.4 Theorem. Let K be a compact finitely connected subset of the complex plane. The non-trivial parts of $R(K)$ are precisely the components of int K.

Proof. By IV.7.5, every representing measure for a point is absolutely continuous with respect to the logmodular measure for that point.

If $z \in bK$, then the point mass at z is the logmodular measure for z. So the point mass is the only representing measure for z. By II.11.2, z is a peak point of $R(K)$. In particular, $\{z\}$ is a one-point part.

By 4.1 and 4.3, logmodular measures for points in distinct components of int K are mutually singular. Consequently each non-trivial part of K must reduce to a component of int K. This proves the theorem.

Let U be a component of int K, and let m be the logmodular measure for a fixed point $z_0 \in U$. If $z \in U$, then there is a constant $c > 0$ such that $m_z/c \leq m \leq cm_z$. In fact, by 1.2 there is a representing measure μ for z and a constant $a > 0$ such that $m \leq a\mu$. Since μ is boundedly absolutely continuous with respect to m_z, we obtain $m \leq cm_z$. By increasing c, if necessary, we also have $m_z \leq cm$.

4.5 Theorem. Let K be a compact finitely connected subset of the complex plane. Let U be a component of int K, let m_z be the logmodular measure for $z \in$ int K, and set $m = m_{z_0}$ for some fixed $z_0 \in U$. If G is a bounded analytic function on U, there is $g \in H^\infty(m)$ such that $\|g\|_\infty = \|G\|_U$ and

$$G(z) = \int g dm_z, \qquad z \in U.$$

Proof. Let $H^\infty(U)$ be the algebra of bounded analytic functions on U. In view of 4.3, every $G \in H^\infty(U)$ determines a function $g \in L^\infty(m)$, via the boundary values of $G \circ f$ on W, such that $\|g\|_\infty = \|G\|_U$ and $G(z) = \int g dm_z$, $z \in U$. Hence we can consider $H^\infty(U)$ as a subalgebra of $L^\infty(m)$.

If $g \in H^\infty(m)$, there is a net $g_\alpha \in R(K)$ which converges weak-star to g. Then $g_\alpha(z) = \int g_\alpha dm_z$ converges to a function $G(z)$ in $H^\infty(U)$ such that $G(z) = \int g dm_z$, $z \in U$. Hence $H^\infty(m) \subset H^\infty(U)$.

In view of 4.3 and the corresponding fact for circle domains, m is multiplicative on $H^\infty(U)$. By the maximality theorem IV.7.7, $H^\infty(m) = H^\infty(U)$. That proves the theorem.

It is customary to identify $H^\infty(m)$ and $H^\infty(U)$. In the same manner, one can identify $H^p(m)$ and $H^p(U)$, where $H^p(U)$ is the set of functions G analytic on U such that $|G|^p$ has a harmonic majorant.

One application of this embedding theorem is the determination of the rational defect.

4.6 Theorem. Let K be a compact finitely connected subset of the complex plane. Let $U_1, U_2, \ldots$ be the components of int K. Let q_j be the number of boundary components of U_j. Then the codimension of the closure of Re $R(K)$ in $C_R(bK)$ is $\sum (q_j - 1)$.

Proof. We remark that $\sum (q_j - 1)$ cannot exceed the number of components of K^c. In particular, bU_j is connected except for finitely many U_j.

Suppose first that $K = \bar{W}$, where W is a circle domain with q boundary circles. Let $z_1, \ldots, z_{q-1}$ be points which belong to distinct bounded components of $\bar{W}^c$. By the proof of IV.8.3, the functions $\log|z - z_j|$, $1 \leq j \leq q - 1$, together with $\operatorname{Re} R(\bar{W})$ span a dense linear subspace of $C_R(bW)$. So the defect of $\operatorname{Re} R(\bar{W})$ in $C_R(bW)$ does not exceed $q - 1$. On the other hand, let $\chi_1, \ldots, \chi_q$ be the characteristic functions of the q boundary circles of W. If $f \in R(\bar{W})$, then every point on the boundary of $f(\bar{W})$ must be assumed by f somewhere on bW. Consequently, it is impossible to approximate $\sum a_j\chi_j$ uniformly by $\operatorname{Re}(f)$, $f \in R(\bar{W})$, unless all the a_j are equal. So the defect of $\operatorname{Re} R(\bar{W})$ in $C_R(bW)$ is exactly $q - 1$.

Let μ be harmonic measure on bW for a fixed $w \in W$. The real measures on bW orthogonal to $R(\bar{W})$ are the measures of the form $h\mu$, where $h \in N^\infty(m)$. Hence the dimension of $N^\infty(m)$ is $q - 1$.

Now suppose K is arbitrary. Fix $z_j \in U_j$, and let m_j be the logmodular measure for z_j. The real measures orthogonal to $R(K)$ are the measures of the form $\sum h_j m_j$, where $h_j \in N^\infty(m_j)$. Transferring the problem to a circle domain, we see that the dimension of $N^\infty(m_j)$ is $q_j - 1$. We sum these dimensions to obtain the theorem.

4.7 Corollary. Suppose K is a compact finitely connected subset of the plane. Then $R(K)$ is dirichlet if and only if every component of the interior of K is simply connected.

An example of a compact set K which is not simply connected, while $R(K)$ is dirichlet, is the "half-moon" obtained from the closed unit disc Δ by deleting an open subdisc which is tangent to $b\Delta$ at one point.

5. Pointwise Bounded Approximation

Let U be a bounded open subset of the complex plane, and let $H^\infty(U)$ be the algebra of bounded analytic functions on U. When is it possible to approximate every $f \in H^\infty(U)$ pointwise boundedly on U by a sequence of polynomials?

Let U^* be the interior of the polynomial hull $\hat{\bar{U}}$ of the closure $\bar{U}$ of U. If a sequence of polynomials converges pointwise boundedly on U, the sequence is bounded on U^*, and so a subsequence must converge pointwise boundedly on U^*. Hence any pointwise bounded limit of polynomials on U extends to be bounded and analytic on U^*.

The converse of this statement was obtained by Farrell, in the case that U^* is connected. Later, Farrell's theorem was generalized to arbitrary bounded open sets by Rubel and Shields. Their theorem is as follows.

5.1 Theorem (Farrell-Rubel-Shields Theorem). Let U be an open subset of the complex plane, and let f be a bounded analytic function on U. There is a sequence $\{p_n\}_{n=1}^\infty$ of polynomials such that $\sup_n \|p_n\|_U < \infty$ and

$p_n(z) \to p(z)$ for all $z \in U$ if and only if f extends to be bounded and analytic on U^*.

The proof of 5.1 will be deferred until the end of the section. First we make some topological remarks.

Let $K = \hat{\bar{U}}$, so that U^* is the interior of K. Any disc centered at a point on the boundary of K must meet U, or that boundary point would not belong to the polynomial hull of $\bar{U}$. So $\bar{U}$ contains bK, and U^* is dense in K. Also, U^* is simply connected, so every component of U^* is conformally equivalent to a disc.

It may occur that U^* has components which do not meet U. Such a set is illustrated by the "cornucopia" (cf. fig. 4). Here U is a ribbon which winds around the outside of the circle $\{|z| = 1\}$ and accumulates on that circle, while U^* is obtained from U by adjoining the interior of the unit disc.

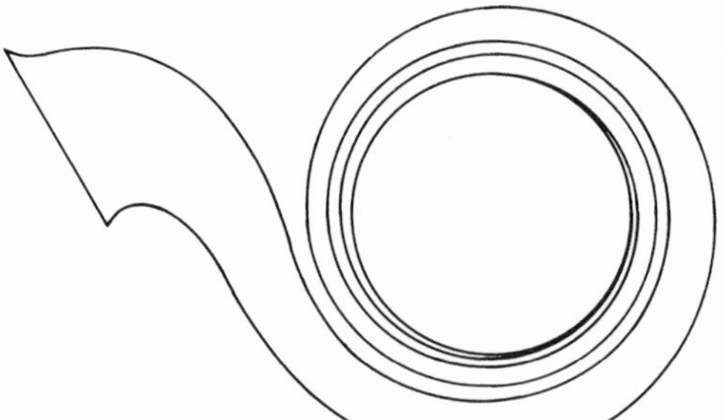

Figure 4. The cornucopia.

Now Farrell's theorem is a consequence of Runge's theorem and an elementary fact about conformal mappings. In fact, suppose that U^* is connected, and that f is bounded and analytic on U^*. Let $\{V_n\}_{n=1}^{\infty}$ be a sequence of open, connected, simply connected sets such that $\bar{V}_{n+1} \subseteq V_n$ and $\bigcap_{n=1}^{\infty} V_n = K$. Fix $z_0 \in U^*$, and let g_n be the conformal map of V_n onto U^* satisfying $g_n(z_0) = z_0$ and $g_n'(z_0) > 0$. It is easy to establish that $g_n(z) \to z$ uniformly on compact subsets of U^*. Since $f \circ g_n$ is analytic on V_n, there is a polynomial p_n such that $\|f \circ g_n - p_n\|_K < 1/n$. Then p_n converges pointwise boundedly to f on U^*.

In order to extend 6.1 to the case in which U^* is disconnected, the notion of parts appears to be decisive. To illustrate this, we prove an abstract version of the Farrell-Rubel-Shields theorem.

5.2 Theorem. Let A be a uniform algebra on X. Let $\{\phi_n\}_{n=1}^{\infty}$ be homomorphisms in M_A, each belonging to a different part of M_A. Suppose each ϕ_n has a dominant representing measure μ_n. Let f be a bounded function on X which belongs to $H^\infty(\mu_n)$ for each n. Then there are $f_k \in A$ such that $\|f_k\|_X \leq \|f\|_X$ and $f_k \to f$ μ_n-almost everywhere, $1 \leq n < \infty$.

Proof. Assume that $\|f\|_X = 1$. Let $\mu = \sum_{n=1}^{\infty} \mu_n/2^n$. We assert that f belongs to the weak-star closure of the unit ball of A in $L_\infty(\mu)$.

Suppose not. By the separation theorem for convex sets, there is $h \in L^1(\mu)$ such that

$$\sup\left\{\operatorname{Re}\int gh\,d\mu : g \in A, \|g\| \leq 1\right\} < \operatorname{Re}\int fh\,d\mu.$$

Consequently

$$(*) \qquad \sup\left\{\left|\int gh\,d\mu\right| : g \in A, \|g\| \leq 1\right\} < \left|\int fh\,d\mu\right|.$$

Let the measure η on X be a norm-preserving extension of the functional

$$L(g) = \int gh\,d\mu, \qquad g \in A,$$

from A to $C(X)$. Then

$$(**) \qquad \|\eta\| = \sup\left\{\left|\int gh\,d\mu\right| : g \in A, \|g\| \leq 1\right\}.$$

If $\eta = k\mu + \eta_s$ is the Lebesgue decomposition of η with respect to μ, then

$$\|\eta\| = \int |k|\,d\mu + \|\eta_s\|.$$

The measure $\eta - h\mu$ is orthogonal to A. By 2.2, the measures μ_n are mutually singular. So the absolutely continuous part of $\eta - h\mu$ with respect to μ_n is $(k - h)\mu_n/2^n$. By Ahern's theorem II.7.8, $(k - h)\mu_n$ is orthogonal to A. Hence $(k - h)\mu$ is orthogonal to A. It follows that $k\mu$ is also an extension of L from A to $C(X)$, of norm $\int |k|\,d\mu$. By the minimality of the norm of η, we obtain $\eta_s = 0$, and $\eta = k\mu$.

Since $f \in H^\infty(\mu_n)$, $(k - h)\mu_n$ is also orthogonal to f. So $(k - h)\mu$ is orthogonal to f, and

$$\left|\int fh\,d\mu\right| = \left|\int fk\,d\mu\right| \leq \int |k|\,d\mu = \|\eta\|.$$

But this contradicts $(*)$ and $(**)$. And the assertion is established.

Now f must also belong to the closure of the unit ball of A in $L^2(\mu)$. Let $\{f_k\}$ be a sequence in A such that $\|f_k\|_X \leq 1$ and $f_k \to f$ in $L^2(\mu)$. Passing to a subsequence, we obtain $f_k \to f$ μ-almost everywhere. This proves the theorem.

5.3 Theorem. Let K be a compact finitely connected subset of the complex plane, and let f be a bounded analytic function on int K. Then there is a sequence $f_n \in R(K)$ such that $\|f_n\|_K \leq \|f\|_\infty$ and $f_n(z) \to f(z)$ for all $z \in \operatorname{int} K$.

Proof. Assume that int K has an infinite number of components $U_1, U_2, \ldots$, the case of a finite number of components being easier. Fix a point $z_j \in U_j$, $1 \leq j < \infty$, and let μ_j be the logmodular (= harmonic) measure on bK for z_j. By 4.5, we can identify $H^\infty(U_j)$ and $H^\infty(\mu_j)$. Now f belongs to $H^\infty(U_j)$ for all j. The hypotheses of 5.2 are met. So there is a sequence of functions $f_n \in R(K)$ such that $\|f_n\|_K \leq \|f\|_\infty$ and f_n converges μ_j-almost everywhere to the function f, regarded as an element of $H^\infty(\mu_j)$, $1 \leq j < \infty$. Since evaluation at points $z \in U_j$ is weak-star continuous on $H^\infty(\mu_j)$, $f_n(z)$ tends to $f(z)$ for all $z \in \operatorname{int} K$. That proves the theorem.

Proof of 5.1. We must show that if f can be extended to be analytic and bounded on U^*, then f can be approximated pointwise boundedly on U by a sequence of polynomials. For this, it suffices to apply 5.3 to the closure K of U^*, and to observe that because K is simply connected, functions in $R(K)$ can be approximated uniformly on K by polynomials.

6. Finitely Generated Ideals

We wish now to consider the question of embedding an analytic structure in M_A. The main result in this direction is Gleason's embedding theorem 6.1. For this, we need the notion of an analytic variety.

Let D be an open subset of C^n. A subset V of D is an **analytic variety** if for every $z \in V$ there is a neighborhood U_z of z and a family $\mathscr{F}_z$ of analytic functions on U_z, such that $U_z \cap V$ is the set of common zeros of the family $\mathscr{F}_z$.

6.1 Theorem (Gleason Embedding Theorem). Let A be a uniform algebra, and let $\phi \in M_A$. Suppose there are $g_1, \ldots, g_n \in A_\phi$ such that every $f \in A_\phi$ can be expressed in the form $f = \sum f_j g_j, f_j \in A$. Then there is a neighborhood $U \subseteq M_A$ of ϕ, a polycylinder $\Delta^n(0;\epsilon)$, an analytic variety $V \subset \Delta^n(0;\epsilon)$, and a homeomorphism $\tau\colon U \to V$, such that $f \circ \tau^{-1}$ extends analytically to $\Delta^n(0;\epsilon)$ for all $f \in A$.

Proof. Let U be the set of $\psi \in M_A$ such that $|\psi(g_j)| < \epsilon$, $1 \leq j \leq n$, ϵ to be specified later. Let $\tau(\psi) = (\psi(g_1), \ldots, \psi(g_n))$. Then τ maps U continuously into $\Delta^n(0;\epsilon)$.

By hypothesis, the multilinear map $(f_1, \ldots, f_n) \to \sum f_j g_j$ maps A^n onto A_ϕ. An application of the principle of uniform boundedness produces a constant $c > 0$ with the following property: For all $g \in A_\phi$, there are $f_1, \ldots, f_n \in A$ such that $g = \sum f_j g_j$ and $\|f_j\| \leq c\|g\|$.

Now suppose that $f \in A$ satisfies $\|f\| \leq 1$. Then $f - \phi(f) \in A_\phi$, and

$\|f - \phi(f)\| \leq 2$. There are $f_1, \ldots, f_n \in A$ such that $f = \phi(f) + \sum f_j g_j$, while $\|f_j\| \leq 2c$. Similarly, there are $f_{jk} \in A$ such that $\|f_{jk}\| \leq 4c^2$ and $f_j = \phi(f_j) + \sum f_{jk} g_k$, that is,

$$f = \phi(f) + \sum_{j=1}^{n} \phi(f_j) g_j + \sum_{j,k} f_{jk} g_k.$$

Proceeding in this manner, we obtain

$$f = \sum_{i_1 + \cdots + i_n < N} c_{i_1 \ldots i_n} g_1^{i_1} \ldots g_n^{i_n} + R_N,$$

where

$$|c_{i_1 \ldots i_n}| \leq (2c)^{i_1 + \cdots + i_n},$$

$$R_N = \sum_{i_1 + \cdots + i_n = N} f_{i_1 \ldots i_n} g_1^{i_1} \ldots g_n^{i_n},$$

$$\|f_{i_1 \ldots i_n}\| \leq (2c)^N.$$

If $\psi \in U$, then $|\psi(R_N)| \leq (2nc\epsilon)^N$. Setting $\epsilon = 1/(4nc)$, we obtain $\|R_N\|_U \leq 1/2^N$. Hence

$$\psi(f) = \sum_0^\infty c_{i_1 \ldots i_n} \psi(g_1)^{i_1} \ldots \psi(g_n)^{i_n}, \qquad \psi \in U.$$

In particular, the values of $\psi(f)$ are determined by $\psi(g_1), \ldots, \psi(g_n)$ when $\psi \in U$. So τ is one-to-one. Evidently τ is a homeomorphism.

The function

$$\tilde{f}(z_1, \ldots, z_n) = \sum_0^\infty c_{i_1 \ldots i_n} z_1^{i_1} \ldots z_n^{i_n}$$

is analytic on $\Delta^n(0;\epsilon)$, and $\tilde{f}|_{\tau(U)} = f \circ \tau^{-1}$. To complete the proof, it suffices to show that $\tau(U)$ is a variety.

Let V be the set of common zeros of the analytic functions on $\Delta^n(0;\epsilon)$ which vanish on $\tau(U)$. Then $V \supseteq \tau(U)$, and V is a variety. Now the series representation $\tilde{f}$ determined by f depends on the choice of the $f_{i_1 \ldots i_n}$. However, if $z \in V$, then the analytic functions $\tilde{f}$ satisfying $\tilde{f}|_{\tau(U)} = f \circ \tau^{-1}$ all must agree at z. So $\theta(f) = \tilde{f}(z)$ is well-defined. It is a homomorphism of A which satisfies $\tau(\theta) = z$. Consequently $\tau(U) = V$, and $\tau(U)$ is a variety. That completes the proof.

Before we proceed, we wish to make some remarks easily obtainable from the parametrization theorem for analytic varieties (cf. Gunning and Rossi).

Let V be a one-dimensional analytic subvariety of $\Delta^n(0;\epsilon)$. Then the set of singular points of V is discrete.

If $z_0 \in V$ is not a singular point, there is a neighborhood of z_0 in V which inherits from Δ^n the structure of an analytic disc. There are functions which extend analytically to Δ^n and which give coordinates on the disc.

If $z_0 \in V$ is an irreducible singular point of the variety, then z_0 again has a neighborhood which is homeomorphic to a disc. There is an analytic function f on Δ^n and an integer $k > 1$ such that $f^{1/k}$ gives coordinates in a neighborhood of z_0, and all g analytic on Δ^n can be expanded in a series of powers of $f^{1/k}$ on the disc.

If $z_0 \in V$ is a reducible singular point, then z_0 has a neighborhood in V which is homeomorphic to the union of a finite number of discs which meet at z_0. Each of the discs represents the germ of an irreducible variety at z_0, and can be given an analytic structure as described above.

Now note the following elementary fact: If $\mathscr{F}$ is any separating family of complex-valued functions on a disc, there is at most one analytic structure on the disc for which all the functions in $\mathscr{F}$ become analytic. This shows that the conformal structures of the discs in V coincide on their intersections.

Hence V can be given the structure of a (possibly disconnected) open Riemann surface, with certain point identifications, so that the functions analytic on Δ^n also become analytic on V. The same fact cited above shows that this analytic structure is unique.

Now suppose that M_A has no isolated points, so the varieties we produce cannot have dimension zero. Suppose also that $F \in A$ is such that FA has finite positive codimension in A (as a vector space over the complex numbers). Then FA is a proper ideal in A, and so there is $\phi \in M_A$ such that $F(\phi) = 0$. Since the homomorphisms in $F^{-1}(0)$ vanish on FA, and since M_A is a linearly independent set of functionals, the number of homomorphisms in $F^{-1}(0)$ cannot exceed dim A/FA.

Each maximal ideal A_ϕ containing FA is finitely generated, by F and any set of functions in A_ϕ which span A_ϕ modulo FA. By 6.1, there are disjoint neighborhoods $U_1, \ldots, U_q$ of the homomorphisms $\phi_1, \ldots, \phi_q$ in $F^{-1}(0)$ which can be given the structure of analytic varieties, on which the functions in A are analytic.

Since F is analytic on U_j, and since $F^{-1}(0) \cap U_j$ consists of just the one point ϕ_j, the dimension of the variety U_j near ϕ_j cannot exceed one. Since M_A is assumed to have no isolated points, the variety cannot have dimension zero. Consequently U_j can be taken to be a connected one-dimensional analytic variety.

There exists $\epsilon > 0$ such that

$$U = F^{-1}(\Delta(0;\epsilon)) \subseteq U_1 \cup \ldots \cup U_q.$$

Indeed, otherwise there is a sequence ψ_n in M_A such that $\psi_n \notin U_1 \cup \ldots \cup U_q$ and $F(\psi_n) \to 0$. But any weak-star adherent point ψ of the ψ_n satisfies $F(\psi) = 0$, and so must coincide with one of the ϕ_j. This contradicts $\psi_n \notin U_1 \cup \ldots \cup U_q$.

We give the set $U = F^{-1}(\Delta(0;\epsilon))$ the natural structure of a Riemann surface with certain point identifications, so that the functions in A become analytic on U.

Suppose $\phi \in U$, and no point identification occurs at ϕ. If F has order k at ϕ, then $(F - F(\phi))^{1/k}$ gives local coordinates at ϕ. Then F is a k-sheeted covering of a neighborhood of ϕ in U over a neighborhood of $F(\phi)$, providing we assign multiplicity k to F at ϕ.

If $\phi \in U$ is the center of analytic discs $V_1, \ldots, V_p$, and $F|_{V_j}$ has order k_j at ϕ, then we assign multiplicity $k = k_1 + \ldots + k_p$ to F at ϕ. Again F is a k-sheeted covering of a neighborhood of ϕ over a neighborhood of $F(\phi)$.

If $z \in \Delta(0;\epsilon)$, and $m(z)$ is the sum of the multiplicities of F at the points in $F^{-1}(z)$, then $m(z)$ is constant in a neighborhood of each $z \in \Delta(0;\epsilon)$. Hence $m(z) = m$ is constant on $\Delta(0;\epsilon)$.

We summarize our results as follows.

6.2 Theorem. Suppose M_A has no isolated points. Suppose $F \in A$ is such that FA has finite codimension in A (as a complex vector space). For sufficiently small $\epsilon > 0$, the subset $U = F^{-1}(\{|z| < \epsilon\})$ of M_A can be given the structure of an open (possibly disconnected) Riemann surface, with a finite number of point identifications, so that the functions in A become analytic on U. There is an integer m, $1 \leq m \leq \dim A/FA$, such that F is an m-sheeted analytic cover of U over $\{|z| < \epsilon\}$.

If F is as above, then no point of $F^{-1}(0)$ belongs to the Shilov boundary ∂_A of A. So we can define W to be the component of $C\backslash F(\partial_A)$ which contains 0.

If $\lambda \in W$, then $F - \lambda$ is bounded away from zero on ∂_A. Consequently $(F - \lambda)A$ is a closed subspace of A. The operator T_λ on A, defined by $T_\lambda(g) = (F - \lambda)g$, has closed range and has null space consisting only of $\{0\}$. In particular, each T_λ is a "semi-Fredholm" operator. By the index theory for Fredholm operators (cf. Gokhberg and Krein [1]), the codimension of the range of T_λ in A must be constant (finite or infinite), for $\lambda \in W$.

Consequently, for each $\lambda \in W$, the codimension of $(F - \lambda)A$ in A is equal to the codimension of FA in A. Applying 6.2 to each $\lambda \in W$, and patching together the resulting analytic structures, we obtain the following global formulation of 6.2.

6.3 Theorem. Let A be a uniform algebra such that M_A has no isolated points. Suppose $f \in A$ is such that FA has finite codimension in A (as a complex vector space). Let W be the connected component of $C\backslash F(\partial_A)$ which contains 0. Then $F^{-1}(W) \subseteq M_A$ can be given the structure of a (possibly disconnected) open Riemann surface, with discrete point identifications, so that the functions in A become analytic on $F^{-1}(W)$. There is an integer m, $1 \leq m \leq \dim A/FA$, such that F is an m-sheeted analytic cover of $F^{-1}(W)$ over W.

7. Extremal Methods

We will now look for functions F such that FH^∞ has finite codimension in H^∞. The first step is the solution of an extremal problem.

7.1 Theorem. Suppose ϕ and ψ belong to the same part of M_A, $\phi \neq \psi$. There exists a representing measure μ for ϕ and $F \in H^\infty(\mu)$ such that:

(i) $|F| = 1$ almost everywhere.

(ii) There is a constant $0 < c < 1$ such that $\nu = ((1 - c^2)/|1 - cF|^2)\mu$ is a representing measure for ψ.

(iii) $\phi(F) = 0$, $\psi(F) = c$.

(iv) $\bar{F}(H^2_\phi(\nu) \cap H^2_\psi(\nu))$ is orthogonal to $\overline{H^2(\nu)}$ in $L^2(\nu)$.

Proof. Let c be the norm of the restriction of ψ to A_ϕ. By 2.1, $0 < c < 1$.

Let the measure η represent a norm-preserving extension of ψ from A_ϕ to $C(X)$. Choose $f_n \in A_\phi$ such that $\|f_n\| < 1$ and $\psi(f_n) \to c$. Let F be a weak-star adherent point of the sequence $\{f_n\}$ in $L^\infty(|\eta|)$. Then $|F| \leq 1$ almost everywhere, while $\int F d\eta = c = \int d|\eta|$. Consequently $F\eta = |\eta|$, and $|F| = 1$ almost everywhere.

If $g \in A_\psi$, then $\int g d|\eta| = \int gF d\eta = \lim \int g f_n d\eta = 0$. Consequently $\nu = |\eta|/c = F\eta/c$ is a representing measure for ψ.

If $h \in A_\phi$, then $\int h|1 - cF|^2 d\nu = \int h(1 - cF)(F - c)d\eta/c = \lim \psi(h(1 - cf_n)(f_n - c))/c = 0$. Since $\int |1 - cF|^2 d\nu = 1 + c^2 - 2c \operatorname{Re} \int F d\nu = 1 - c^2$, we see that $\mu = (|1 - cF|^2/(1 - c^2))\nu$ is a representing measure for ϕ.

Note that the measures ν and μ are comparable, so their Hardy spaces are identical.

Properties (i), (ii), and (iii) are now clear. To verify (iv), it suffices to observe that $\int g f \bar{F} d\nu = \int g f d\eta/c = 0$ whenever $f \in A$ and $g \in A_\Psi \cap A_\phi$. This completes the proof.

We are now in a position to give another proof of Wermer's embedding theorem, which was treated in V.7.

7.2 Theorem (Wermer Embedding Theorem). Let A be a uniform algebra on X. Suppose that $\phi \in M_A$ has a unique representing measure on X, and that the part P of ϕ consists of more than one point. Then there is a one-to-one continuous map σ of the open unit disc $\{|z| < 1\}$ into M_A, such that the image of σ coincides with P, and $f \circ \sigma$ is analytic on $\{|z| < 1\}$ for all $f \in A$.

Proof. Let μ be the representing measure for ϕ. Suppose $\psi \neq \phi$ belongs to P. By 1.3, ψ also has a unique representing measure, which must coincide with the measure ν of 7.1. From IV.6.2 we obtain the direct sum decomposition

$$L^2(\nu) = H^2_\psi(\nu) \oplus \overline{H^2(\nu)}.$$

From 7.1 (iv) we obtain

$$\bar{F}(H^2_\phi(\nu) \cap H^2_\psi(\nu)) \subseteq H^2_\psi(\nu).$$

It follows that $FH^2_\psi(\nu)$ contains $H^2_\phi(\nu) \cap H^2_\psi(\nu)$, and so has codimension two, at most, in $H^2(\nu)$. Hence $FH^2(\nu)$ has codimension at most one in $H^2(\nu)$. Since $F \in H^\infty_\phi(\nu)$, we see that $FH^2(\nu) = H^2_\phi(\nu)$. Intersecting with L^∞, we obtain $FH^\infty(\mu) = H^\infty_\phi(\mu)$.

Now apply 6.3 to the algebra $H^\infty(\mu)$. Since $|F| = 1$ on ∂_{H^∞}, the set W of 6.3 is the open unit disc $\{|z| < 1\}$. Since F is a one-sheeted analytic cover of $F^{-1}(W)$ over W, F^{-1} embeds W homeomorphically as a subset of M_{H^∞}. Since $\|F\|_\infty = 1$, the set $F^{-1}(W) \subseteq M_{H^\infty}$ coincides with the part of ϕ, regarded as a homomorphism of H^∞.

For $z \in W$, define $\sigma(z) \in M_A$ to be the restriction of the homomorphism $F^{-1}(z)$ to A. Then σ is a one-to-one continuous map of W into M_A, such that $\sigma(W) \subseteq P$, and $f \circ \sigma$ is analytic for all $f \in A$. Since every $\theta \in P$ has a weak-star continuous extension to $H^\infty(\mu)$ which belongs to the same part as ϕ, the image of σ must coincide with P. That proves the theorem.

It is more difficult to handle the case of a finite dimensional set of representing measures. We will give some results in that direction.

7.3 Lemma. Let ϕ, ψ, F, μ, and ν be as in 7.1. Suppose the set of representing measures for ϕ is finite dimensional. Then $FH^\infty(\mu)$ has finite codimension in $H^\infty(\mu)$.

Proof. Note that $H^2(\nu) \subseteq \bar{F}H^2(\nu)$, and that $\bar{F}H^2(\nu) \ominus H^2(\nu)$ is orthogonal to $H^2(\nu)$ and to $\overline{H^2_\phi(\nu)} \cap \overline{H^2_\psi(\nu)}$. Since $H^2(\nu) \oplus \overline{H^2_\psi(\nu)}$ has finite codimension in $L^2(\nu)$, $\bar{F}H^2(\nu) \ominus H^2(\nu)$ is finite dimensional. So $H^2(\nu) \ominus FH^2(\nu)$ is finite dimensional, and $FH^2(\mu)$ has finite codimension in $H^2(\mu)$. Hence $F(H^2(\mu) \cap L^\infty)$ has finite codimension in $H^2(\mu) \cap L^\infty$.

To complete the proof, it will suffice to show that $H^\infty(\mu)$ has finite codimension in $H^2(\mu) \cap L^\infty$.

Let m be a core measure for ϕ, regarded as a homomorphism of $H^\infty(\mu)$. Then μ and m are mutually absolutely continuous, and $d\mu/dm$ is bounded. Since $N^\infty(m)$ is finite dimensional, it suffices to show that every $f \in H^2(\mu) \cap L^\infty$ which is orthogonal to $N^\infty(m)$ in $L^2(m)$ belongs to $H^\infty(\mu)$.

If $f = u + iv \in H^2(\mu) \cap L^\infty$, and $f \perp N^\infty(m)$, then $u \perp N^\infty(m)$, so $u + i{*u} \in H^2(m) \subseteq H^2(\mu)$. Consequently ${*u} - v$ is constant, and $u + i{*u} \in H^2(m) \cap L^\infty = H^\infty(m) = H^\infty(\mu)$. This completes the proof.

7.4 Theorem. Let ϕ, ψ, F, μ, and ν be as in 7.1. Suppose the set of representing measures for ϕ is finite dimensional. Then the part P of ϕ in $M_{H^\infty(\mu)}$ can be given the structure of a (possibly disconnected) finite open

Riemann surface, with a finite number of point identifications, such that the functions in $H^\infty(\mu)$ become analytic on P. Also, F is an m-sheeted analytic cover of P over $\{|z| < 1\}$, and $P = F^{-1}(\{|z| < 1\})$. Moreover, F has a finite number of ramification points on P.

Proof. Let $W = \{|z| < 1\}$. Since $|F| = 1$ on ∂_{H^∞}, 6.3 shows that $F^{-1}(W)$ is a Riemann surface with point identifications, and F is an m-sheeted analytic cover of $F^{-1}(W)$ over W.

Also, $(\lambda - F)H^\infty$ has constant finite codimension in H^∞, for $|\lambda| < 1$. If $|F(\theta)| < 1$, then $(F(\theta) - F)H^\infty$ is a weak-star closed subspace of H^∞ which has finite codimension in H^∞, and which is contained in H^∞_θ. So H^∞_θ is weak-star closed, and θ is weak-star continuous. Hence θ has a representing measure which is absolutely continuous with respect to μ, and θ belongs to the same part as ϕ. If $|F(\theta)| = 1$, evidently $\theta \notin P$. This shows that $P = F^{-1}(W)$.

If $\theta \in P$ is such that only one analytic disc passes through θ, the ramification index of F at θ will be that integer $k \geq 0$ such that $F^{(1)}(\theta) = \ldots = F^{(k)}(\theta) = 0$, while $F^{(k+1)}(\theta) \neq 0$. If q analytic discs pass through $\theta \in P$, the ramification index of F at θ will be $q - 1$ plus the sum of the ramification indices of F on each disc through θ.

Let S be the subset of P at which F has a positive ramification index. Then S is discrete. We must show that S is finite.

Let $\{r_n\}$ be a sequence of real numbers such that $r_n \uparrow 1$, and $F(S)$ does not meet the circle $\{|z| = r_n\}$. Let $P_n = F^{-1}(\{|z_n| \leq r_n\})$, and let V_n be the sum of the ramification indices at points $\theta \in P_n$. We must show that the increasing sequence $\{V_n\}$ is bounded. To do this, we will need some cohomology theory and a formula of Hurwitz.

Let χ be the Euler characteristic. The Hurwitz formula reads

$$\chi(P_n) = m - V_n.$$

To prove the formula, we triangulate the disc $\{|z| \leq r_n\}$ so that the points of $F(S)$ are vertices of the triangles, and so that the triangulation lifts via F to a triangulation of P_n. The number of triangles in P_n is m times the number of triangles in $\{|z| \leq r_n\}$. The same goes for the sides. The number of vertices in P_n is V_n less than m times the number of vertices on $\{|z| \leq r_n\}$. Hence $\chi(P_n) = m\chi(\{|z| \leq r_n\}) - V_n$. Since the Euler characteristic of a disc is one, we obtain the Hurwitz formula.

Let $T_n = M_{H^\infty} \backslash \operatorname{int} P_n$. Then $M_{H^\infty} = T_n \cup P_n$, and $T_n \cap P_n = bP_n$. From the Mayer-Vietoris sequence of the triad (M_{H^∞}, T_n, P_n), we obtain the following exact sequence:

$$H^1(M_{H^\infty}, Z) \to H^1(T_n, Z) \oplus H^1(P_n, Z) \to H^1(bP_n, Z).$$

Now bP_n is homeomorphic to q_n circles, where $q_n \leq m$. So $H^1(bP_n, Z) \cong Z^{q_n}$. From IV.6.4, $H^1(M_{H^\infty}, Z) \cong D' \oplus Z^k$, where D' is a completely divisible group. Finally $H^1(P_n,Z) \cong Z^{p_n}$ for some integer p_n. The above sequence now reads

$$D' \oplus Z^k \to H^1(T_n,Z) \oplus Z^{p_n} \to Z^{q_n}.$$

Since D' is completely divisible, the image of D' is contained in $H^1(T_n,Z)$. The exactness condition now easily yields $p_n \leq q_n + k \leq m + k$. Since the p_n are bounded, the $\chi(P_n)$ are bounded. By the Hurwitz formula, the V_n are bounded. Hence S is finite.

Choose $r < 1$ such that $|F(S)| < r$. Let Y be a component of $F^{-1}(\{r < |z| < 1\})$. The restriction of F to Y is a k-sheeted covering of the annulus, for some integer k. A suitable branch of $F^{1/k}$ is then a conformal map of Y and the annulus $\{r^{1/k} < |z| < 1\}$. This allows us to adjoin a boundary circle to each component Y of $F^{-1}(\{r < |z| < 1\})$. Hence P is the interior of a (possibly disconnected) finite bordered Riemann surface with a finite number of point identifications. This completes the proof.

At this point the picture becomes hazy. However, if we make the additional assumption that ϕ have a unique logmodular measure, the analysis can be carried a good deal further. We state the results without proof.

7.5 Theorem. Let A be a uniform algebra on X. Suppose that the set of representing measures for ϕ on X is finite dimensional, and that ϕ has a unique logmodular measure m on X. Suppose, moreover, that the part P of ϕ contains more than one point. Then P can be given the structure of a (connected) finite open Riemann surface, on which the functions in A are analytic. There is a natural isometric embedding of the algebra $H^\infty(P)$ of bounded analytic functions on P into $H^\infty(m)$, so that $H^\infty(m)$ is the direct sum of $H^\infty(P)$ and the ideal of functions in $H^\infty(m)$ which vanish identically on P.

NOTES

Parts were introduced by Gleason [1], who took 2.1(ii) as his definition. The equivalence of Gleason's definition to the one given in section 1 was noted by Bishop [9], who obtained 1.1 and 1.2. Theorems 3.1 through 3.3 are due to Wilken [2, 4]. Section 4 is based on Ahern and Sarason [2], and is close in spirit to Bishop [1, 4]. For another approach, see Glicksberg [1]. The "abstract" proof of the Farrell-Rubel-Shields theorem is due to Wermer [17]. The extended version 5.2 is due to Ahern and Sarason [2]. The Gleason embedding theorem is in Gleason [2]. Theorem 7.1 is due to Bishop (cf. O'Neill [1]). The Wermer embedding theorem is in Wermer [12]. Theorem 7.5 is in Gamelin [4]. Various partial results had been obtained earlier by Wermer [19], O'Neill [1], and Wermer and O'Neill [1].

EXERCISES

1. The metric log $b(z, w)$, which int Δ is endowed with as a part of $P(\Delta)$, coincides with the Poincaré metric. If $z, w \in$ int Δ, then

$$\|\phi_z - \phi_w\| = 2\left(1 - \left(1 - \left|\frac{z-w}{1-\bar{w}z}\right|^2\right)^{1/2}\right)$$

2. The Borel sets of 2.2 can be taken to be disjoint F_σ-sets.

3. Suppose that $f \in A$ has an nth root in A, for all integers n.

(a) f vanishes identically on any part on which it has a zero.

(b) If A is logmodular, then the subset of M_A on which f vanishes either is empty or contains a one-point part.

(c) $H^\infty(\Delta)$ has a one-point part lying off its Shilov boundary. *Hint:* Take $f(z) = \exp((z-1)/(z+1))$.

4. Let A be the algebra of continuous functions on the torus $b\Delta \times b\Delta$ such that for every fixed $w \in b\Delta$, $f(z, w)$ extends to be analytic in z for $|z| < 1$, and $f(0, w)$ and $f(\frac{1}{2}, w)$ extend to be analytic in w for $|w| < 1$.

(a) Identify M_A.

(b) M_A has a disconnected part.

(c) Using an algebra with a one-point part off the Shilov boundary, modify the example to obtain a two-point part.

5. Let $K_n = \{(z, w)\colon\ |z| \leq 1,\ \ w = z/n\}$, $1 \leq n < \infty$. Let $K = \{(z, 0)\colon |z| \leq 1\} \cup \bigcup_{n=1}^{\infty} K_n$.

(a) K is polynomially convex.

(b) $M_A \backslash \partial_A$ is a part of $P(K)$.

(c) If $0 < |z_0| < 1$ and $\epsilon > 0$ is sufficiently small, then the set $\{(z, w) \in K\colon \|\phi_{(z,w)} - \phi_{(z_0,0)}\| < \epsilon\}$ is not connected.

6. Let K be a compact subset of the plane.

(a) Every part of $A(K)$ meets bK in a set of zero area. *Hint:* Consider $\iint_E \frac{dxdy}{z-\zeta}$, where E is appropriately chosen.

(b) Every part of $A(K)$ which is not a peak point must contain a component of int K.

(c) If E is a peak set of $A(K)$ such that $E \subseteq bK$, then every point of E is a peak point.

7. The closure of every part of $R(K)$ is connected.

8. If U is a component of int K, and $z \in bU$, then either z is a peak point of $R(K)$, or z belongs to the same part as U in $R(K)$.

9. If Δ^n is a polydisc in C^n with center at 0, then the ideal of functions in $P(\Delta^n)$ which vanish at 0 is generated (algebraically) by the coordinates $z_1, \ldots, z_n$.

10. If a Banach algebra A is generated by $f_1, \ldots, f_n$, and if a maximal ideal J of A is (algebraically) finitely generated, then it is generated (algebraically) by

$f_1 - \hat{f}_1(J), \ldots, f_n - \hat{f}_n(J)$. *Hint:* If $g_1, \ldots, g_m$ generate J, and $h_1, \ldots, h_m$ are near $g_1, \ldots, g_m$, then also $h_1, \ldots, h_m$ generate J.

11. If M_ϕ is finite dimensional, and ϕ has a unique logmodular measure m, then no point identification can occur on the Riemann surface constructed in section 7. *Hint:* Use the maximality theorem IV.7.7 for H^∞, and exercise 9 of chapter III.

CHAPTER VII

GENERALIZED ANALYTIC FUNCTIONS

In this chapter, we will study in detail algebras of bounded analytic almost periodic functions in the upper half-plane, of the form

$$f(z) = \sum c_j e^{it_j z}, \qquad \operatorname{Im}(z) \geq 0,$$

where the t_j belong to the positive half of a subgroup Γ of the real numbers. If Γ is taken to be the integers, the functions f become periodic in the upper half-plane, and the algebra we obtain is the standard disc algebra.

In sections 4 through 12, these algebras are studied under the standing assumption that Γ be a dense subgroup of the reals. Section 2 is devoted to a more general situation, but the results of sections 2 and 3 are not used later.

A rudimentary knowledge of the theory of locally compact abelian groups and their character groups will be assumed.

1. Preliminaries

In this chapter, Γ will always denote a discrete abelian group, and G will be its compact character group. The normalized Haar measure on G will be denoted by σ.

Every $a \in \Gamma$ gives rise to a continuous character χ_a of G, defined by setting $\chi_a(x) = x(a)$, $x \in G$. The Pontryagin duality theorem tells us that every continuous character of G is of the form χ_a, for some $a \in \Gamma$. It is customary to identify Γ with the character group of G.

The character group of a compact subgroup H of G is denoted by $\hat{H}$. Every character in $\hat{H}$ can be extended to be a character of G. That is, $\hat{H} = \Gamma|_H$.

The characters in Γ which are identically one on H are denoted by $H^\perp$. Then $\hat{H} = \Gamma/H^\perp$.

Finite linear combinations of characters form a separating self-adjoint subalgebra of $C(G)$. By the Stone-Weierstrass theorem, linear combinations of characters are dense in $C(G)$.

The characters χ_a are orthogonal in $L^2(\sigma)$, that is,

$$\int \chi_a \bar{\chi}_b d\sigma = \int \chi_{a-b} d\sigma = 0, \qquad a \neq b.$$

Since continuous functions are approximable by linear combinations of characters, the χ_a form an orthonormal basis for $L^2(\sigma)$.

The formal Fourier series of a function $f \in L^1(\sigma)$ is given by

$$f \sim \sum_{a \in \Gamma} c(a)\chi_a,$$

where

$$c(a) = \int_G f\bar{\chi}_a d\sigma.$$

More generally, every finite measure μ on G has a formal Fourier-Stieltjes series given by

$$\mu \sim \sum_{a \in \Gamma} c(a)\chi_a,$$

where

$$c(a) = \int_G \bar{\chi}_a d\mu.$$

If the Fourier-Stieltjes coefficients of μ all vanish, then μ is the zero measure.

The convolution of two measures μ and ν is defined on Borel sets E by

$$(\mu * \nu)(E) = \int_G \nu(E - x) d\mu(x).$$

Convolution is commutative. The Fourier-Stieltjes coefficients of $\mu * \nu$ are obtained by multiplying the respective coefficients of μ and ν.

2. Algebras Associated with Groups

We will be studying uniform algebras A on G which contain enough characters to order Γ.

Let Γ_+ be a semigroup in Γ which satisfies $\Gamma_+ \cup (-\Gamma_+) = \Gamma$. Define "$a \geq b$" to mean $a - b \in \Gamma_+$. This relation satisfies the axioms of an ordering:

(i) If $a, b, c \in \Gamma$ and $a \geq b$, then $a + c \geq b + c$.
(ii) If $a \in \Gamma$, then either $a \geq 0$ or $0 \geq a$ (or both).

Conversely, if "$\geq$" is an ordering of Γ, and Γ_+ is the set of non-negative elements in Γ, then Γ_+ is a semigroup, by (i), and $\Gamma_+ \cup (-\Gamma_+) = \Gamma$, by (ii).

Consequently there is a bijective correspondence between orderings of Γ and subsemigroups Γ_+ of Γ satisfying $\Gamma = \Gamma_+ \cup (-\Gamma_+)$. Such a semigroup Γ_+ is said to **order** Γ.

The semigroup Γ_+ **totally orders** Γ if Γ_+ orders Γ and $\Gamma_+ \cap (-\Gamma_+) = \{0\}$. In this case, $a \geq 0$ and $0 \geq a$ imply that $a = 0$.

The semigroup Γ_+ determines an **archimedean ordering** of Γ if whenever $a, b \in \Gamma_+$ and $-a \notin \Gamma_+$, there is an integer $m > 0$ such that $ma + b \in \Gamma_+$. This occurs if and only if $\Gamma_+/(\Gamma_+ \cap (-\Gamma_+))$ induces a total archimedean ordering of $\Gamma/(\Gamma_+ \cap (-\Gamma_+))$. And this occurs if and only if there is an order-preserving homomorphism of Γ into the reals with kernel $\Gamma_+ \cap (-\Gamma_+)$.

2.1 Theorem. Let A be a uniform algebra on G. If $A \cap \Gamma$ totally orders Γ, then A is generated by $A \cap \Gamma$. Moreover, σ is multiplicative on A, and A is antisymmetric.

Proof. Let Λ be the set of characters obtained by deleting 1 from $A \cap \Gamma$. First we show that Λ is contained in a maximal ideal of A.

If not, there are $\chi_1, \ldots, \chi_m \in \Lambda$ and $f_1, \ldots, f_m \in A$ such that $\sum f_j \chi_j = 1$. Choose k so that χ_k precedes each of the χ_j, $1 \leq j \leq m$, in the ordering induced by $A \cap \Gamma$. Then $\bar{\chi}_k = \sum f_j \chi_j \bar{\chi}_k \in A$. This contradicts the totality of the ordering. Our assertion is established.

Any representing measure for a maximal ideal containing Λ is orthogonal to Λ and $\bar{\Lambda}$, and so must be the Haar measure σ of G. So σ is multiplicative on A. Every real-valued function in any uniform algebra is constant on the support of any representing measure. Since the closed support of σ is G, A is antisymmetric.

If $f \in A$, then $\int f \chi d\sigma = 0$ for all $\chi \in \Lambda$. Consequently f is uniformly approximable by linear combinations of characters not in $\bar{\Lambda}$, that is, of characters in $A \cap \Gamma$. That completes the proof.

2.2 Theorem. Suppose that $A \cap \Gamma$ orders Γ. Then every maximal set of antisymmetry of A is a coset $x_0 + H$ of some compact subgroup H of G. The restriction of A to $x_0 + H$ is generated by $A \cap \Gamma$.

In particular, the restriction algebra $A|_{x_0+H}$ arises from a total ordering of the dual group $\hat{H} = \Gamma/H^{\perp}$ of H.

Proof. Let K be a maximal set of antisymmetry of A. Let $x_0 \in K$. Let H be the smallest closed subgroup of G containing $K - x_0$. And let B be the uniform algebra on H of $f \in C(H)$ such that $g(y) = f(x_0 + y)$ belongs to A on K.

Now $B \cap H$ orders H. If $\chi \in \hat{H}$ and both $\chi \in B$ and $\bar{\chi} \in B$, then χ is constant on $K - x_0$, so χ is constant on H. Hence $B \cap \hat{H}$ totally orders $\hat{H}$, and 2.1 applies to B. Since B is antisymmetric, $K - x_0$ coincides with H. Since the restriction of $A \cap \Gamma$ to H is $B \cap \hat{H}$, and since this generates B, we see that $(A \cap \Gamma)|_K$ generates $A|_K$. That completes the proof.

2.3 Theorem. Suppose that $A \cap \Gamma$ totally orders Γ, while $A \neq C(G)$. Then A is a maximal subalgebra of $C(G)$ if and only if the ordering induced by $A \cap \Gamma$ is archimedean.

Proof. Suppose first the ordering induced by $A \cap \Gamma$ is archimedean. Let B be a uniform algebra on G such that $A \subseteq B \subseteq C(G)$. By the archimedean property, either $B \cap \Gamma = \Gamma$ or $B \cap \Gamma = A \cap \Gamma$. If $B \cap \Gamma = \Gamma$, then $B = C(G)$. If $B \cap \Gamma = A \cap \Gamma$, then $B \cap \Gamma$ totally orders Γ. By 2.1, $B = A$. This shows that A is a maximal subalgebra of $C(G)$.

Now suppose the ordering is not archimedean. Choose characters $\chi_1, \chi_2 \in A \cap \Gamma$ such that $\bar{\chi}_1 \notin A \cap \Gamma$ and $\chi_1^m \bar{\chi}_2 \notin A \cap \Gamma$ for all $m \geq 1$. Let B be the uniform algebra generated by A and $\bar{\chi}_1$. Then $B \neq A$. By 2.1, σ is multiplicative on A. Since $\overline{\chi_1^m}\chi_2 \in A \cap \Gamma$ for all $m \geq 0$, and $\overline{\chi_1^m}\chi_2$ cannot be 1, we obtain $\int f \overline{\chi_1^m} \chi_2 d\sigma = 0$ for all $f \in A$. Hence $\int g \chi_2 d\sigma = 0$ for all $g \in B$. So $B \neq C(G)$, and A is not maximal. That completes the proof.

2.4 Theorem. Suppose that $A \cap \Gamma$ orders Γ, while $A \neq C(G)$. Then A is contained in a maximal subalgebra of $C(G)$ if and only if there is a non-zero order-preserving homomorphism of Γ into the reals.

Proof. Suppose first that A is contained in a maximal subalgebra of $C(G)$. We can assume that A itself is maximal. Let $K = x_0 + H$ be the non-trivial maximal set of antisymmetry of A. Let B be the translate of $A|_K$ to H. Then $B \cap \hat{H}$ totally orders $\hat{H}$. By 2.3, the ordering is archimedean. Hence there is an order-preserving isomorphism of $\hat{H}$ and a subgroup of the reals. Composing with the quotient homomorphism of Γ onto $\Gamma|_H = \hat{H}$, we obtain the desired homomorphism.

Conversely, suppose that φ is a non-zero order-preserving homomorphism of Γ into the reals. Let $H = (\text{kernel } \varphi)^{\perp}$.

We assert that every maximal set of antisymmetry of A is either a point or a coset of some subgroup H_0 of G which contains H. In fact, suppose that $x_0 + H_0$ is a maximal set of antisymmetry which contains at least two points. There is then $\chi_1 \in A \cap \Gamma$ such that χ_1 is not constant on H_0. Choose $\chi_2 \in A \cap \Gamma$ such that $\varphi(\chi_2) > 0$. Then also $\varphi(\chi_1\chi_2) > 0$, and either χ_2 or $\chi_1\chi_2$ is not constant on H_0. This shows that there is $\chi_3 \in A \cap \Gamma$ such that $\varphi(\chi_3) > 0$ while χ_3 is not constant on H_0. Suppose $\chi \in A \cap \Gamma$ satisfies $\varphi(\chi) > 0$. If m is sufficiently large, then $\varphi(\chi^m \bar{\chi}_3) > 0$, and $\chi^m \bar{\chi}_3 \in A$. Since $x_0 + H_0$ is a set of antisymmetry, $\bar{\chi}_3$ cannot belong to A on $x_0 + H_0$. Hence χ cannot be constant on H_0, that is, $\chi \notin H_0^{\perp}$. It follows that $H_0^{\perp} \subseteq \text{kernel } \varphi = H^{\perp}$. Consequently, $H \subseteq H_0$. This proves the assertion.

Now let B be the uniform algebra of functions in $C(G)$ which are uniformly approximable on H by $A \cap \Gamma$. The ordering induced by $(B|_H) \cap \hat{H}$ is total and archimedean, so B is maximal. If we translate B to any non-point maximal set of antisymmetry of A, we obtain a maximal subalgebra of $C(G)$ which contains A. That completes the proof.

The dual group of the infinite torus T^∞, the countable product of circles, is the countable direct sum Z^∞ of the integers. Let Γ_+ be the set of $(n_1, n_2, \ldots) \in Z^\infty$ whose last non-zero entry is positive, together with 0. Let A

be the uniform algebra on T^∞ generated by Γ_+. Then A is a dirichlet algebra. Since there are no non-trivial order-preserving homomorphisms of Z^∞ into the reals, A is not contained in any maximal subalgebra of $C(T^\infty)$.

3. A Theorem of Bochner

Let Γ be the set of integral lattice points in the plane. The character group of Γ is the torus T^2, the product of two circles. The lattice point $(m,n) \in \Gamma$ corresponds to the character χ_{mn} of T^2 given by

$$\chi_{mn}(\theta,\psi) = e^{im\theta}e^{in\psi},$$

where θ and ψ are the angular parameters of T^2. The Fourier-Stieltjes coefficients of a finite measure μ are given by

$$\hat{\mu}(j,k) = \int_0^{2\pi}\int_0^{2\pi} e^{-ij\theta}e^{-ik\psi}d\mu(\theta,\psi).$$

Let γ be a fixed irrational number. Let Γ_+ be the set of lattice points (m,n) in Γ which satisfy $n \geq \gamma m$. Then Γ_+ induces a total archimedean ordering of Γ. Projection onto the reals parallel to the line $y = \gamma x$ is an order-preserving isomorphism of Γ and a subgroup of the reals.

Let A_γ be the uniform algebra generated by the characters χ_{mn}, where $n \geq \gamma m$. Then A_γ is a dirichlet algebra, and the normalized Haar measure $\sigma = d\theta d\psi/4\pi^2$ is multiplicative on A_γ.

If $\gamma < 0$, the algebra A_γ contains the functions $e^{i\theta}$ and $e^{i\psi}$, and so it contains the bicylinder algebra on T^2. In fact, the bicylinder algebra is the intersection of the A_γ, for $\gamma < 0$. One reason for the importance of the A_γ is that, by studying them, information concerning the more mysterious bicylinder algebra can be obtained.

An example of a theorem which can be proved by considering the algebras A_γ is the following theorem of Bochner. The idea involved is that a measure in A_γ which is singular to Haar measure cannot have a "first" non-vanishing Fourier-Stieltjes coefficient.

3.1 Theorem. Let μ be a finite complex measure on the torus T^2. If the Fourier-Stieltjes coefficients of μ vanish off a sector in the plane whose aperture is less than π, then μ is absolutely continuous with respect to Haar measure on T^2.

Proof. The Fourier-Stieltjes coefficients of $\chi_{mn}\mu$ are obtained from those of μ by translating them by (m,n). Hence by multiplying μ by an appropriate χ_{mn}, and perhaps by widening the aperture slightly, we can assume that there are irrational numbers $\gamma_1 < \gamma_2$ such that $\hat{\mu}(m,n) = 0$ whenever $n \leq \gamma_1 m$ or $n \leq \gamma_2 m$.

Now μ is orthogonal to A_{γ_1} and to A_{γ_2}. Since $d\sigma = d\theta d\psi/4\pi^2$ is a representing measure for both algebras, the abstract F. and M. Riesz theorem

II.7.6 tells us that the singular piece μ_s of μ with respect to σ is orthogonal to A_{γ_1} and to A_{γ_2}. So $\hat{\mu}_s(n,m) = 0$ whenever $n \leq \gamma_1 m$ or $n \leq \gamma_2 m$.

Choose γ irrational so that $\gamma_1 < \gamma < \gamma_2$. Then μ_s is orthogonal to A_γ.

Moreover, if $\mu_s \neq 0$, there is a lattice point (m,n) nearest to the line $y = \gamma x$ for which $\hat{\mu}_s(m,n) \neq 0$. Then $\overline{\chi_{mn}}\mu_s$ is orthogonal to the functions $f \in A_\gamma$ satisfying $\int f d\sigma = 0$. By II.7.9, $\overline{\chi_{mn}}\mu_s$ is orthogonal to A_γ. In particular, $\hat{\mu}_s(m,n) = \int \overline{\chi_{mn}} d\mu_s = 0$. This contradiction establishes the theorem.

3.2 Corollary. Let ν be a finite complex measure on the plane, whose Fourier transform

$$\hat{\nu}(s,t) = \int_{-\infty}^{\infty}\int_{-\infty}^{\infty} e^{-i(sx+ty)} d\nu(x,y)$$

vanishes outside a sector of aperture less than π. Then ν is absolutely continuous with respect to $dxdy$.

Proof. Write $\nu = hdxdy + \nu_s$, where ν_s is singular with respect to $dxdy$. Suppose $\nu_s \neq 0$. We will obtain a contradiction.

Identify the torus T^2 with the square $\{-\pi < x \leq \pi, -\pi < y \leq \pi\}$. Replacing ν by its dilation $\nu^c(E) = \nu(cE)$, for c appropriately large, we can assume that $|\nu_s|(T^2) > \|\nu_s\|/2$.

Now define a measure μ on Borel subsets E of T^2 by setting

$$\mu(E) = \sum_{n,m=-\infty}^{\infty} \nu((2n\pi, 2m\pi) + E).$$

The assumption on ν_s guarantees that μ is not absolutely continuous with respect to $dxdy$ on T^2. However, $\hat{\mu}(m,n) = \hat{\nu}(m,n)$ when m and n are integers. This contradicts 3.1.

4. Generalized Analytic Functions

For the remainder of this chapter, we will restrict ourselves to the totally ordered archimedean case. *So we assume henceforth that* Γ *is a subgroup of the reals, endowed with the discrete topology. We assume also that* Γ *is dense in the reals,* although many of the proofs will remain valid when Γ is isomorphic to the integers.

The uniform algebra on G generated by the characters χ_a, $a \in \Gamma$, $a \geq 0$, will be denoted henceforth by A. Evidently A is a dirichlet algebra on G. By 2.3, A is a maximal subalgebra of $C(G)$.

For each real number s, the character $\boldsymbol{e}_s \in G$ is defined by

$$e_s(a) = e^{ias}, \qquad a \in \Gamma.$$

The mapping $s \to e_s$ is an isomorphism of the real line onto a dense subgroup of G. This subgroup will be denoted by $\boldsymbol{R}$.

If $f = \sum c_j \chi_{a_j}$ is a finite linear combination of characters, then

$f(e_s) = \sum c_j e^{ia_js}$ can be extended analytically to the "upper half-plane above R," by setting

$$\hat{f}(e_s + it) = \sum c_j e^{ia_j(s+it)}.$$

If $f \in A$, then $a_j \geq 0$ for all j, and $\hat{f}$ is bounded. Hence all $f \in A$ extend to be analytic and bounded in the upper half-plane above R. The extension $\hat{f}$ can be expressed by means of the integral formula

$$\hat{f}(e_s + it) = \frac{t}{\pi} \int_{-\infty}^{\infty} f(e_s) \frac{ds}{s^2 + t^2}.$$

More generally, if $x \in G$, and $f = \sum c_j \chi_{a_j}$, then $f(x + e_s) = \sum c_j \chi_{a_j}(x) e^{ia_js}$ can be extended analytically to the "upper half-plane above $x + R$" by setting

$$\hat{f}(x + e_s + it) = \sum c_j \chi_{a_j}(x) e^{ia_j(s+it)}$$
$$= \frac{t}{\pi} \int_{-\infty}^{\infty} f(x + e_s) \frac{ds}{s^2 + t^2}.$$

Define a measure μ_r on R by

$$\mu_r = \frac{t}{\pi} \frac{ds}{s^2 + t^2}, \qquad 0 < r < 1,$$

where $r = e^{-t}$ and $t > 0$. If f_r is defined by

$$f_r(x + e_s) = \hat{f}(x + e_s + it),$$

then

$$f_r = \mu_r * f.$$

The measures μ_r, $0 < r < 1$, are pairwise mutually absolutely continuous, with derivatives bounded and bounded away from zero. The Fourier-Stieltjes coefficients of μ_r are given by

$$\int \bar{\chi}_a d\mu_r = r^{|a|}.$$

If $f \in A$ has Fourier series

$$f \sim \sum_{a \geq 0} c(a) \chi_a,$$

then

$$f_r \sim \sum_{a \geq 0} c(a) r^a \chi_a.$$

As $r \uparrow 1$, the measures μ_r concentrate at the identity e_0 of G. Consequently, $f_r \to f$ uniformly as $r \to 1$. As $r \downarrow 0$, the μ_r tend weak-star to σ, as can be seen by checking the Fourier-Stieltjes coefficients. Hence

$$\lim_{r \to 0} f_r = \int f d\sigma.$$

Evaluation at the point $x + it$ of the upper half-plane above x is a homomorphism of A which we label ϕ_{rx}, where $r = e^{-t}$. We also set $\phi_0(f) = \int f d\sigma$.

4.1 Theorem. The correspondence $\phi_{rx} \leftrightarrow (r,x)$ is a homeomorphism of M_A and the cone obtained from $[0,1] \times G$ by identifying the slice $\{0\} \times G$ to a point. The representing measure for ϕ_0 is σ. The representing measures for the ϕ_{rx}, $0 < r < 1$, are the measures μ_{rx} defined by

$$\mu_{rx}(E) = \mu_r(x + E), \qquad E \text{ a Borel set.}$$

Proof. Suppose that $\phi \in M_A$. If $\phi(\chi_a) = 0$ for all positive $a \in \Gamma_+$, then $\phi = \phi_0$.

Otherwise, $\phi(\chi_a) \neq 0$ for all $a \in \Gamma_+$. In this case, $\theta(a) = \log|\phi(\chi_a)|$ satisfies $\theta(a + b) = \theta(a) + \theta(b)$ and $\theta(a) \leq 0$, for $a,b \in \Gamma_+$. Since θ is monotone, it extends to be a continuous linear function on the real line. Consequently $\theta(a) = ua$ for some $u < 0$. And $|\phi(\chi_a)| = r^a$, $a \in \Gamma_+$, for some $0 < r \leq 1$. Set

$$x(a) = \begin{cases} r^{-a}\phi(\chi_a), & a \in \Gamma_+ \\ r^a\overline{\phi(\chi_{-a})}, & a \notin \Gamma_+. \end{cases}$$

One verifies that x is a character of Γ. From

$$\phi(\chi_a) = r^a\chi_a(x) = \phi_{rx}(\chi_a), \qquad a \in \Gamma_+,$$

it follows that $\phi = \phi_{rx}$.

The remainder of the proof follows easily from the previous discussion.

The one-point parts of A are the points of G, together with ϕ_0. This provides the simplest known example of a one-point part which is not a peak point. The remaining parts of M_A are the half-planes above the lines $x + R$ on G. These are all analytic discs, in accordance with Wermer's embedding theorem VI.7.2.

5. Analytic Measures

Suppose H is any compact abelian group, and ν is a non-zero measure on H. Then ν is mutually absolutely continuous with respect to Haar measure τ on H if and only if the sets of $|\nu|$-measure zero are invariant under translation by elements of H. This observation follows immediately from the commutativity of convolution:

$$\int_H |\nu|(-x + E)d\tau(x) = \int_H \tau(-x + E)d|\nu|(x),$$

E a Borel set.

Applying this remark to the circle group, we obtain the following version of the F. and M. Riesz theorem: If ν is a measure on $b\Delta$ which is orthogonal to the disc algebra $P(\Delta)$ then the sets of $|\nu|$-measure zero are invariant under rotations of $b\Delta$.

The direct analogue of this theorem is not true for our algebra A. Every line $x + R$ on G supports non-zero measures orthogonal to A, while $x + R$ has Haar measure zero. In other words, there are measures ν in $A^\perp$ such that the null sets of $|\nu|$ are not invariant under translation by arbitrary elements of G. However, the following weaker assertion is true.

5.1 Theorem (deLeeuw–Glicksberg Theorem). Suppose $\nu \in A^\perp$, E is a Borel subset of G, and t is real. Then $|\nu|(E) = 0$ if and only if $|\nu|(E + e_t) = 0$.

Proof. Suppose that ν is a non-zero measure in $A^\perp$. Let L^2 stand for $L^2(|\nu|)$.

For each Borel subset E of G, the projection $Q(E)$ of L^2 is defined by

$$Q(E)f = h_E f, \qquad f \in L^2,$$

where h_E is the characteristic function of E. The Fourier transform of the spectral measure Q is a unitary group operating in L^2, defined by

$$V_a = \int_G \chi_a(x) dQ(x), \qquad a \in \Gamma.$$

By approximating the integral by Riemann sums, one sees that

$$V_a f = \chi_a \cdot f, \qquad f \in L^2, a \in \Gamma.$$

Set $|\nu| = k\nu$, where k is a Borel function of unit modulus. For each real a, let M_a be the closed linear span of the functions $\{\chi_b k : b \in \Gamma, b < a\}$ in L^2. Then

(i) $$M_a \subset M_b \text{ if } -\infty < a < b < \infty.$$

Since $\bigcup_{a<\infty} M_a$ contains all $\chi_a k$, $a \in \Gamma$, and the χ_a span L^2, we obtain

(ii) $$\bigcup_{a<\infty} M_a \text{ is dense in } L^2.$$

If $b,c \in \Gamma$, $c < a$ and $b \geq a$, then

$$\int \chi_b \bar{\chi}_c k d|\nu| = \int \chi_{b-c} d\nu = 0.$$

So M_a is orthogonal in L^2 to all characters χ_b with $b \geq a$. Consequently,

(iii) $$\bigcap_{a>-\infty} M_a = \{0\}.$$

Let P_a be the orthogonal projection of L^2 onto M_a. By (i), (ii), and (iii), the family $\{P_a\}$ is a resolution of the identity. Its Fourier transform is

$$U_t = \int_{-\infty}^{\infty} e^{ita} dP_a.$$

If $a \in \Gamma$, then $\chi_a M_b = M_{b+a}$. Hence one verifies easily that $\chi_a P_b \chi_{-a} = P_{b+a}$, that is, that $V_a P_b V_{-a} = P_{b+a}$. So $V_a U_t = \int_{-\infty}^{\infty} e^{itb} d(V_a P_b) = \int_{-\infty}^{\infty} e^{itb} d(P_{b+a} V_a) = e^{-iat} U_t V_a$.

Now fix a real number t. We define another spectral measure Q_t on Borel sets E by

$$Q_t(E) = U_t Q(E + e_t) U_{-t}.$$

The Fourier transform of Q_t is given, for $a \in \Gamma$, by

$$\begin{aligned}\int_G \chi_a(x) dQ_t(x) &= U_t \int_G \chi_a(x) dQ(x + e_t) U_{-t} \\ &= U_t \int_G \chi_a(x - e_t) dQ(x) U_{-t} \\ &= e^{-iat} U_t V_a U_{-t} \\ &= V_a.\end{aligned}$$

Since a spectral measure is determined by its Fourier transform, we obtain $Q_t = Q$, i.e.,

$$Q(E + e_t) = U_{-t} Q(E) U_t, \qquad E \text{ Borel.}$$

Since U_t is unitary, $Q(E) = 0$ if and only if $Q(E + e_t) = 0$. On the other hand, $Q(E) = 0$ if and only if $|\nu|(E) = 0$. That does it.

5.2 Corollary. Let E be a Borel subset of G. Then $|\nu|(E) = 0$ for all measures $\nu \in A^\perp$ if and only if E meets every line $x + R$ in a set of linear measure zero.

Proof. If E meets the line $x + R$ in a set of positive linear measure, one constructs easily a measure ν in $A^\perp$ such that $|\nu|(E) \neq 0$. In fact, if $a \in \Gamma$, $a > 0$, and $0 < r < 1$, then $(\chi_a - \phi_{rx}(\chi_a))\mu_{rx}$ will do.

Conversely, suppose E meets every line $x + R$ in a set of linear measure zero. Let $\nu \in A^\perp$. By the commutativity of convolution, we have

$$\begin{aligned}0 &= \int \mu_r(-x + E) d|\nu|(x) \\ &= \int |\nu|(-x + E) d\mu_r(x).\end{aligned}$$

Since μ_r is supported on R, $|\nu|(-e_t + E) = 0$ for at least one $e_t \in R$. By 5.1, $|\nu|(E) = 0$. That proves the corollary.

6. Local Product Decomposition

There is a local product decomposition which is very useful for understanding the group G. We suppose for simplicity that $2\pi \in \Gamma$.

Let K be the subgroup of $y \in G$ such that $\chi_{2\pi}(y) = 1$. Then K is a closed subgroup of G which contains e_n, for $n = 0, \pm 1, \ldots$. Let $\tilde{G} = K \times R$, and let ρ be the continuous map from $\tilde{G}$ to G defined by

$$\rho(y, s) = y + e_s, \qquad y \in K, s \in R.$$

Then ρ is a homomorphism of $\tilde{G}$ onto G. The kernel of ρ is the discrete subgroup $\{(-e_n,n)\}_{n=-\infty}^{\infty}$ of G. A fundamental domain for ρ is $K \times [0,1)$; that is, ρ is one-to-one on $K \times [0,1)$, and $\rho(K \times [0,1)) = G$. The group G is obtained topologically from $K \times [0,1]$ by identifying $(y,1)$ with $(y - e_1,0)$, $y \in K$.

Let τ be the Haar measure of K. Then the transplant of Haar measure σ on G to the fundamental domain $K \times [0,1)$ is $\tau \times dt$. It will be convenient to identify G with $K \times [0,1)$ and σ with $\tau \times dt$.

A Borel function f on $\tilde{G}$ is **automorphic** if $f(y,s+1) = f(y + e_1,s)$ for $\tau \times dt$—almost all $(y,s) \in \tilde{G}$. An automorphic function is determined by its values on the fundamental domain $K \times [0,1)$. Conversely, any Borel function on $K \times [0,1)$ can be extended uniquely (modulo null functions, of course) to be automorphic on $\tilde{G}$.

If $f \in L^p(\tau \times (dt/(1 + t^2)))$, then the restriction of f to $K \times [0,1)$ belongs to $L^p(\sigma)$. Conversely, if $f \in L^p(\sigma)$, then the automorphic extension of f to $\tilde{G}$ belongs to $L^p(\tau \times (dt/(1 + t^2)))$. The closed subspace of automorphic functions in $L^p(\tau \times (dt/(1 + t^2)))$ is equivalent to the Banach space $L^p(\sigma)$.

7. The Hardy Spaces

The conformal map $w = i(1 - z)/(1 + z)$ of the unit disc onto the upper half-plane establishes an isomorphism of the bounded analytic functions on $\{|z| < 1\}$ and the bounded analytic functions on $\{\text{Im}\,(w) > 0\}$. The representing measure $d\theta/2\pi$ for $z = 0$ is carried onto the representing measure $dt/\pi(1 + t^2)$ for $w = i$. Using Fatou's theorem, we identify the algebra of bounded analytic functions in the upper half-plane with a weak-star closed subalgebra $\boldsymbol{H^\infty(dt/(1 + t^2))}$ of $L^\infty(dt/(1 + t^2))$. The closure of $H^\infty(dt/(1 + t^2))$ in $L^p(dt/(1 + t^2))$ is denoted by $\boldsymbol{H^p(dt/(1 + t^2))}$, $0 < p < \infty$. Since these spaces are isomorphic to the corresponding spaces $H^p(d\theta/2\pi)$, the results developed in chapters IV and V for the unique representing measure situation apply to $H^p(dt/(1 + t^2))$.

If f is a Borel function on G, and $x \in G$, the function f_x is defined on the real line by

$$f_x(t) = f(x + e_t), \qquad -\infty < t < \infty.$$

A Borel function f **belongs to $\boldsymbol{H^p}$ on the line through $\boldsymbol{x}$** if $f_x \in H^p(dt/(1 + t^2))$. We say that f **belongs to $\boldsymbol{H^p}$ on almost all lines** if f_x belongs to H^p on the line through x for σ-almost all $x \in G$, or equivalently, for τ-almost all $x \in K$.

By **almost all lines** we mean a family of lines of full σ-measure, or equivalently, a family of lines which meets K in a subset of full τ-measure.

If $f \in A$, then f belongs to H^∞ on all lines. From the local product decomposition, we deduce the following.

7.1 Theorem. Let $0 < p \leq \infty$. If $f \in H^p(\sigma)$, then f belongs to H^p on almost all lines.

Proof. If $f_n \in A$, and $f_n \to f$ in $L^p(\sigma)$, then the automorphic extensions of f_n to the covering group $\tilde{G}$ converge in $L^p(\tau \times (dt/(1 + t^2)))$ to the automorphic extension of f. Hence $(f_n)_x$ converges to f_x in $L^p(dt/(1 + t^2))$ for τ-almost all $x \in K$. That proves the theorem, for $0 < p < \infty$. The same proof works for $p = \infty$, providing we replace (as we may) L^p-convergence with pointwise bounded convergence.

7.2 Corollary. If $f \in H^1(\sigma)$ and $0 < r < 1$, then

$$\log|\mu_r * f| \leq \mu_r * \log|f| \qquad \sigma\text{-almost everywhere.}$$

Proof. This is just Jensen's inequality for the algebras of analytic functions on the upper half-plane, in view of the description of representing measures given in section 4.

The converse of 7.1 is also true, but the proof requires some preliminary effort. We begin with the following lemma.

7.3 Lemma. Let ν be a non-negative finite measure on the compact abelian group H whose closed support is H. If E is a Borel subset of H of positive Haar measure, then $\nu(x - E) > 0$ for Haar-almost all $x \in H$.

Proof. The assertion of the lemma is equivalent to the assertion that $\nu * h > 0$ η-almost everywhere, where h is the characteristic function of E, and η is the Haar measure of H.

Let F be the set where $\nu * h = 0$, and let g be the characteristic function of $-F$. Then $h * g$ is a continuous non-negative function on G. Since

$$\begin{aligned}\int (h * g)(x)d\nu(x) &= (\nu * h * g)(0)\\ &= \int g(-x)(\nu * h)(x)d\eta(x)\\ &= 0,\end{aligned}$$

and since the support of ν is dense, we have $h * g = 0$. In particular,

$$0 = \int h * g d\eta = \eta(E)\eta(F).$$

This shows that $\eta(F) = 0$, proving the lemma.

7.4 Corollary. If u is a Borel function on G which is constant dt-almost everywhere on almost all lines, then u is constant σ-almost everywhere.

Proof. We can assume that u is real. It suffices to show that for each real t, the set $E_t = \{x\colon u(x) \geq t\}$ either has zero σ-measure or full σ-measure.

Suppose that $\sigma(E_t) > 0$. Applying 7.3 to the group $H = G$ and the measure $\nu = \mu_r$, we find that $\mu_r(x - E_t) > 0$ for almost all $x \in G$. Consequently $\mu_r(x - E_t) = 1$ for almost all $x \in G$, that is, E_t contains almost all points of almost all lines. In view of the local product decomposition, $\sigma(E_t) = 1$. That proves the corollary.

7.5 Corollary. If E is a Borel subset of G which contains every line which it meets, then either $\sigma(E) = 0$ or $\sigma(E) = 1$.

7.6 Corollary. Suppose $0 < p \leq \infty$, $f \in H^p(\sigma)$, and $f \neq 0$. Then $\mu_r * \log|f| > -\infty$ almost everywhere. In particular, f cannot vanish on a set of positive measure.

Proof. We can assume that f belongs to H^p on all lines. Then $\mu_r * \log|f| = -\infty$ precisely on the set of lines on which f vanishes dt-almost everywhere. By 7.5, this set has measure zero or measure one. The latter case is excluded, or else f would vanish σ-almost everywhere.

Using 7.4, we can now prove the converse of 7.1.

7.7 Theorem. Suppose $0 < p \leq \infty$, and $f \in L^p(\sigma)$. If f belongs to H^p on almost all lines, then $f \in H^p(\sigma)$.

Proof. Suppose first that f is as above, with $p = 2$. Since $L^2(\sigma) = H^2(\sigma) \oplus \overline{H_0^2(\sigma)}$, we can write $f = g + \bar{h}$, where $g \in H^2(\sigma)$ and $h \in H_0^2(\sigma)$. Since both h and $\bar{h} = f - g$ belong to H^2 on almost all lines, h must be constant on almost all lines. By 7.4, h is constant. Hence $f \in H^2(\sigma)$.

Once known for $p = 2$, the theorem follows immediately for $2 \leq p \leq \infty$, in view of the inclusions $H^p(dt/(1 + t^2)) \subseteq H^2(dt/(1 + t^2))$, and $H^2(\sigma) \cap L^p(\sigma) \subseteq H^p(\sigma)$. The first of these inclusions is obvious, the second comes from IV.6.2.

Now suppose that $0 < p < 2$, and that f is as above. A simple application of Fubini's theorem shows that $\log|t + f|$ belongs to $L^1(\sigma)$ for almost all real t. Replacing f by $t + f$ for such a t, we can assume that $\log|f| \in L^1(\sigma)$.

Let $u = \max(0, \log|f|)$. As in V.5, $e^{-(u+i^*u)} \in H^\infty(\sigma)$. Also, $e^{-(u+i^*u)}f$ belongs to H^∞ on almost all lines, and $e^{-(u+i^*u)}f$ is bounded. By what we have already proved, $e^{-(u+i^*u)}f \in H^\infty(\sigma)$. By V.5, $f \in H^p(\sigma)$. That completes the proof.

7.8 Theorem. Suppose $f \in H^1(\sigma)$. The following are equivalent:

(i) f_x is outer for almost all x.
(ii) $\log|\mu_r * f| = \mu_r * \log|f| > -\infty$ almost everywhere.
(iii) $A_0 f$ is dense in $H_0^1(\sigma)$.
(iv) f is an extremal function in $H^1(\sigma)$.

Proof. Assertions (i) and (ii) are equivalent, by the definition of outer function (cf. V.5). Assertions (iii) and (iv) are equivalent, by V.9.5. We will show that (i) implies (iv) and that (iii) implies (ii).

Suppose that f_x is outer for almost all x. Suppose $k \in L_R^\infty(\sigma)$ satisfies $kf \in H^1(\sigma)$. Then kf belongs to H^1 on almost all lines. Since f is outer on almost all lines, k belongs to H^∞ on almost all lines. By 7.7, $k \in H^\infty(\sigma)$. Since k is real, k must be constant. Hence f is extremal. This shows that (i) implies (iv).

Now suppose that $A_0 f$ is dense in $H_0^1(\sigma)$. For $\epsilon > 0$ fixed, let E be the set where $\log|\mu_r * f| \leq \mu_r * \log|f| - \epsilon$. We must show that $\sigma(E) = 0$.

Choose $s \in \Gamma$ such that $s > 0$ and $-\epsilon < s \log r$. There is a sequence $g_n \in A$ such that $g_n f \to \chi_s$ in $L^1(\sigma)$. Then $(\mu_r * g_n)(\mu_r * f) = \mu_r * (g_n f)$ tends to $\mu_r * \chi_s = r^s \chi_s$ in $L^1(\sigma)$. Almost everywhere on E we obtain

$$\begin{aligned} s \log r &= \log|\mu_r * \chi_s| \\ &\leq \log|\mu_r * f| + \limsup_{n\to\infty} \log|\mu_r * g_n| \\ &\leq \mu_r * \log|f| - \epsilon + \limsup_{n\to\infty} (\mu_r * \log|g_n|) \\ &\leq -\epsilon + \limsup_{n\to\infty} (\mu_r * \log|g_n f|) \\ &\leq -\epsilon. \end{aligned}$$

Consequently $\sigma(E) = 0$. That shows that (iii) implies (ii), completing the proof.

It is not known whether there exists an $f \in H_0^1(\sigma)$ such that $A_0 f$ is dense in $H_0^1(\sigma)$.

8. Weak-star Maximality of H^∞

We return momentarily to an arbitrary uniform algebra B.

8.1 Lemma. Suppose $\phi \in M_B$ has a unique representing measure m. The following statements are equivalent:

(i) $H^\infty(m)$ is a maximal weak-star closed subalgebra of $L^\infty(m)$.

(ii) If $f \in H^1(m)$, $f \neq 0$, $h \in L^\infty(m)$, and $h^n f \in H^1(m)$ for $n \geq 1$, then $h \in H^\infty(m)$.

(iii) If M is a closed invariant subspace of $L^1(m)$ which is not of the form $h_E L^1(m)$, h_E the characteristic function of a set E, and if $h \in L^\infty(m)$ satisfies $hM \subseteq M$, then $h \in H^\infty(m)$.

Proof. Assume (i) is true. Let M be a closed invariant subspace of L^1. Let H be the set of all $h \in L^\infty$ such that $hM \subseteq M$. Then H is a weak-star closed subalgebra of L^∞. If $H = L^\infty$, then M must be of the form $h_E L^1$ for some measurable set E. Otherwise, $H = H^\infty$. It follows that (iii) is true.

Assertion (ii) is obtained from (iii) by considering the invariant subspace generated by the functions $h^n f$, $n \geq 0$.

Now suppose (ii) is true. Suppose $h \in L^\infty$, $h \notin H^\infty$, and let H be the algebra generated by H^∞ and h. If $f \in L^1$ is orthogonal to H, then $h^n f \perp B$, and $h^n f \in H_0^1$ for all $n \geq 0$, by IV. 6.2 Applying (ii), we obtain

$f = 0$. Hence H is weak-star dense in L^∞. That proves (i), completing the proof of the lemma.

8.2 Theorem. The algebra $H^\infty(d\theta/2\pi)$ of bounded analytic functions on the unit disc is a maximal weak-star closed subalgebra of $L^\infty(d\theta/2\pi)$.

Proof. This can be proved just as was Wermer's maximality theorem II.5.1. It can also be proved using the classification of weak-star closed invariant subspaces of L^∞ given in V.6. In fact, suppose B is a weak-star closed subalgebra of L^∞ containing H^∞. Then B is an invariant subspace of L^∞. If B is not simply invariant, V.6.4 shows that $B = L^\infty$. If B is simply invariant, then V.6.2 states that $B = FH^\infty$ for some unitary function $F \in L^\infty$. Now F, and hence F^2, belong to B. Since $H^\infty = \bar{F}B$, we obtain $F = \bar{F}F^2 \in H^\infty$. Consequently $B = H^\infty$. That proves the theorem.

This theorem shows that the algebra $H^\infty(dt/(1 + t^2))$ of bounded analytic functions in the upper half-plane is a maximal weak-star closed subalgebra of $L^\infty(dt/(1 + t^2))$. Using this fact, and 8.1, we can now prove the following.

8.3 Theorem. $H^\infty(\sigma)$ is a maximal weak-star closed subalgebra of $L^\infty(\sigma)$.

Proof. We will verify property (ii) of 8.1. Suppose $f \in H^1(\sigma)$, $f \neq 0$, $h \in L^\infty(\sigma)$, and $h^n f \in H^1(\sigma)$ for $n \geq 1$. Then $h^n f$ belongs to H^1 on almost all lines, for all $n \geq 1$. Since 8.1(ii) is valid for the algebra $H^\infty(dt/(1 + t^2))$, we see that h belongs to H^∞ on almost all lines. By 7.7, $h \in H^\infty$. That proves the theorem.

Our knowledge of $H^\infty(\sigma)$ is limited. Virtually nothing is known, for instance, about the maximal ideal space of $H^\infty(\sigma)$.

9. Weight Functions

In this section, we will be considering the closures of A and A_0 in $L^2(wd\sigma)$, where w is a non-negative weight function in $L^1(\sigma)$. From Szegö's theorem V.8.3, the distance from 1 to A_0 in $L^2(wd\sigma)$ is $e^{\int \log wd\sigma}$. So the closures of A_0 and A in $L^2(wd\sigma)$ coincide if and only if $\int \log wd\sigma = -\infty$. In 9.1 we will derive more precise information.

First, note that $\log w$ belongs to L^1 on the line through $x \in G$ if and only if $(\mu_r * \log w)(x) > -\infty$. Now the set where $\mu_r * \log w = -\infty$ contains every line which it meets. By 7.5, either $\mu_r * \log w > -\infty$ almost everywhere or $\mu_r * \log w = -\infty$ almost everywhere. That is, either $\log w$ belongs to L^1 on almost all lines, or on almost all lines $\log w$ does not belong to L^1.

9.1 Theorem. Let w be a non-negative function in $L^1(\sigma)$. Let M be the closure of A in $L^2(wd\sigma)$.

(i) If $\log w \in L^1(\sigma)$, then A_0 is not dense in M.

(ii) If $\log w \notin L^1(\sigma)$, while $\log w$ belongs to L^1 on almost all lines, then A_0 is dense in M, while $\chi_s A$ is not dense in M for each $s > 0$, $s \in \Gamma$.

(iii) If on almost all lines $\log w$ does not belong to L^1, then $\chi_s A$ is dense in M for each $s \in \Gamma$, $s > 0$.

Proof. We have already noted that (i) and the first assertion in (ii) are consequences of Szegö's theorem.

Fix $s \in \Gamma$, $s > 0$. Let d_ϵ denote the distance from 1 to the closure M_s of $\chi_s A$ in $L^2((w + \epsilon)d\sigma)$, $\epsilon \geq 0$. Let g_ϵ be the outer function in $H^2(\sigma)$ such that $|g_\epsilon|^2 = w + \epsilon$. Since the χ_t form an orthonormal basis for $L^2(\sigma)$, we can write

$$g_\epsilon = \sum_{t \geq 0, t \in \Gamma} c_\epsilon(t)\chi_t$$

where

$$\int (w + \epsilon)d\sigma = \int |g_\epsilon|^2\, d\sigma = \sum_t |c_\epsilon(t)|^2.$$

The theorem will follow from the following assertions, where we assume $\epsilon > 0$ and $0 < r < 1$:

(a) d_ϵ decreases to d_0 as ϵ decreases to 0.

(b) $d_\epsilon = \sum_{0 \leq t < s} |c_\epsilon(t)|^2$.

(c) $\| \mu_r * g_\epsilon \|^2_{L^2(\sigma)} = \sum_{t \geq 0} |c_\epsilon(t)|^2 r^{2t}$.

(d) $\log |\mu_r * g_\epsilon|^2 = \mu_r * \log (w + \epsilon)$ almost everywhere.

First we will verify these assertions.

For (a), note that d_ϵ decreases as ϵ decreases, and $d_0 \leq \lim_{\epsilon \to 0} d_\epsilon$. For fixed $f \in A$,

$$\int |1 - \chi_s f|^2 w d\sigma = \lim_{\epsilon \to 0} \int |1 - \chi_s f|^2 (w + \epsilon) d\sigma \geq \lim_{\epsilon \to 0} d_\epsilon.$$

Consequently $d_0 \geq \lim_{\epsilon \to 0} d_\epsilon$, which proves (a).

For (b), we have

$$\begin{aligned} d_\epsilon &= \inf_{f \in A} \int |1 - \chi_s f|^2 (w + \epsilon) d\sigma \\ &= \inf_{f \in A} \int |g_\epsilon - \chi_s f g_\epsilon|^2 d\sigma \\ &= \inf_{h \in A} \int |g_\epsilon - \chi_s h|^2 d\sigma. \end{aligned}$$

Hence d_ϵ is the square of the distance from g_ϵ to $\chi_s H^2(\sigma)$ in $L^2(\sigma)$. The component of g_ϵ orthogonal to $\chi_s H^2(\sigma)$ is $\sum_{0 \leq t < s} c_\epsilon(t)\chi_t$. Hence $d_\epsilon = \sum_{0 \leq t < s} |c_\epsilon(t)|^2$, proving (b).

Assertion (c) follows from the expansion

$$\mu_r * g_\epsilon = \sum_t c_\epsilon(t) r^t \chi_t.$$

Assertion (d) follows from 7.8.

Now suppose that on almost all lines $\log w$ does not belong to L^1. Then $\mu_r * \log w = -\infty$ almost everywhere. By (d), $|\mu_r * g_\epsilon|$ decreases pointwise almost everywhere to zero. Hence $\mu_r * g_\epsilon \to 0$ in $L^2(\sigma)$. By (c) and (b), $\lim_{\epsilon \to 0} d_\epsilon = 0$. By (a), $d_0 = 0$. Consequently $\chi_s A$ is dense in M.

If, on the other hand, $\log w$ belongs to L^1 on almost all lines, then $\mu_r * \log w > -\infty$ almost everywhere. By (d), $\mu_r * g_\epsilon$ cannot tend to zero in $L^2(\sigma)$. By (c) and (b), d_ϵ cannot tend to zero. By (a), $d_0 \neq 0$, and $\chi_s A$ is not dense in M. This concludes the proof.

Next we note that case (ii) of 9.1 is not vacuous.

9.2 Lemma. There exists a non-negative function $w \in L^\infty(\sigma)$ such that $\log w \notin L^1(\sigma)$, while $\log w$ belongs to L^1 on almost all lines.

Proof. Since $\sigma(R) = 0$, there is a closed set $E \subset G$ such that $\sigma(E) > 0$ while E does not meet R. Let h be the characteristic function of $-E$. Since most of the mass of μ_r is concentrated on compact subsets of R, which remain outside E when translated slightly, we see that $\lim_{x \to 0} (\mu_r * h)(x) = 0$.

Let u be a continuous function on $G \backslash \{0\}$ such that $u \geq 0$, $\int u d\sigma = \infty$, and $(\mu_r * h * u)(0) = \int (\mu_r * h)(x) u(-x) d\sigma(x) < \infty$. Then $(\mu_r * u * h)(0) = \int_E (\mu_r * u)(x) d\sigma(x) < \infty$, so $\mu_r * u < \infty$ on a set of positive measure. Applying 7.5, we find that $\mu_r * u < \infty$ almost everywhere.

The function $w = e^{-u}$ has the properties required by the lemma.

9.3 Theorem. Suppose w is a non-negative function in $L^1(\sigma)$ such that $\log w$ is integrable on almost all lines. Then there is $f \in H^1(\sigma)$ such that $0 < |f| \leq w$ almost everywhere.

Proof. We can assume that $0 < w \leq 1$.

From 9.2, A is not dense in $L^2(w d\sigma)$. Choose $g \in L^2(w d\sigma)$, $g \neq 0$, such that g is orthogonal to A in $L^2(w d\sigma)$. Since $\int |g|^2 w^2 d\sigma \leq \int |g|^2 w d\sigma < \infty$, $gw \in L^2(\sigma)$. Since $\int h g w d\sigma = 0$ for all $h \in A$, $gw \in H^2(\sigma)$.

Let k be the outer function in $H^2(\sigma)$ such that $|k| = \max(1, |g|\sqrt{w})$. Then $|gw/k| \leq \sqrt{w}$, and $gw/k \in H \cap L^\infty = H^\infty$. The function $f = (gw/k)^2$ does the job.

It is not known whether one can choose f in 9.3 so that $|f| = w$ almost everywhere.

Combining 9.2 and 9.3, we see that there are functions $f \in H^\infty(\sigma)$ such that $f \neq 0$ while $\log |f| \notin L^1(\sigma)$. In contrast, there is the following result, which shows that no such f can be continuous on G.

9.4 Theorem. If $f \in A$ and $f \neq 0$, then $\log |f| \in L^1(\sigma)$.

Proof. $\mu_r * f$ is continuous and not identically zero. Let E be an open set on which $\mu_r * f$ is bounded away from zero. Choose $h \in C(G)$ such that $h \geq 0$, $h = 0$ off $-E$, and $h > 0$ somewhere on $-E$. Since every line meets $-E$ on a set of positive linear measure, $\mu_r * h > 0$. Since $\mu_r * h$ is continuous, $\mu_r * h \geq \epsilon$ for some $\epsilon > 0$. Hence

$$\begin{aligned} -\infty &< \int h(-x) \log |(\mu_r * f)(x)| \, d\sigma(x) \\ &= (h * \log |\mu_r * f|)(0) \\ &\leq (h * \mu_r * \log |f|)(0) \\ &\leq \epsilon \int \log |f| \, d\sigma. \end{aligned}$$

That proves the theorem.

10. Invariant Subspaces

An invariant subspace M of $L^2(\sigma)$ is **doubly invariant** if $\chi_s M \subseteq M$ for all $s \in \Gamma$.

10.1 Theorem. Suppose M is a doubly invariant subspace of $L^2(\sigma)$. There is a Borel subset E of G with characteristic function h_E, such that $M = h_E L^2(\sigma)$.

Proof. This can be proved as was V.6.4, by considering the projection h of 1 onto M. Since $h(1 - h)$ is orthogonal to all the χ_s, $h(1 - h) = 0$. Consequently h is the characteristic function of a Borel set E. It follows easily that $M = hL^2(\sigma)$.

10.2 Theorem. Suppose M is an invariant subspace of $L^2(\sigma)$ which is not doubly invariant. If $f \in M$ and $f \neq 0$, then $\log |f|$ is integrable on almost all lines. In particular, f cannot vanish on a set of positive measure.

Proof. Suppose that $\log |f|$ is not integrable on almost all lines. We must show that M is doubly invariant.

Let N be the invariant subspace of $L^2(\sigma)$ generated by f. By 9.1(iii), we can approximate every χ_s by functions in A in the norm of $L^2(|f|^2 \, d\sigma)$. Hence $\chi_s f \in N$ for all $s \in \Gamma$. So N is doubly invariant. If E is the set where f does not vanish, then $N = h_E L^2(\sigma)$.

Now $M \ominus N$ is also invariant, and all the functions in $M \ominus N$ vanish on E. By the same argument as used above, every invariant subspace of $L^2(\sigma)$ generated by a single function $g \in M \ominus N$ must be doubly invariant. So $M \ominus N$ must be doubly invariant. Hence M is doubly invariant, as was to be shown.

An invariant subspace M is **right continuous** if M is not doubly invariant, while $\bigcup \{\chi_s M: s \in \Gamma, s > 0\}$ is dense in M. An invariant subspace M is **left continuous** if M is not doubly invariant, while $M = \bigcap \{\chi_s M: s \in \Gamma, s < 0\}$. Finally, M is **continuous** if it is both left and right continuous.

10.3 Theorem. If M is a left (respectively right) continuous invariant subspace of $L^2(\sigma)$ which is not continuous, then there is a measurable function F on G such that $|F| = 1$ almost everywhere and $M = FH^2(\sigma)$ (respectively $M = FH_0^2(\sigma)$).

Proof. F is the function of unit norm in $\bigcap_{s<0} \chi_s M$ which is orthogonal to $\bigcup_{s>0} \chi_s M$. The proof that F has the desired properties proceeds as in V.6.2. Suppose, for instance, that M is right continuous. Since $F \perp \chi_s F$ for all $s > 0$, $|F|^2\, d\sigma$ is orthogonal to A_0. Hence $|F| = 1$ almost everywhere. Since $\chi_s F \in M$ for all $s > 0$, M contains $FH_0^2(\sigma)$. Suppose $g \in M$ is orthogonal to $FH_0^2(\sigma)$. Then $g\bar{F} \perp H_0^2(\sigma)$. Since F is orthogonal to $g\chi_s$ for all $s > 0$, $g\bar{F} \perp \overline{H_0^2(\sigma)}$. Since F is orthogonal to M, $g\bar{F} \perp 1$. Consequently $g\bar{F} = 0$, and $g = 0$. So $M = FH_0^2(\sigma)$, as asserted.

11. Structure of Cocycles

Complex-valued functions of modulus one will be called unitary functions. A **cocycle on G** is a unitary Borel function B on $G \times R$ which satisfies the cocycle identity

$$B(x + e_s, t) = \overline{B(x,s)}B(x, s + t)$$

for all $x \in G$ and $s, t \in R$. In particular, $B(x, 0) = 1$ for all $x \in G$. We identify any two cocycles which differ only on a set of measure zero on $G \times R$.

The cocycles on G form a group, under pointwise multiplication. The identity of the group is the cocycle which is identically 1. The inverse of a cocycle B is its complex conjugate $\bar{B}$.

If F is a unitary Borel function on G, then F determines a cocycle via the formula

$$B(x,t) = \overline{F(x)}F(x + e_t).$$

Cocycles which arise in this manner from unitary functions on G are called **coboundaries**. The coboundaries form a subgroup of the group of cocycles. We will say that two cocycles are **cohomologous** if their quotient is a coboundary.

Let B be a cocycle. To every point $x \in G$ there corresponds the unitary Borel function $x + e_t \to B(x,t)$ on the line through x. The cocycle relation shows that the functions on the line through x corresponding to other points on the line are all constant multiples of each other, normalized so that the function determined by the point $x + e_s$ assumes the value 1 at $x + e_s$.

If two cocycles A,B determine the same unitary functions, modulo null functions, on almost every line, then on almost all lines, $A\bar{B}$ determines the function which is 1 dt-almost everywhere. Consequently $A\bar{B} = 1$ almost everywhere on $G \times R$, and A and B are the same cocycle.

This leads us to the following important observation. Each cocycle can be regarded as a rule which assigns, in a measurable way, a unitary function to each line in G, each function specified up to constant multiples. Two such rules determine the same cocycle if the functions agree dt-almost everywhere on almost all lines. If B is the cocycle determined by a unitary Borel function F, the corresponding function on a line in G is the restriction of F to that line.

Every cocycle B determines the invariant subspace of $L^2(\sigma)$, denoted by $\boldsymbol{BH^2(\sigma)}$, consisting of all functions $f \in L^2(\sigma)$ such that $\overline{B(x,t)}f(x + e_t) \in H^2(dt/(1 + t^2))$ for almost all $x \in G$. Equivalently, $BH^2(\sigma)$ consists of all $f \in H^2(\sigma)$ such that on almost all lines $x + R$ of G, f_x belongs to $B_xH^2(dt/(1 + t^2))$, where B_x is the unitary function on the line $x + R$ determined by B.

If B is the coboundary of the unitary function F on G, then the invariant subspace $BH^2(\sigma)$ coincides with $FH^2(\sigma)$. This follows from 7.1 and 7.7.

We will show later in 12.6 that $B \leftrightarrow BH^2(\sigma)$ is a bijective correspondence of cocycles and left-continuous invariant subspaces of $L^2(\sigma)$. First we will examine cocycles in more detail.

Let K be the compact subgroup of G introduced in section 6, and let $\boldsymbol{\mathscr{U}(K)}$ denote the multiplicative group of unitary Borel functions on K, modulo null functions. For each $\beta \in \mathscr{U}(K)$, we define a cocycle B_β by prescribing unitary functions on the lines in G as follows: The unitary function on a line is to be constant on the open intervals of length one with endpoints in K, and each time the line passes through the point $y \in K$ the function is multiplied by $\beta(y - e_1)$. The cocycle $B_\beta(x,t)$ is given explicitly for $t \geq 0$ by

$$B_\beta(y,t) = \begin{cases} 1, & y \in K,\ 0 \leq t < 1 \\ \prod_{j=0}^{[t]-1} \beta(y + e_j), & y \in K,\ 1 \leq t < \infty \end{cases}$$

$$B_\beta(y + e_s,t) = \overline{B_\beta(y,s)}B_\beta(y,s + t), \qquad y \in K,\ 0 \leq s < 1,\ 0 \leq t < \infty,$$

where $[t]$ is the largest integer not exceeding t. Note that the cocycle B_β is normalized to be continuous from the right on each line. Also,

$$B_\beta(y,1) = \beta(y), \qquad y \in K.$$

11.1 Theorem. Every cocycle B on G has the factorization

$$B = B_\beta C,$$

where β is a unitary Borel function on K, and C is a coboundary.

Proof. Every $x \in G$ can be expressed uniquely in the form $x = y + e_s$, $y \in K$, $0 \leq s < 1$. Consequently the formula

$$F(y + e_s) = B(y,s), \qquad y \in K, 0 \leq s < 1,$$

defines a Borel function F on G. Set $C(x, t) = \overline{F(x)}F(x + e_t)$, and define $\beta(y) = B(y,1)$. We wish to show that $B = B_\beta C$.

Suppose $y \in K$ and $t \geq 0$. Write $t = [t] + s$, where $0 \leq s < 1$. Then $F(y) = 1$ and $F(y + e_t) = B(y + e_{[t]},s)$. So

$$\begin{aligned} B_\beta(y,t)C(y,t) &= \beta(y + e_{[t]-1}) \ldots \beta(y + e_1)\beta(y)\overline{F(y)}F(y + e_t) \\ &= B(y + e_{[t]-1},1) \ldots B(y + e_1,1)B(y,1)B(y + e_{[t]},s) \\ &= B(y,[t])B(y + e_{[t]},s) \\ &= B(y,t), \end{aligned}$$

where we have used the cocycle identity to collapse the product. The equality $B_\beta(y,t)C(y,t) = B(y,t)$ continues to hold for $y \in K$ and $t < 0$. It follows that the cocycles $B_\beta C$ and B determine the same unitary functions on each line, so they coincide. That proves the theorem.

Let $\mathcal{Q}(K)$ denote the subgroup of $\mathcal{U}(K)$ of functions β which can be expressed in the form

$$\beta(y) = \overline{f(y)}f(y + e_1), \qquad \tau\text{-almost all } y \in K,$$

for some $f \in \mathcal{U}(K)$. If β is such a function, then B_β is the coboundary of the function F on G defined by

$$F(y + e_s) = f(y), \qquad y \in K, 0 \leq s < 1.$$

Conversely, suppose that B_β is the coboundary of F. Then B_β and F determine the same functions on almost all lines. Altering F on a set of measure zero, we can assume that B_β and F determine the same functions on all lines. We can also assume that F is continuous from the right on each line. With F so normalized, $F(y + e_s)$ must be constant for $0 \leq s < 1$, whenever $y \in K$. Clearly, $\beta(y) = \overline{F(y)}F(y + e_1)$ now for all $y \in K$. So β belongs to $\mathcal{Q}(K)$.

The correspondence $\beta \rightarrow B_\beta$ is a monomorphism of the group $\mathcal{U}(K)$ into the group of cocycles on G. We have just shown B_β is a coboundary if and only if $\beta \in \mathcal{Q}(K)$. Combining this observation with 11.1, we obtain the following theorem.

11.2 Theorem. The monomorphism $\beta \rightarrow B_\beta$ induces an isomorphism of $\mathcal{U}(K)/\mathcal{Q}(K)$ and the quotient group of cocycles on G modulo coboundaries.

If we restrict ourselves to real cocycles, we can carry this isomorphism one step further. First note that if $\beta \in \mathscr{Q}(K)$ is real, then there is a real function $f \in \mathscr{U}(K)$ such that $\beta(y) = \overline{f(y)}f(y + e_1)$, $y \in K$. In fact, let $g \in \mathscr{U}(K)$ satisfy $\beta(y) = \overline{g(y)}g(y + e_1)$. Then the function f, defined to be 1 where $0 \leq \arg g < \pi$, and -1 where $\pi \leq \arg g < 2\pi$, does the trick.

Let $\mathscr{U}_r(K)$ and $\mathscr{Q}_r(K)$ denote the real-valued functions in $\mathscr{U}(K)$ and $\mathscr{Q}(K)$ respectively. The preceding remark shows that the inclusion $\mathscr{U}_r(K) \to \mathscr{U}(K)$ induces an isomorphism of $\mathscr{U}_r(K)/\mathscr{Q}_r(K)$ and a subgroup of $\mathscr{U}(K)/\mathscr{Q}(K)$. We proceed to identify $\mathscr{U}_r(K)/\mathscr{Q}_r(K)$.

Let $\mathscr{M}(K)$ be the family of Borel subsets of K, modulo null sets. The operation

$$E_0 \Delta E_1 = (E_0 \cap E_1^c) \cup (E_1 \cap E_0^c)$$

makes $\mathscr{M}(K)$ into a group. Subsets of $\mathscr{M}(K)$ of the form $E \Delta (E + e_1)$, $E \in \mathscr{M}(K)$, form a subgroup of $\mathscr{M}(K)$, denoted by $\mathscr{N}(K)$. The Borel sets in $\mathscr{N}(K)$ are also called **coboundaries**.

For each $E \in \mathscr{M}(K)$, define the function $\beta_E \in \mathscr{U}_r(K)$ to be -1 on E and $+1$ off E. The following theorem is easy to verify.

11.3 Theorem. The correspondence $E \to \beta_E$ is an isomorphism of $\mathscr{M}(K)$ and $\mathscr{U}_r(K)$ which induces an isomorphism of $\mathscr{M}(K)/\mathscr{N}(K)$ and $\mathscr{U}_r(K)/\mathscr{Q}_r(K)$.

In order to find cocycles on G which are not coboundaries, it now suffices to find Borel subsets of K which are not coboundaries. If K is connected, then K itself is not a coboundary. More generally, the following is true.

11.4 Theorem. The set K is a coboundary if and only if there is a closed subgroup H of K of index two, such that $K = H \Delta (H + e_1)$. This occurs if and only if $\pi \in \boldsymbol{\Gamma}$.

Proof. Recall that 2π belongs to $\boldsymbol{\Gamma}$, and that K is the set of $y \in G$ such that $\chi_{2\pi}(y) = 1$.

Suppose that $\pi \in \boldsymbol{\Gamma}$. Then $K = H \Delta (H + e_1)$, where H is the closed subgroup of K consisting of all $y \in K$ such that $\chi_\pi(y) = 1$.

Conversely, suppose that K is a coboundary. Choose a Borel set F such that $\tau(F \cup (F + e_1)) = 1$, while $\tau(F \cap (F + e_1)) = 0$. Evidently $\tau(F) = \tau(K \backslash F) = \frac{1}{2}$. Also, for almost all $x \in F$, $x + e_2$ also belongs to F.

We assert that the subgroup $\Lambda = \{e_{2n}: -\infty < n < \infty\}$ is not dense in K. For suppose the contrary. Let ν be a finite measure on Λ which assigns positive mass to every point of Λ. Applied to the group K and the measure ν, lemma 7.3 states the following: If E is a Borel subset of K such that $\tau(E) > 0$, then $x - E$ meets Λ for τ-almost all $x \in K$, that is, E meets $x + \Lambda$ for τ-almost all $x \in K$. However, for almost all $x \in K \backslash F$, $x + \Lambda$ does not meet $K \backslash F$. This contradiction establishes the assertion that Λ is not dense in K.

The closure H of Λ is a closed subgroup of K, such that $K = H \cup (H + e_1)$, and $H \cap (H + e_1)$ is empty, that is, such that $K = H \Delta (H + e_1)$. The character of K which is 1 on H and -1 on $H + e_1$ extends to be a character χ_a of G, for some $a \in \Gamma$. Since $\chi_a(e_n) = e^{\pi i n}$, $-\infty < n < \infty$, evidently $a = \pi$. That completes the proof.

11.5 Corollary. If there is some $a \in \Gamma$ such that $a/2 \notin \Gamma$, then there exist real cocycles on G which are not coboundaries.

Proof. Multiplying Γ by $2\pi/a$ only changes the parametrization of the line $s \to e_s$. Consequently we can assume that $2\pi \in \Gamma$ and $\pi \notin \Gamma$, so that 11.4 is applicable.

12. Cocycles and Invariant Subspaces

In this section, we classify left continuous invariant subspaces in terms of cocycles. First some measure theoretic difficulties must be attended to.

12.1 Lemma. If B is a cocycle on G, then the functions $B(\cdot,t)$ move continuously with t in $L^2(\sigma)$.

Proof. Let K be the compact subgroup of G constructed in section 6. Then

$$\|B(\cdot,t) - B(\cdot,t_0)\|_2^2 = \int_G |B(x,t) - B(x,t_0)|^2 \, d\sigma(x)$$
$$= \int_K \int_0^1 |B(y + e_s,t) - B(y + e_s,t_0)|^2 \, ds d\tau(y).$$

By the cocycle relation, this integral is

$$\int_K \int_0^1 |B(y,s + t) - B(y,s + t_0)|^2 \, ds d\tau(y).$$

Now the inner integral tends to zero as $t \to t_0$, for each fixed $y \in K$. By the pointwise bounded convergence theorem, the double integral tends to zero as $t \to t_0$.

12.2 Lemma. Suppose f_t moves continuously in $L^2(\sigma)$ with $t \in R$. Then there is a Borel function $A(x,t)$ on $G \times R$ such that for each $t \in R$, $A(x,t) = f_t(x)$ σ-almost everywhere.

Proof. We can restrict our attention to values of t in the interval $[0,1]$. Also, we can assume that each f_t is a Borel function.

For each n, let $A_n(x,t)$ be the Borel function on $G \times [0,1]$ which linearly interpolates $f_{m/n}$ and $f_{(m+1)/n}$ on the interval $m/n \leq t \leq (m+1)/n$. That is, if $t = m/n + a/n$, $0 \leq a \leq 1$, and if $x \in G$, then

$$A_n(x,t) = af_{m/n}(x) + (1-a)f_{(m+1)/n}(x).$$

Since $t \to f_t$ is a uniformly continuous map of [0,1] into $L^2(\sigma)$, we obtain

$$\lim_{n\to\infty} \sup_{0\leq t\leq 1} \| A_n(\cdot,t) - f_t \|_2 = 0.$$

Let $\{n_k\}_{k=1}^{\infty}$ be a subsequence such that

$$\sum_{k=1}^{\infty} \sup_{0\leq t\leq 1} \| A_{n_{k+1}}(\cdot,t) - A_{n_k}(\cdot,t) \|_2 < \infty.$$

From the convergence of this series, it follows that for each t, $\lim_{k\to\infty} A_{n_k}(x,t)$ exists for almost all x. This limit determines a Borel function with the required properties.

12.3 Lemma. Suppose $A(x,t)$ is a unitary Borel function on $G \times R$ such that

$$(*) \qquad A(x + e_s,t) = \overline{A(x,s)}A(x,s+t), \qquad \text{a.a. } (x,s,t) \in G \times R \times R.$$

Then there is a cocycle B on G such that $B(x,t) = A(x,t)$ for almost all $(x,t) \in G \times R$.

Proof. The assumption is that

$$A(y + e_r + e_s,t) = \overline{A(y + e_r,s)}A(y + e_r,s + t)$$

for almost all $(y,r,s,t) \in K \times [0,1] \times R \times R$. Fix r, $0 \leq r < 1$, so that this identity holds for almost all $(y,s,t) \in K \times R \times R$. Define a unitary Borel function A_1 on $G \times R$ by

$$A_1(x,t) = A(x - e_r,t).$$

Then A_1 satisfies $(*)$ and the additional identity

$$(**) \qquad A_1(y + e_s,t) = \overline{A_1(y,s)}A_1(y,s+t),$$

for almost all $(y,s,t) \in K \times R \times R$.

Let J be a Borel subset of K of τ-measure zero such that for each $y \notin J$, $(**)$ holds for almost all (s,t). Replacing J by $\bigcup_{n=\infty}^{\infty} (J + e_n)$, we can assume that $J \pm e_1 = J$. Define the unitary Borel function A_2 on $G \times R$ by

$$A_2(y + e_s,t) = \begin{cases} 1, & y \in J \\ A_1(y + e_s,t), & y \notin J. \end{cases}$$

Then $A_2 = A_1$ almost everywhere, and A_2 satisfies $(*)$. Moreover, for each $y \in K$, A_2 satisfies $(**)$ for almost all $(s,t) \in R \times R$.

Now define a unitary Borel function F on G by setting

$$F(y + e_s) = A_2(y,s), \qquad y \in K, 0 \leq s < 1.$$

Let $C(x,t) = \overline{F(x)}F(x + e_t)$ be the coboundary of F, and let $A_3 = \bar{C}A_2$. Then A_3 is a unitary Borel function on $G \times R$ which satisfies (*). For all $y \in K$, A_3 satisfies (**) for almost all $(s,t) \in R \times R$. Moreover,

$$A_3(y,s) = 1, \qquad y \in K, 0 \leq s < 1.$$

Fix $y \in K$. For almost all $1 \leq s < 2$, we have

$$\begin{aligned} A_3(y + e_s,t) &= \overline{A_3(y + e_1,s - 1)}A_3(y + e_1,s - 1 + t) \\ &= A_3(y + e_1,s - 1 + t), \qquad \text{a.a. } t \in R, \end{aligned}$$

and

$$A_3(y + e_s,t) = \overline{A_3(y,s)}A_3(y,s + t), \qquad \text{a.a. } t \in R.$$

Regarding s as fixed and setting $u = s + t - 1$, we obtain

$$A_3(y + e_1,u) = \overline{A_3(y,s)}A_3(y,u + 1), \qquad \text{a.a. } u \in R,$$

for almost all $1 \leq s < 2$. In particular, $\overline{A_3(y,s)}$ is constant almost everywhere for $1 \leq s < 2$. That constant is given by

$$\beta(y) = \int_1^2 \overline{A_3(y,s)}ds, \qquad y \in K.$$

Evidently β is a unitary Borel function on K. For each fixed $y \in K$,

$$A_3(y + e_1,u) = \beta(y)A_3(y,u + 1), \qquad \text{a.a. } u \in R.$$

Let B_β be the cocycle constructed in section 11, and set $A_4 = \bar{B}_\beta A_3$. Then A_4 satisfies (*), and for all $y \in K$, A_4 satisfies (**) for almost all $(s,t) \in R \times R$. Moreover,

$$A_4(y,s) = 1, \qquad 0 \leq s < 1$$

$$A_4(y + e_1,u) = A_4(y,u + 1), \qquad \text{a.a. } u \in R.$$

It follows that $A_4 = 1$ almost everywhere on $G \times R$.

If $B(x,t) = B_\beta(x + e_r,t)C(x + e_r,t)$, then B is a cocycle on G which coincides almost everywhere with A. That completes the proof.

12.4 Lemma. A function $f \in L^2(dt/(1 + t^2))$ belongs to $H^2(dt/(1 + t^2))$ if and only if

$$\int_{-\infty}^{\infty} f(t)g(t)\frac{dt}{1 + t^2} = 0$$

for all $g \in H^2(dt/(1 + t^2))$ satisfying $g(i) = 0$.

Proof. As remarked in section 7, the H^p-theory developed in chapter IV applies to the representing measure $dt/[\pi(1 + t^2)]$ for i. By IV.6.2,

$$L^2(dt/(1 + t^2)) = H^2(dt/(1 + t^2)) \oplus H_i^2(dt/(1 + t^2))$$

where H_i^2 is the set of functions in H^2 which vanish at i. That proves 12.4.

12.5 Lemma. Let ν be a finite complex measure on the reals, and define

$$f(t) = \int_{-\infty}^{\infty} e^{it\lambda} d\nu(\lambda), \qquad -\infty < t < \infty.$$

Then $f \in H^2(dt/(1 + t^2))$ if and only if $|\nu|(-\infty,0) = 0$.

Proof. If $|\nu|(-\infty,0) = 0$, and $g \in H_i^2(dt/(1 + t^2))$, then

$$\int_{-\infty}^{\infty} f(t)g(t)\frac{dt}{1+t^2} = \int_0^{\infty}\int_{-\infty}^{\infty} e^{it\lambda}g(t)\frac{dt}{1+t^2}d\nu(\lambda).$$

If $\lambda \geq 0$, then $e^{it\lambda}g(t) \in H_i^2(dt/(1 + t^2))$, so the inner integral vanishes. By 12.4, $f \in H^2(dt/(1 + t^2))$.

Conversely, suppose $f \in H^2(dt/(1 + t^2))$. Since

$$\int_0^{\infty} e^{it\lambda} d\nu(\lambda) \in H^2(dt/(1 + t^2)),$$

also

$$h(t) = \int_{-\infty}^{0} e^{it\lambda} d\nu(\lambda) \in H^2(dt/(1 + t^2)).$$

By what we have already shown,

$$\overline{h(t)} = -\int_0^{\infty} e^{it\lambda} d\bar{\nu}(-\lambda) \in H^2(dt/(1 + t^2)).$$

It follows that the continuous function h must be identically constant, say $h = c$. If δ_0 is the point mass at 0, then

$$\int_{-\infty}^{0} e^{it\lambda} d(\nu - c\delta_0)(\lambda) = 0, \qquad -\infty < t < \infty.$$

This can only happen when $|\nu|(-\infty,0) = 0$. That proves the lemma.

Recall that every cocycle B determines the invariant subspace, denoted by $BH^2(\sigma)$, of all $f \in L^2(\sigma)$ such that the function $t \to \overline{B(x,t)}f(x + e_t)$ belongs to $H^2(dt/(1 + t^2))$ for almost all $x \in G$. If B is the coboundary of the unitary Borel function F, then $BH^2(\sigma)$ coincides with $FH^2(\sigma)$.

12.6 Theorem. The invariant subspace $BH^2(\sigma)$ arising from a cocycle B is left continuous. Every left continuous invariant subspace of $L^2(\sigma)$ is of the form $BH^2(\sigma)$ for a unique cocycle B.

Proof. The proof divides into three parts. First the Fourier transform establishes a bijective correspondence between left continuous invariant subspaces and continuous unitary groups on $L^2(\sigma)$ which satisfy the Weyl commutation relation

$$U_t V_a = e^{ita} V_a U_t, \qquad -\infty < t < \infty,\ a \in \Gamma,$$

where V_a is the multiplication operator

$$V_a f = \chi_a f, \qquad f \in L^2(\sigma),\ a \in \Gamma.$$

Next we establish a bijective correspondence between such continuous unitary groups and cocycles, by showing that every such group is of the form

$$U_t = B(\cdot, t)T_t,$$

where B is a cocycle on G, and T_t is the continuous unitary group of translations

$$(T_t f)(x) = f(x + e_t), \qquad -\infty < t < \infty, x \in G.$$

Finally, we show that if B is the cocycle determined in this manner by the left continuous invariant subspace M, then $M = \bar{B}H^2(\sigma)$.

Let M be a left continuous invariant subspace of $L^2(\sigma)$. Define the closed subspace M_a of $L^2(\sigma)$ by

$$M_a = \cap \{\chi_b M : b \in \Gamma, b < a\}, \qquad -\infty < a < \infty.$$

Since M is left continuous, $M = M_0$. In fact, $M_a = \chi_a M$ whenever $a \in \Gamma$. Also,

(i) $$M_b \supseteq M_a, \qquad -\infty < b < a < \infty.$$

(ii) $$M_a = \bigcap_{b<a} M_b, \qquad -\infty < a < \infty.$$

The closure of $\bigcup_{a>-\infty} M_a$ is a doubly invariant subspace of $L^2(\sigma)$. By 10.1, it must coincide with the set of all $f \in L^2(\sigma)$ which vanish off some fixed Borel set E. By 10.2, no $f \in M$ can vanish on a set of positive measure without vanishing identically. Since $M = M_0 \subset \bigcup_{a\in\Gamma} M_a$, E must have full measure. Consequently

(iii) $$\bigcup_{a>-\infty} M_a \text{ is dense in } L^2(\sigma).$$

Applying 10.1 and 10.2 to the doubly invariant subspace $\bigcap_{a<\infty} M_a$, we find also that

(iv) $$\bigcap_{a<\infty} M_a = \{0\}.$$

Let P_a be the orthogonal projection of $L^2(\sigma)$ onto M_a. If $a,b \in \Gamma$, then

$$P_b(\chi_a f) = \chi_a P_{b-a}(f), \qquad f \in L^2(\sigma).$$

Since the M_b are continuous from the left, this relation continues to hold for all real b. Hence

(v) $$P_b V_a = V_a P_{b-a}, \qquad a \in \Gamma, -\infty < b < \infty.$$

On account of (i), (iii), and (iv), the projections $\{I - P_a : -\infty < a < \infty\}$ form a resolution of the identity. The Fourier transform of this spectral resolution is the continuous unitary group $\{U_t\}$ operating in $L^2(\sigma)$, defined by

$$U_t = \int_{-\infty}^{\infty} e^{ita} d(I - P_a) = -\int_{-\infty}^{\infty} e^{ita} dP_a.$$

If $f \in L^2(\sigma)$, then

$$U_t V_a f = -\int_{-\infty}^{\infty} e^{itb} d(P_b V_a f)$$

$$= -\int_{-\infty}^{\infty} e^{itb} d(V_a P_{b-a} f)$$

$$= -e^{ita} V_a \int_{-\infty}^{\infty} e^{it(b-a)} dP_{b-a} f$$

$$= e^{ita} V_a U_t f.$$

Consequently,

(vi) $$U_t V_a = e^{ita} V_a U_t, \qquad a \in \Gamma,\ -\infty < t < \infty.$$

Conversely, let $\{U_t\}$ be a continuous unitary group in $L^2(\sigma)$ satisfying the commutation relation (vi). Then there are projections P_a, $-\infty < a < \infty$, such that $I - P_a$ is a resolution of the identity whose Fourier transform is the group $\{U_t\}$. The subspaces $M_a = P_a L^2(\sigma)$ satisfy (i), (iii), and (iv). The projections P_a are uniquely determined by the group $\{U_t\}$, once we specify that the M_a satisfy (ii). The commutation relation forces the projections to satisfy (v). Together (ii) and (v) show that $M = M_0$ is a left continuous invariant subspace. And $\{U_t\}$ is the unitary group that corresponds to M.

This completes the first step of the proof. Now we turn to the description of unitary groups satisfying the commutation relation.

Let B be a cocycle on G. Let U_t be the unitary operator on $L^2(\sigma)$ defined by

$$(U_t f)(x) = B(x,t) f(x + e_t).$$

In other words,

(vii) $$U_t = B(\cdot,t) T_t.$$

Applying 12.1, we see that $U_t f$ moves continuously in $L^2(\sigma)$, for each $f \in L^2(\sigma)$. The cocycle identity shows that $U_{s+t} = U_s U_t$. Consequently $\{U_t\}$ is a strongly continuous group. One verifies that the U_t satisfy the commutation relation with the V_a. Moreover, distinct cocycles give rise to distinct unitary groups.

Conversely, suppose that $\{U_t\}$ is a continuous unitary group satisfying the commutation relation (vi). Then

$$U_t(\chi_a) = e^{ita} \chi_a U_t(1) = U_t(1) T_t(\chi_a), \qquad a \in \Gamma.$$

It follows that

$$U_t(f) = U_t(1) T_t(f)$$

for all linear combinations f of the χ_a, and hence for all $f \in L^2(\sigma)$. That is,

$$U_t = U_t(1) \cdot T_t.$$

Since U_t is unitary, we can choose $U_t(1)$ to be a unitary Borel function, for each fixed t. Since $U_t(1)$ moves continuously in $L^2(\sigma)$ with t, 12.2 produces a unitary Borel function A on $G \times R$ such that for each t,

$$(U_t 1)(x) = A(x,t), \qquad \text{a.a. } x \in G.$$

Applying the group relation $U_{s+t} = U_s U_t$ to the function 1, we obtain for each fixed s and t,

$$A(x + e_s,t) = \overline{A(x,s)}A(x,s + t), \qquad \text{a.a } x \in G.$$

By 12.3, there is a cocycle B which coincides almost everywhere with A. The operators $B(\cdot,t)T_t$ coincide with the unitary group U_t for almost all t. So the continuous unitary groups must coincide.

This shows that (vii) yields a bijective correspondence between cocycles on G and continuous unitary groups $\{U_t\}$ satisfying (vi). The second step of the proof is complete.

For the third step of the proof, let M be a left continuous invariant subspace of $L^2(\sigma)$, let $\{U_t\}$ be its unitary group, and let B be the cocycle satisfying (vii). We will show that $M = \bar{B}H^2(\sigma)$.

Now $f \in M$ if and only if $P_0 f = f$. This occurs if and only if $(I - P_a)f = 0$ for all $a < 0$. This happens if and only if for all $g \in L^2(\sigma)$, the measure $d\nu(a) = d((I - P_a)f,g)$ is supported on $[0,\infty)$. By 12.5, this occurs if and only if

$$(U_t f,g) = \int_{-\infty}^{\infty} e^{ita} d((I - P_a)f,g)$$

belongs to $H^2(dt/(1 + t^2))$ for all $g \in L^2(\sigma)$. By 12.4, this occurs if and only if

$$\begin{aligned} 0 &= \int_{-\infty}^{\infty} (U_t f,g)F(t)\frac{dt}{1 + t^2} \\ &= \int_G \overline{g(x)} \int_{-\infty}^{\infty} B(x,t)f(x + e_t)F(t)\frac{dt}{1 + t^2}\,d\sigma(x) \end{aligned}$$

for all $g \in L^2(\sigma)$ and all $F \in H_i^2(dt/(1 + t^2))$. This happens if and only if for each $F \in H_i^2(dt/(1 + t^2))$,

$$\int_{-\infty}^{\infty} B(x,t)f(x + e_t)F(t)\frac{dt}{1 + t^2} = 0, \qquad \text{a.a. } x \in G.$$

The null subset of G involved in this last condition depends on F. However, $H_i^2(dt/(1 + t^2))$ is separable, and the integrals vary continuously with $F \in H_i^2(dt/(1 + t^2))$. Hence this condition will be satisfied if and only if

$$\int_{-\infty}^{\infty} B(x,t)f(x + e_t)F(t)\frac{dt}{1 + t^2} = 0 \quad \text{for all} \quad F \in H_i^2\Big(\frac{dt}{1 + t^2}\Big),$$

except for x belonging to some countable union of null sets. By 12.4, this occurs if and only if $B(x,t)f(x + e_t) \in H^2(dt/(1 + t^2))$ for almost all $x \in G$, that is, if and only if $f \in \bar{B}H^2(\sigma)$. That completes the proof.

NOTES

This chapter is based primarily on the series of papers of Helson and Lowdenslager [1, 2, 3] and Helson [2]. Most of the results of sections 7–10 and 12 are taken from this brilliant collection. The study of analytic almost periodic functions from the point of view of uniform algebras was initiated by Arens and Singer [2]. Theorem 2.3 is due to Hoffman and Singer [1], and 2.4 is due to Kaufman [1]. Bochner's theorem is in Bochner [1], and this proof is in Helson and Lowdenslager [1]. The deLeeuw-Glicksberg theorem is in deLeeuw and Glicksberg [1], this proof being due to Mandrekar and Nadkarni [1]. The local product decomposition was introduced by Hoffman [1]. The pieces of 7.8 were known to Forelli and Yale, before V.9.5 appeared. Theorem 9.4 is due to Arens [1], and this proof is from Helson and Lowdenslager [2]. Section 11 comes from Gamelin [3].

EXERCISES

1. The discrete abelian group Γ can be totally ordered if and only if Γ has no elements of finite order (except the identity). This occurs if and only if $\hat{\Gamma} = G$ is connected. *Hint:* If G is not connected, then G has a proper open-closed subgroup.

2. If Γ_+ orders Γ, then the maximal sets of antisymmetry of A_Γ are the cosets of the subgroup $[\Gamma_+ \cap (-\Gamma_+)]^\perp$ of G.

3. If Γ is totally ordered, and $\phi \in M_{A_\Gamma}$, then $\phi = \rho\phi_x$, where ϕ_x is the evaluation homomorphism at some $x \in G$, and $\rho \in M_{A_\Gamma}$ satisfies $\rho(\chi) \geq 0$ for all $\chi \in \Gamma_+$. This "polar decomposition" is unique if and only if the ordering of Γ is archimedean.

4. Let G be the torus T^2, and let $\Gamma = Z^2$ have the lexicographic ordering, that is, define Γ_+ to be the pairs of integers (m,n) such that either $n > 0$, or $n = 0$ and $m \geq 0$.

(a) Identify M_A.
(b) Identify the parts of A.
(c) A is triply generated. *Hint:* Take $g \in C(b\Delta)$ such that $P(\Delta)g$ is dense in the functions in $C(b\Delta)$ which vanish at $z = 1$, and consider z, w, and $g(z)w$.

5. If ν is a non-zero complex measure on R such that $\int_{-\infty}^{\infty} e^{isx}\,d\nu(x) = 0$ for $s \leq 0$, then ν is mutually absolutely continuous with respect to dx.

6. Let A be the bicylinder algebra on the torus T^2.

(a) If $a \geq 0$, and if $f \in A$ vanishes on a subset of the line $\{(e^{i\theta}, e^{i\psi}): \psi = a\theta, -\infty < \theta < \infty\}$ of positive linear measure, then $f = 0$.
(b) If $a < 0$, then every compact subset of the line $\{(e^{i\theta}, e^{i\psi}): \psi = a\theta, -\infty < \theta < \infty\}$ is a peak interpolation set for A.

7. Let Γ be a subgroup of the reals, with the discrete topology, and let $A = A_\Gamma$. The following are equivalent, for a closed G_δ-set $E \subseteq G$.

(i) E is a peak interpolation set.
(ii) E meets every line $x + R$ in a set of linear measure zero.
(iii) There is $f \in A$ which vanishes precisely on E (considered as a function on M_A).
(iv) There is a non-zero $f \in A$ such that $f^{-1}(0) \cap G \supseteq E$.

8. Let p be a fixed prime number, and let Γ be the set of numbers of the form

$2\pi m/p^n$, where m and n are integers. Each p-adic integer $x = \sum_{j=0}^{\infty} a_j p^j$ determines a character of Γ via the pairing $\langle x, 2\pi m/p^n \rangle = \exp(2\pi i m x/p^n)$. This correspondence establishes an isomorphism of the group of p-adic integers and the compact group K of section 6.

9. Let $w \in L^1$, $w \geq 0$, and let M_t be the closure of $\chi_t A$ in $L^2(wd\sigma)$, $t \in \Gamma$. There exists $g \in H_0^2(\sigma)$ such that $|g|^2 = w$ and $A_0 g$ is dense in $H_0^2(\sigma)$ if and only if $\bigcup_{t>0} M_t$ is dense in M_0, while $\bigcap_{t<0} M_t \neq M_0$.

10. The group of cocycles modulo coboundaries is completely divisible.

11. Let B and C be cocycles. Products of functions in $BH^\infty(\sigma)$ with functions in $CH^\infty(\sigma)$ generate an invariant subspace of $L^2(\sigma)$ (possibly continuous from the wrong side) whose cocycle is BC.

12. A cocycle B is **analytic** if B belongs to $H^\infty(\sigma)$ on almost all lines.

(a) If B and C are cocycles, then $BH^2(\sigma) \supseteq CH^2(\sigma)$ if and only if $\bar{B}C$ is analytic.

(b) Part (a) is true if we relax the measurability assumption on C, and assume only that $t \rightarrow C(x,t)$ is measurable for each $x \in G$.

(c) If the invariant subspace M_f generated by $f \in L^2(\sigma)$ is not doubly invariant, then $M_f = BH^2(\sigma)$, where B is that uniquely determined cocycle such that $\bar{B}f$ is an outer function in H^2 of almost all lines. *Hint:* Use (b).

CHAPTER VIII

ANALYTIC CAPACITY AND RATIONAL APPROXIMATION

The main goal of this chapter is to prove Vitushkin's theorems which reduce the problem of rational approximation to the study of analytic capacity. Vitushkin's theorem characterizing those compact sets K for which $A(K) = R(K)$ is given in section 8. Along the way, a constructive proof of Mergelyan's theorem is obtained in section 7.

Vitushkin's characterization of those compact sets K for which $R(K) = C(K)$ is given in section 5. The reader familiar which Chapter II can pass to section 5 immediately after reading the definition of analytic capacity and theorem 4.1. In turn, the results of section 5 should provide sufficient motivation for a more detailed study of analytic capacity.

1. Analytic Capacity

Let K be a compact subset of the complex plane. By $\boldsymbol{\Omega}(\boldsymbol{K})$ we will denote the open subset of the Riemann sphere S^2 obtained by adjoining the point at infinity to the unbounded component of K^c. In other words,

$$\Omega(K) = \hat{K}^c \cup \{\infty\}.$$

A function f is **admissible for K** if f is defined and analytic on $\Omega(K)$, and if f satisfies

$$\|f\|_{\Omega(K)} \leq 1$$
$$f(\infty) = 0.$$

A function is admissible for an arbitrary set E if it is admissible for some compact subset of E. The set of admissible functions for a set E will be denoted by $\mathscr{A}(E)$.

Any function f which is analytic at ∞ can be represented near ∞ as a Laurent series in descending powers of z,

$$f(z) = a_0 + \frac{a_1}{z} + \frac{a_2}{z^2} + \dots.$$

Here $a_0 = f(\infty)$. We define

$$f'(\infty) = a_1 = \lim_{z\to\infty} z[f(z) - f(\infty)].$$

The **analytic capacity** of a set E is defined by

$$\gamma(E) = \sup\{|f'(\infty)| : f \in \mathscr{A}(E)\}.$$

Some of the immediate properties of analytic capacity are contained in the following lemmas.

1.1 Lemma. The analytic capacity is a monotone set function, that is, if $E \subseteq F$, then $\gamma(E) \leq \gamma(F)$.

1.2 Lemma. If z_0 and a are complex, then

$$\gamma(z_0 + aE) = |a|\gamma(E).$$

1.3 Lemma. The analytic capacity of a compact set K depends only on the outer boundary of K. That is,

$$\gamma(K) = \gamma(bK) = \gamma(\hat{K}) = \gamma(b\hat{K}).$$

Proof. $\Omega(K) = \Omega(bK) = \Omega(\hat{K}) = \Omega(b\hat{K})$.

1.4 Theorem. Let K be a compact connected subset of the complex plane which contains more than one point. Let g be the conformal map of $\Omega(K)$ onto the interior of the unit disc, satisfying $g(\infty) = 0$ and $g'(\infty) > 0$. Then $\gamma(K) = g'(\infty)$.

Proof. Since g is admissible, $g'(\infty) \leq \gamma(K)$. If h is also admissible, then $h \circ g^{-1}$ is an analytic function on $\{|w| < 1\}$ satisfying $(h \circ g^{-1})(0) = 0$ and $|h \circ g^{-1}| \leq 1$. By Schwarz's lemma, $|(h \circ g^{-1})(w)| \leq |w|$, that is, $|h(z)| \leq |g(z)|$ for all $z \in \Omega(K)$. Consequently $|h'(\infty)| \leq g'(\infty)$, and $\gamma(K) \leq g'(\infty)$.

1.5 Corollary. The analytic capacity of a disc is its radius. The analytic capacity of a straight line segment of length L is $L/4$.

Proof. The mapping function for the exterior of the disc $\Delta(z_0;r)$ with center z_0 and radius r is

$$g(z) = \frac{r}{z - z_0}.$$

Since $g'(\infty) = r$, we obtain $\gamma(\Delta(z_0;r)) = r$.

The second assertion is obtained by considering the function

$$z = \left(w + \frac{1}{w}\right)\frac{L}{4},$$

which maps the interior of the unit disc $\{|w| < 1\}$ onto the Riemann sphere with the line segment $K = [-L/2, L/2]$ removed. The analytic capacity of K is

$$\gamma(K) = \lim_{z\to\infty} zw(z) = \lim_{z\to\infty} (w(z)^2 + 1)\frac{L}{4} = \frac{L}{4}.$$

Since analytic capacity is translation invariant, the same result holds for any straight line segment.

1.6 Theorem. Let K be compact. There is a unique admissible function g for K satisfying $g'(\infty) = \gamma(K)$.

The function g will be called the **Ahlfors function** for K. If K is connected and has more than one point, then the Ahlfors function for K is the normalized conformal map of $\Omega(K)$ onto the interior of the unit disc.

Proof of 1.6. The existence of such a function is always assured by the fact that $\mathscr{A}(K)$ is a normal family of analytic functions on $\Omega(K)$.

Suppose that $g_0, g_1 \in \mathscr{A}(K)$ both have derivatives at ∞ equal to $\gamma(K)$. Then the function $g = (g_0 + g_1)/2 \in \mathscr{A}(K)$ also satisfies $g'(\infty) = \gamma(K)$. Let $h = (g - g_0)/2$. Then $g_1 = g + h$, and $g_2 = g - h$. It suffices to show that $h = 0$.

From the inequality

$$|g \pm h|^2 = |g|^2 + |h|^2 \pm 2\,\mathrm{Re}\,(g\bar{h}) \leq 1$$

we obtain

$$|g|^2 + |h|^2 \leq 1.$$

Let $k = h^2/2$. Then

$$|k| \leq \frac{1 - |g|^2}{2} = \frac{(1 - |g|)(1 + |g|)}{2}$$
$$\leq 1 - |g|,$$

so

$$|g| + |k| \leq 1.$$

If k is not zero, we can write

$$k(z) = \frac{a_n}{z^n} + \frac{a_{n+1}}{z^{n+1}} + \cdots,$$

where $a_n \neq 0$. Since $h(\infty) = 0 = h'(\infty)$, also $k(\infty) = 0 = k'(\infty)$, and $n > 1$. Consequently for $\epsilon > 0$ small enough, $\epsilon |a_n||z|^{n-1} \leq 1$ in a neighborhood of K. Let

$$f = g + \epsilon \bar{a}_n z^{n-1} k.$$

Then $f(\infty) = 0$. Since $|f| \leq |g| + |k| \leq 1$ in a neighborhood of K, also $|f| \leq 1$ on $\Omega(K)$, and f is admissible. However,

$$f'(\infty) = g'(\infty) + \epsilon |a_n|^2 > \gamma(K).$$

This contradiction shows that $k = 0$. Hence also $h = 0$, and the theorem is established.

Now we turn to some continuity properties of analytic capacity.

1.7 Theorem. Let $\{K_n\}_{n=1}^{\infty}$ be a sequence of compact sets such that $K_n \supseteq K_{n+1}$ and $\bigcap_{n=1}^{\infty} K_n = K$. Then the Ahlfors functions g_n of K_n converge uniformly on compact subsets of $\Omega(K)$ to the Ahlfors function g of K. In particular, $\gamma(K_n) \to \gamma(K)$ as $n \to \infty$.

Proof. Since the g_n are bounded, they form a normal family of analytic functions. Let h be any uniform limit on compacta of a subsequence of the g_n. Then h is admissible for K, so $h'(\infty) \leq \gamma(K)$. On the other hand, $\gamma(K) \leq \liminf \gamma(K_n) = \liminf g_n'(\infty) \leq h'(\infty)$. Hence $h'(\infty) = \gamma(K)$, and $h = g$ is the Ahlfors function for K. It follows that the sequence $\{g_n\}$ itself must converge to g.

1.8 Corollary. If K is compact, then

$$\gamma(K) = \inf\{\gamma(U)\colon U \text{ open, } U \supseteq K\}.$$

The compact set K is a **Painlevé null set** if $\Omega(K)$ supports no non-constant bounded analytic function. A Painlevé null set coincides with its outer boundary. Painlevé null sets are totally disconnected, since any connected component J of K consisting of more than one point would give rise to a non-zero bounded analytic function on $\Omega(K)$, namely, the conformal map of $\Omega(J)$ onto the interior of the unit disc.

1.9 Theorem. The Painlevé null sets coincide with the compact sets of zero analytic capcity.

Proof: If K is a Painlevé null set, then evidently $\gamma(K) = 0$. On the other hand, suppose that $\Omega(K)$ supports a non-constant bounded analytic function f. If $\epsilon > 0$ is small, and if n is the order of $f - f(\infty)$ at ∞, then $\epsilon z^{n-1}[f - f(\infty)]$ is admissible for K and has a non-zero derivative at ∞. So $\gamma(K) > 0$. That proves the theorem.

Compact sets of analytic capacity zero can also be characterized as the compact sets which are removable singularities for bounded analytic functions.

1.10 Theorem. Let D be open, and let K be a compact subset of D. Every function which is bounded and analytic on $D \backslash K$ extends to be analytic on D if and only if $\gamma(K) = 0$.

Proof. If the Ahlfors function g of K extends analytically across K, then $g = 0$, and $\gamma(K) = g'(\infty) = 0$.

Conversely, suppose $\gamma(K) = 0$. Let f be bounded and analytic on $D \backslash K$. Using the Cauchy integral formula, we can write $f = f_1 + f_2$, where f_1 is bounded and analytic off K, and f_2 is analytic on D. By 1.9, $f_1 = 0$, and so f_2 is the desired analytic extension of f.

2. Estimates of Analytic Capacity

We have already calculated (in 1.5) the analytic capacity of discs and of straight line segments. Now we wish to show that the analytic capacity of any compact continuum is comparable to its diameter.

2.1 Theorem. If K is compact and connected, then

$$\gamma(K) \leq \text{diam}\,(K) \leq 4\gamma(K).$$

Proof. Any bounded set E is contained in a disc D of radius diam (E). So the inequality $\gamma(E) \leq \gamma(D) = \text{diam}\,(E)$ holds for arbitrary sets E.

We can assume that K contains at least two points. Let g be the Ahlfors function of K, and let h be the inverse of g, mapping the interior of the unit disc onto $\Omega(K)$. Let $z_1 \in K$ be fixed. Define $f(z) = \gamma(K)/(h(z) - z_1)$. Then f is univalent on the unit disc, $f(0) = 0$, and $f'(0) = 1$. If $z_2 \in K$, then $\gamma(K)/(z_2 - z_1)$ does not belong to the range of f. By the $\frac{1}{4}$-theorem of Koebe, $\gamma(K)/|z_2 - z_1| \geq \frac{1}{4}$, or $|z_2 - z_1| \leq 4\gamma(K)$. Since $z_1, z_2 \in K$ are arbitrary, we obtain diam $(K) \leq 4\gamma(K)$.

Pommerenke has shown that the analytic capacity of any compact linear set of length L is $L/4$. We will content ourselves with the following estimate.

2.2 Theorem. Let K be a compact subset of a line, and let L be the length of K. Then

$$\frac{L}{4} \leq \gamma(K) \leq \frac{L}{\pi}.$$

Proof. We can assume that K lies on the real axis. Let

$$f(z) = \frac{1}{2} \int_K \frac{ds}{z - s}.$$

If $z = x + iy$, where $y \neq 0$, then

$$\begin{aligned} |\text{Im}\, f(z)| &= \frac{|y|}{2} \int_K \frac{ds}{(x - s)^2 + y^2} \\ &< \frac{|y|}{2} \int_{-\infty}^{\infty} \frac{ds}{s^2 + y^2} \\ &= \frac{\pi}{2}. \end{aligned}$$

Consequently the range of f lies in the vertical strip $-\pi/2 < \text{Im}\,(w) < \pi/2$. If we compose f with the conformal map $w \to (e^w - 1)/(e^w + 1)$ of this vertical strip onto the interior of the unit disc, we obtain a function

$$h(z) = \frac{e^{f(z)} - 1}{e^{f(z)} + 1}$$

which satisfies $|h| \leq 1$. Also, $h(\infty) = 0$, so h is admissible. Since $f'(\infty) = L/2$, we obtain $h'(\infty) = L/4$. Consequently $\gamma(K) \geq L/4$.

For the other inequality, let g be the Ahlfors function of K. Given $\epsilon > 0$, there exists a finite union Γ of smooth closed Jordan curves in $\Omega(K)$, such that Γ surrounds K in the usual sense of contour integration, and such that the length of Γ does not exceed $2L + \epsilon$. Then

$$\gamma(K) = g'(\infty) = \frac{1}{2\pi i}\int_\Gamma g(z)\,dz \leq \frac{2L+\epsilon}{2\pi}.$$

Letting ϵ tend to zero, we obtain $\gamma(K) \leq L/\pi$, as was required.

The analytic capacity is not commensurate with area, as we have just seen. However, it is not difficult to show that for any bounded measurable set E, the inequality

$$\text{Area}\,(E) \leq \pi\gamma(E)^2$$

obtains (cf. Ahlfors and Beurling [1]). This inequality is sharp, since equality holds when E is a disc. Again we content ourselves with a slightly cruder estimate.

2.3 Theorem. If E is a bounded measurable subset of the complex plane, then

$$\text{Area}\,(E) \leq 4\pi\gamma(E)^2.$$

Proof. It suffices to prove the theorem for compact sets K. For this, we consider the convolution

$$f(\zeta) = \iint_K \frac{dxdy}{z-\zeta}$$

of the locally integrable function $1/z$ and the characteristic function of K. The function f is continuous on the Riemann sphere, and f is analytic off K. Also, $f(\infty) = 0$, and $f'(\infty) = \text{Area}\,(K)$.

Let $a = (\text{Area}\,(K)/\pi)^{1/2}$. For ζ fixed, the disc $\Delta(\zeta;a)$ has the same area as K. Since $1/|z-\zeta|$ is a decreasing function of $|z-\zeta|$, we obtain

$$|f(\zeta)| \leq \iint_K \frac{dxdy}{|z-\zeta|} \leq \iint_{\Delta(\zeta;a)} \frac{dxdy}{|z-\zeta|} = 2\pi a.$$

It follows that $f/2\pi a$ is an admissible function, and

$$\gamma(K) \geq \frac{f'(\infty)}{2\pi a} = \frac{\sqrt{\text{Area}\,(K)}}{2\sqrt{\pi}}.$$

This is the desired inequality.

Now we turn to estimates for admissible functions in terms of analytic capacity.

2.4 Theorem. Let K be compact, and let f be admissible for K. If $z \in \Omega(K)$, then

$$|f(z)| \leq \frac{\gamma(K)}{d(z,K)},$$

$$|f'(z)| \leq \frac{4\gamma(K)}{d(z,K)^2},$$

where $d(z, K)$ is the distance from z to K.

Proof. For z fixed, and $a < d(z, K)$, let

$$h(\zeta) = \left(\frac{a}{\zeta - z}\right)\left(\frac{f(\zeta) - f(z)}{1 - \overline{f(z)}f(\zeta)}\right).$$

Now $|f(\zeta) - f(z)|/|1 - \overline{f(z)}f(\zeta)| < 1$ for all $\zeta \in \Omega(K)$. So $|h(\zeta)| < 1$ whenever $\zeta \in \Omega(K)$ satisfies $|\zeta - z| \geq a$. Since h is analytic in the disc $\{|\zeta - z| \leq a\}$, we obtain $|h(\zeta)| < 1$ for all $\zeta \in \Omega(K)$, and $h \in \mathscr{A}(K)$. Hence $|h'(\infty) = a|f(z)| \leq \gamma(K)$. Letting a tend to $d(z, K)$, we obtain the first estimate.

Again fix $z \in \Omega(K)$, and let $d = d(z, K)$. By the first estimate, $|f(\zeta)| \leq 2\gamma(K)/d$ when $|\zeta - z| = d/2$. Hence

$$|f'(z)| = \left|\frac{1}{2\pi i}\int_{|\zeta - z| = d/2} \frac{f(\zeta)}{(\zeta - z)^2}\, d\zeta\right| \leq \frac{4\gamma(K)}{d^2}.$$

2.5 Theorem. Let K be a compact subset of the disc $\Delta(0; R)$, and let $f \in \mathscr{A}(K)$. If

$$f(z) = \frac{f'(\infty)}{z} + \frac{a_2}{z^2} + \frac{a_3}{z^3} + \cdots$$

represents f near ∞, then

$$|a_n| < enR^{n-1}\gamma(K), \qquad n \geq 2.$$

Proof. Fix R_0 so that $R_0 > R$. By 2.4, $|f(z)| \leq \gamma(K)/(R_0 - R)$ if $|z| = R_0$. Hence

$$|a_n| = \left|\frac{1}{2\pi i}\int_{|z| = R_0} f(z) z^{n-1}\, dz\right| \leq \frac{\gamma(K) R_0^n}{R_0 - R}.$$

Setting $R_0 = nR/(n - 1)$, and noting that $(n/(n - 1))^{n-1} \leq e$, we obtain the desired inequality.

Now we wish to prove a more technical result, which will be used in the proof of Vitushkin's theorem. First we need a lemma.

2.6 Lemma. Let $p > 0$ be an integer. Let $\{c_{n,k}: 1 \leq k \leq pn, 1 \leq n < \infty\}$ be a double sequence such that $0 \leq c_{n,k} \leq 1$. Then

$$\left(\sum_{n,k} \frac{c_{n,k}}{n}\right)^2 \leq 2p \sum_{n,k} c_{n,k}.$$

Proof. Let $a_n = \sum_{k=1}^{pn} c_{nk}/n$. Using the estimate $0 \le a_n \le p$, one verifies easily by induction that

$$\left(\sum_{n=1}^{N} a_n\right)^2 \le 2p \sum_{n=1}^{N} na_n, \qquad N \ge 1.$$

Now let N tend to ∞.

2.7 Theorem. Let $p > 0$ be an integer, and let $\{E_j\}$ be a family of subsets of the bounded set E, such that every disc $\Delta(z;\gamma(E))$ meets at most p of the sets $\{E_j\}$. Then

$$\sum \gamma(E_j) \le 100p\gamma(E).$$

If f_j is an admissible function for E_j, then

$$\sum |f_j(z)| \le 100p,$$

the series being summed over those indices j for which f_j is defined at z.

Proof. Fix z_0. The annulus $\{n\gamma(E) \le |z - z_0| \le (n+1)\gamma(E)\}$ can be covered by $20n$ discs of radius $\gamma(E)$. Hence there are at most $20np$ of the sets $\{E_j\}$ which meet $\Delta(z_0; (n+1)\gamma(E))$ but not $\Delta(z_0; n\gamma(E))$. We renumber these with double indices $E_{n1}, E_{n2}, \ldots$. Then

$$n \le \frac{d(z_0;E_{nk})}{\gamma(E)} \le n + 1.$$

If f_{nk} is admissible for E_{nk}, then

$$|f_{nk}(z_0)| \le \frac{\gamma(E_{nk})}{d(z_0,E_{nk})} \le \frac{\gamma(E_{nk})}{n\gamma(E)},$$

by 2.5. Using the estimate $|f_{0k}| \le 1$, and 2.6, we obtain

$$\sum |f_{nk}(z_0)| \le p + \sum \frac{\gamma(E_{nk})}{n\gamma(E)}$$
$$\le p + \left[40p \sum \frac{\gamma(E_{nk})}{\gamma(E)}\right]^{1/2}.$$

Let $a = \sum \gamma(E_j)/\gamma(E)$. The estimate becomes

$$\sum |f_j(z_0)| \le p + \sqrt{40pa}.$$

If h_j is an admissible function for E_j such that $h_j'(\infty) = \gamma(E_j)/2$, then $h = \sum h_j/(p + \sqrt{40pa})$ is admissible for E. So

$$\sum \frac{\gamma(E_j)}{2(p + \sqrt{40pa})} = h'(\infty) \le \gamma(E),$$

or

$$a \le 2(p + \sqrt{40pa}).$$

A little elementary reasoning shows that $a \le 100p$. This proves the first inequality. Also,

$$\sum |f_j(z_0)| \leq p + \sqrt{4000p} \leq 100p,$$

which proves the second inequality.

3. Continuous Analytic Capacity

Let E be a subset of the complex plane. By $\mathscr{AC}(E)$ will be denoted the family of functions f such that f is continuous on the Riemann sphere S^2, f is analytic off some compact subset of E, $\|f\|_{S^2} \leq 1$, and $f(\infty) = 0$. The **continuous analytic capacity** of E is

$$\alpha(E) = \sup \{|f'(\infty)|: f \in \mathscr{AC}(E)\}.$$

Evidently

$$\alpha(E) = \sup \{\alpha(K): K \text{ compact}, K \subseteq E\}.$$

The continuous analytic capacity α enjoys some of the properties of the analytic capacity γ. For instance, α is a monotone set function which satisfies

$$\alpha(z_0 + aE) = |a|\,\alpha(E).$$

However, the continuous analytic capacity of a compact set K no longer depends only on the outer boundary of K. For instance, if D is any disc, then $\alpha(D)$ is the radius of D, while $\alpha(bD) = 0$.

3.1 Lemma. For any set E, it is true that

$$\alpha(E) \leq \gamma(E).$$

If U is open, then

$$\alpha(U) = \gamma(U).$$

Proof. The first statement reflects the fact that $\mathscr{AC}(E) \subseteq \mathscr{A}(E)$. The second statement reflects the fact that for any compact $K \subseteq U$, there is a compact set $J \subseteq U$ such that every function in $\mathscr{A}(K)$ agrees with some function in $\mathscr{AC}(J)$ near ∞. In fact, any compact set $J \subseteq U$ satisfying $K \subseteq \operatorname{int} J$ will do.

The continuous analytic capacity of any line segment is zero. In particular, if K is a line segment, then

$$\begin{aligned}\gamma(K) &= \inf \{\gamma(U): U \text{ open}, U \supseteq K\} \\ &= \inf \{\alpha(U): U \text{ open}, U \supseteq K\} \\ &\neq \alpha(K).\end{aligned}$$

So α does not share the continuity properties 1.7 and 1.8 of γ.

The estimates 2.1 and 2.2 are no longer valid for α. However, the function constructed in the proof of 2.3 actually belongs to $\mathscr{AC}(K)$. And so we obtain the following sharpened version of 2.3.

3.2 Theorem. If E is a bounded measurable subset of the complex plane, then

$$\text{Area}\,(E) \leq 4\pi\alpha(E)^2.$$

Theorems 2.4 and 2.5 remain valid when $\gamma(K)$ is replaced by $\alpha(K)$, providing $f \in \mathscr{AC}(K)$. Theorem 2.7 also remains true, with γ replaced by α.

4. Peaking Criteria

Roughly speaking, a point z_0 is a peak point for $R(K)$ if K^c is sufficiently thick at z_0, and z_0 is a peak point for $A(K)$ if $(\text{int } K)^c$ is sufficiently thick at z_0. In this section we give some explicit criteria to formalize this assertion.

4.1 Theorem (Curtis's Criterion). Let K be compact, and let $z_0 \in K$. If

$$\limsup_{r\to 0+} \frac{\gamma(\Delta(z_0;r)\backslash K)}{r} > 0,$$

then z_0 is a peak point of $R(K)$. If

$$\limsup_{r\to 0+} \frac{\alpha(\Delta(z_0;r)\backslash \text{int } K)}{r} > 0,$$

then z_0 is a peak point of $A(K)$.

Proof. We will prove only the first assertion. The proof of the second follows exactly the same pattern.

Choose $r_n \to 0+$ and $b > 0$ so that $\gamma(\Delta(z_0; r_n)\backslash K) > br_n$. From the definition of analytic capacity, there is a compact set $E_n \subseteq \Delta(z_0; r_n)\backslash K$ and a function f_n analytic in $\Omega(E_n)$ such that $|f_n| \leq 1$, $f_n(\infty) = 0$, and $f_n'(\infty) > br_n$. Define an analytic function g_n on $S^2\backslash E_n$ by setting

$$g_n(z) = \begin{cases} \dfrac{(z - z_0)f_n(z)}{f_n'(\infty)}, & z \in \Omega(E_n) \\ \quad 0 \quad , & z \in E_n^c\backslash\Omega(E_n). \end{cases}$$

For $z \in E_n^c$ satisfying $|z - z_0| \leq r_n$ we have $|g_n(z)| \leq 1/b$. By the maximum modulus principle, $|g_n| \leq 1/b$ on $S^2\backslash E_n$. In particular, the g_n form a normal family of analytic functions. Passing to a subsequence, we can assume that the g_n converge to some function g uniformly on compact subsets of $S^2\backslash\{z_0\}$. Since g is analytic and bounded, g is constant. Since $g_n(\infty) = 1$, $g \equiv 1$.

Now let $h_n = 1 - g_n$. Then h_n is a bounded sequence in $R(K)$, $h_n(z_0) = 1$, and h_n tends uniformly to zero on compact subsets of $K\backslash\{z_0\}$. By II.11.1, z_0 is a peak point of $R(K)$. This proves the theorem.

In view of 2.3 and 3.2, we obtain the following corollary.

4.2 Corollary. Let K be compact, and let $z_0 \in K$. If

$$\limsup_{r\to 0+} \frac{\text{Area}\,(\Delta(z_0;r)\backslash K)}{r^2} > 0,$$

then z_0 is a peak point for $R(K)$. If

$$\limsup_{r\to 0+} \frac{\text{Area}\,(\Delta(z_0;r)\backslash \text{int}\, K)}{r^2} > 0,$$

then z_0 is a peak point for $A(K)$.

From the estimate 2.1, we obtain the following criterion.

4.3 Corollary (Gonchar's Criterion). Let K be compact, and let $z_0 \in K$. For each $r > 0$, let $d(r)$ denote the supremum of the diameters of the components of int $\Delta(z_0; r)\backslash K$. If

$$\limsup_{r\to 0+} \frac{d(r)}{r} > 0,$$

then z_0 is a peak point for $R(K)$.

4.4 Corollary. Let K be compact. Suppose there is $d > 0$ such that every component of K^c has diameter larger than d. Then every point of bK is a peak point of $R(K)$.

Melnikov [1] has given a necessary and sufficient criterion for a point $z_0 \in K$ to be a peak point for $R(K)$. His criterion is identical in form to the celebrated necessary and sufficient criterion of Wiener that a boundary point of a domain in Euclidean 3-space be a regular point for the dirichlet problem. The criterion, together with its analogue for $A(K)$, is as follows.

4.5 Theorem (Melnikov's Criterion). Let K be compact, and let $z_0 \in K$. Let $0 < a < 1$, and let E_n be the annulus $\{a^{n+1} < |z - z_0| < a^n\}$. Then z_0 is a peak point for $R(K)$ if and only if

$$\sum_{n=1}^{\infty} \frac{\gamma(E_n\backslash K)}{a^n} = +\infty.$$

Also, z_0 is a peak point for $A(K)$ if and only if

$$\sum_{n=1}^{\infty} \frac{\alpha(E_n\backslash \text{int}\, K)}{a^n} = +\infty .$$

Proof. The proof of the necessity of Melnikov's criterion will be postponed to section 12. It follows immediately from 12.7.

For the sufficiency, we treat only the algebra $A(K)$. The proof for $R(K)$ is identical.

So assume that $\sum \alpha(E_n\backslash \text{int}\, K)/a^n = +\infty$, and assume, for simplicity, that $z_0 = 0$. We must show that 0 is a peak point for $A(K)$.

Choose $f_n \in \mathscr{AC}(E_n \backslash \text{int } K)$ such that $f_n'(\infty) \geq \alpha(E_n \backslash \text{int } K)/2$. Then

$$\sum_{n=1}^{\infty} \frac{f_n'(\infty)}{a^n} = +\infty.$$

Since $f_n'(\infty) \leq \alpha(E_n \backslash \text{int } K) \leq \alpha(\Delta(0;a^n)) = a^n$, none of the terms of this series exceeds 1. Hence for each integer m, there is an integer $p = p(m)$ such that

$$1 \leq \sum_{n=m}^{p} \frac{f_n'(\infty)}{a^n} \leq 2.$$

Let

$$g_m(z) = z \sum_{n=m}^{p} \frac{f_n(z)}{a^n}.$$

Then g_m is continuous on S^2, g_m is analytic on int (K), g_m is analytic outside the disc $\Delta(0; a^m)$, $1 \leq g_m(\infty) \leq 2$, and $g_m(0) = 0$. We wish to show that the g_m are uniformly bounded.

Suppose $0 < |z| \leq 1$. Choose an integer j such that $z \in \bar{E}_j$, that is, such that $a^{j+1} \leq |z| \leq a^j$. If $n \neq j-1, j, j+1$, the distance from z to E_n is at least $a^{j+1} - a^{j+2}$. The analogue of 2.4 for continuous analytic capacity yields the estimate

$$|f_n(z)| \leq \frac{\alpha(E_n \backslash \text{int } K)}{a^{j+1} - a^{j+2}}, \qquad n \neq j-1, j, j+1.$$

Hence

$$\left| \frac{z f_n(z)}{a^n} \right| \leq \frac{\alpha(E_n \backslash \text{int } K)}{a^n(a - a^2)}$$
$$\leq \frac{2 f_n'(\infty)}{a^n(a - a^2)}, \qquad n \neq j-1, j, j+1.$$

Since f_{j-1}, f_j, and f_{j+1} are each bounded by 1, we have

$$\left| \frac{z f_n(z)}{a^n} \right| \leq \frac{a^j}{a^n} \leq \frac{1}{a}, \qquad n = j-1, j, j+1.$$

Summing these estimates, we obtain

$$|g_m(z)| \leq \frac{3}{a} + \sum_{n=m}^{p} \frac{2 f_n'(\infty)}{a^n(a - a^2)} \leq \frac{3}{a} + \frac{4}{a - a^2}.$$

By the maximum modulus principle, this estimate holds for all $z \in S^2$.

It follows that $\{g_m\}_{m=1}^{\infty}$ is a normal family on $S^2 \backslash \{0\}$. Let $\{g_{m_j}\}_{j=1}^{\infty}$ be a subsequence which converges uniformly on compact subsets of $S^2 \backslash \{0\}$ to g. The g is bounded and analytic, so g is constant. Since $1 \leq g_m(\infty) \leq 2$, also $1 \leq g(\infty) \leq 2$.

Now $\{1 - g_{m_j}/g\}_{j=1}^{\infty}$ is a bounded sequence in $A(K)$ which is 1 at $z = 0$, and which converges uniformly to zero on compact subsets of $K \backslash \{0\}$. By II.11.1, 0 is a peak point of $A(K)$.

5. Criteria for $R(K) = C(K)$

Using Curtis's criterion 4.1, together with some results from chapter II, we can reduce the problem of rational approximation for compact sets with no interior to a problem involving analytic capacity.

5.1 Theorem (Vitushkin's Theorem). The following assertions are equivalent, for a compact set K.

(i) $R(K) = C(K)$.

(ii) For every open set D, $\gamma(D\backslash K) = \gamma(D)$.

(iii) For $dxdy$-almost all $z \in K$, it is true that

$$\limsup_{r\to 0} \frac{\gamma(\Delta(z;r)\backslash K)}{r} > 0.$$

Proof. That (ii) implies (iii) follows from 1.5, which states that $\gamma(\Delta(z;r)) = r$.

That (iii) implies (i) follows from 4.1, together with Bishop's peak point criterion for rational approximation II.11.4.

That (i) implies (ii) follows from the localization theorem II.10.3. In fact, suppose $R(K) = C(K)$, and let D be an open set. We must show that $\gamma(D\backslash K) \geq \gamma(D)$.

Let $\epsilon > 0$. Choose f analytic in a neighborhood of $S^2\backslash D$, so that $\|f\|_{S^2\backslash D} < 1$ and $|f'(\infty)| > \gamma(D) - \epsilon$. By modifying f off a neighborhood of $S^2\backslash D$, and extending f to S^2, we can assume, in addition, that $f \in C(S^2)$ and that $\|f\|_{S^2} < 1$. Now $J = (S^2\backslash D) \cup K$ is a compact subset of S^2. Since f remains analytic in a neighborhood of $S^2\backslash D$, there is a neighborhood of U of $S^2\backslash D$ such that $f \in R(U \cap J)$. Since $R(K) = C(K)$ and $K \cap D = J \cap D$, also $f \in R(D \cap J)$. The open sets U and D cover J, so $f \in R(J)$, by the localization theorem II.10.3. Consequently f can be approximated uniformly on J by functions admissible for $D\backslash K$. So $\gamma(D\backslash K) \geq |f'(\infty)| \geq \gamma(D) - \epsilon$. Letting $\epsilon \to 0$, we obtain the desired result.

Vitushkin [1] shows that for any compact set K, and for almost all complex z, either

$$\lim_{r\to 0} \frac{\gamma(\Delta(z;r)\backslash K)}{r} = 1$$

or

$$\lim_{r\to 0} \frac{\gamma(\Delta(z;r)\backslash K)}{r^2} = 0.$$

Consequently the validity of the condition

$$\limsup_{r\to 0} \frac{\gamma(\Delta(z;r)\backslash K)}{r^2} > 0$$

at almost all points $z \in K$ is sufficient to guarantee that $R(K) = C(K)$.

This result is sharp, in the sense that r^2 cannot be replaced by $r^{2+\epsilon}$ for any $\epsilon > 0$. However, the validity of this condition at a single point of K does not necessarily imply that that point is a peak point of $R(K)$.

6. Analytic Diameter

To give Vitushkin's characterization of compact sets K for which $R(K) = A(K)$, we must study not only the coefficient of $(z - z_0)^{-1}$ but also the coefficient of $(z - z_0)^{-2}$ in the Laurent expansion of admissible functions. This leads to the concepts of analytic diameter and analytic center.

It z_0 is complex and f is analytic at ∞, we can express f near ∞ in the form

$$f(z) = f(\infty) + \frac{f'(\infty)}{z - z_0} + \frac{\beta(f, z_0)}{(z - z_0)^2} + \cdots.$$

If $R > 0$ is large, then

$$\beta(f,z_0) = \frac{1}{2\pi i}\int_{|z|=R} f(z)(z - z_0)\, dz.$$

A simple calculation shows that

$$(*) \qquad \beta(f,z_1) = \beta(f,z_0) + (z_0 - z_1)f'(\infty).$$

The coefficient estimate 2.5 shows that

$$(**) \qquad |\beta(f,z_0)| \le 6R\gamma(E)$$

whenever f is admissible for E, and $E \subseteq \Delta(z_0; R)$.

For each bounded set E such that $\gamma(E) > 0$, define

$$\beta(E,z) = \frac{\sup|\beta(f,z)|}{\gamma(E)},$$

where the supremum is taken over all admissible functions for E. If $\gamma(E) = 0$, we set $\beta(E,z) = 0$. It is clear that

$$\beta(E, z) = \sup\{\beta(K,z) \colon K \text{ compact}, K \subseteq E\}.$$

The estimate (**) shows that

$$\beta(E,z) \le 6 \operatorname{diam}(E), \qquad z \in E.$$

From (*) it follows that for admissible functions f,

$$|\beta(f,z_1)| \le \beta(E,z_0)\gamma(E) + |z_0 - z_1|\gamma(E).$$

Taking the supremum, we obtain

$$\beta(E,z_1) \le \beta(E, z_0) + |z_0 - z_1|.$$

Since z_0 and z_1 are interchangeable, we obtain

$$|\beta(E,z_1) - \beta(E,z_0)| \le |z_0 - z_1|.$$

In particular, $\beta(E,z)$ is a continuous function of z.

Suppose $\gamma(E) \neq 0$. Let f be an admissible function for E such that $f'(\infty) > 0$. Using (*), we obtain

$$(***) \qquad \gamma(E)\beta(E,z_1) \geq |\beta(f,z_1)| \geq |z_0 - z_1||f'(\infty)| - |\beta(f,z_0)|.$$

Fixing z_0 and letting z_1 tend to ∞, we find that

$$\lim_{z\to\infty} \beta(E,z) = \infty$$

whenever $\gamma(E) \neq 0$. Consequently there is at least one finite point at which the function $\beta(E,z)$ assumes its minimum value.

The **analytic diameter** $\beta(E)$ of E is the minimum value of $\beta(E, z)$, that is,

$$\beta(E) = \inf_z \beta(E, z).$$

Any point w_0 for which $\beta(E) = \beta(E,w_0)$ is called the **analytic center of** E.

6.1 Lemma. If E is bounded, then

$$\gamma(E) \leq \beta(E) \leq 6 \operatorname{diam}(E).$$

Proof. Since $\beta(E, z) \leq 6 \operatorname{diam}(E)$ whenever $z \in E$, also $\beta(E) \leq 6 \operatorname{diam}(E)$. For $\epsilon > 0$, choose an admissible function f for E such that $f'(\infty) > \gamma(E) - \epsilon$. Then $\beta(f^2, z) > (\gamma(E) - \epsilon)^2$ for all z. Consequently $\beta(E, z) \geq \gamma(E)$ for all z, and $\beta(E) \geq \gamma(E)$.

The following crude estimate is easy to sharpen.

6.2 Lemma. If $z_0 \in E$, and w_0 is the analytic center of E, then $|z_0 - w_0| \leq 24 \operatorname{diam}(E)$.

Proof. Substituting in (***) an admissible function f such that $f'(\infty) > \gamma(E)/2$, and using the estimate $|\beta(f, z_0)| \leq 6 \operatorname{diam}(E)\, \gamma(E)$, we obtain

$$|z_0 - z_1| \leq 12 \operatorname{diam}(E) + 2\beta(E, z_1).$$

Now set $z_1 = w_0$, and use 6.1.

6.3 Lemma. Let E be bounded, and let w_0 be the analytic center of E. If a and b are complex numbers such that $|a| \leq \gamma(E)$ and $|b| \leq \beta(E)\gamma(E)$, there is a function f such that $f/20$ is admissible for E, $f'(\infty) = a$, and $\beta(f,w_0) = b$.

Proof. Let h be an admissible function for E, such that $h'(\infty) = \gamma(E)/2$. Let $z_0 = w_0 + 2\beta(h,w_0)/\gamma(E)$. By (*), $\beta(h,z_0) = 0$. Also, $|w_0 - z_0| \leq 2\beta(E)$.

Let f_0 be an admissible function for E, such that $\beta(f_0, z_0) = \gamma(E)\beta(E)/2$. Let $f_1 = 2f_0 - 4f_0'(\infty)h/\gamma(E)$. Then $f_1(\infty) = 0$, $f_1'(\infty) = 0$, $\beta(f_1, w_0) = \beta(f_1,z_0) = \gamma(E)\beta(E)$, and $|f_1| \leq 2|f_0| + 4|h| \leq 6$.

Set $f_2 = 2h + (w_0 - z_0)f_1/\beta(E)$. Then $f_2(\infty) = 0$, $f_2'(\infty) = \gamma(E)$, $\beta(f_2,w_0) = 0$, and $|f_2| \leq 2|h| + |f_1||w_0 - z_0|/\beta(E) \leq 2 + 6\cdot 2 = 14$. Now the function $f = af_1/\gamma(E) + bf_2/\beta(E)\gamma(E)$ does the trick.

7. A Scheme for Approximation

Suppose K is compact, and $f \in A(K)$. We wish to develop an approximation procedure which will allow us to approximate f by functions which extend analytically beyond the interior of K.

We can assume that $f \in C(S^2)$. Let g be a continuously differentiable function on the complex plane with compact support. The integrals used in II.1 to prove Arens' theorem will play a decisive role:

$$G(\zeta) = \frac{1}{\pi}\iint \frac{f(z)-f(\zeta)}{z-\zeta}\frac{\partial g}{\partial \bar{z}}\,dxdy$$
$$= f(\zeta)g(\zeta) + \frac{1}{\pi}\iint \frac{f(z)}{z-\zeta}\frac{\partial g}{\partial \bar{z}}\,dxdy.$$

Recall that G is continuous on S^2, G is analytic where f is, and G is analytic off the closed support of g. Also, $f - G$ is analytic on the interior of the set on which g is 1.

7.1 Lemma. If the continuously differentiable function g is supported on the disc $\Delta(z_0;\delta)$, if $f \in C(S^2)$ is analytic on the open set U, and if

$$G(\zeta) = \frac{1}{\pi}\iint \frac{f(z)-f(\zeta)}{z-\zeta}\frac{\partial g}{\partial \bar{z}}\,dxdy,$$

then $G \in C(S^2)$, G is analytic on U, and G is analytic off the disc $\Delta(z_0;\delta)$. Also,

$$G(\infty) = 0$$

$$G'(\infty) = -\frac{1}{\pi}\iint f(z)\frac{\partial g}{\partial \bar{z}}\,dxdy$$

$$\beta(G,w) = -\frac{1}{\pi}\iint f(z)(z-w)\frac{\partial g}{\partial \bar{z}}\,dxdy$$

$$\| G \|_\infty \le 2\delta\omega(f;2\delta)\left\|\frac{\partial g}{\partial \bar{z}}\right\|_\infty$$

$$| G'(\infty) | \le 2\delta\omega(f;2\delta)\left\|\frac{\partial g}{\partial \bar{z}}\right\|_\infty \alpha(\Delta(z_0;\delta)\backslash U)$$

$$\frac{1}{\pi}\left|\iint f(z)(z-z_0)\frac{\partial g}{\partial \bar{z}}\,dxdy\right| \le 4\delta^2\omega(f;2\delta)\left\|\frac{\partial g}{\partial \bar{z}}\right\|_\infty \alpha(\Delta(z_0;\delta)\backslash U),$$

where $\omega(f;\cdot)$ is the modulus of continuity of f.

Proof. We have already discussed the qualitative behavior of G. The estimate for $\| G \|_\infty$ is straightforward. The second estimate follows from the first and the inequality $| G'(\infty) | \le \| G \| \alpha(\Delta(z_0;\delta)\backslash U)$. The third estimate follows from the second estimate, by replacing f everywhere by $(z - z_0)f$, and noting that $\omega(f;\delta)$ can be replaced by the oscillation of $(z - z_0)f$ on the disc $\Delta(z_0;\delta)$.

Now we begin describing the approximation procedure.

For each $\delta > 0$, choose discs $\Delta_{k,\delta} = \Delta(z_k; \delta)$ which cover the complex plane, and continuously differentiable functions $g_{k\delta}$ such that

(i) $g_{k\delta}$ is supported on $\Delta_{k\delta}$,

(ii) $\sum g_{k\delta}(z) = 1$ for all complex z,

(iii) $\left\| \frac{\partial g_{k\delta}}{\partial \bar{z}} \right\|_\infty \leq \frac{4}{\delta}$.

(iv) No point z is contained in more than M of the discs $\Delta_{k\delta}$, where M is a universal constant.

One way to construct the $g_{k\delta}$ is as follows. Let g be a continuously differentiable function supported on the disc $\Delta(0; \frac{1}{2})$ such that $g \geq 0$, $\iint g(z)dxdy = 1$, and $|\partial g/\partial \bar{z}| \leq 16$. Let $\{E_k\}_{k=1}^\infty$ be a partition of the plane into squares whose sides have length $\frac{1}{2}$, and set

$$g_{k1}(\zeta) = \iint_{E_k} g(\zeta - z)dxdy.$$

Then each g_{k1} is supported on the disc $\Delta_{k1} = \Delta(z_k; 1)$, where z_k is the center of E_k. And the g_{k1} and Δ_{k1} satisfy (i) through (iv), with $\delta = 1$. For arbitrary $\delta > 0$, one sets $g_{k\delta}(z) = g_{k1}(z/\delta)$, and $\Delta_{k\delta} = \Delta(z_k/\delta; \delta)$.

In the sequel, we will find it convenient often to drop the subscript δ, and ambiguously write $g_k = g_{k\delta}$, $\Delta_k = \Delta_{k\delta}$, etc.

Now suppose $f \in C(S^2)$ is analytic on an open set U containing ∞. Define

$$G_k(\zeta) = \frac{1}{\pi} \iint \frac{f(\zeta) - f(z)}{\zeta - z} \frac{\partial g_k}{\partial \bar{z}} dxdy.$$

Then G_k is analytic on U, and G_k is analytic off Δ_k. So $G_k = 0$ except for those indices k, only finite in number, for which Δ_k meets U^c. Also

$$\| G_k \|_{S^2} \leq 8\omega(f; 2\delta).$$

Now

$$f(\zeta) - \sum G_k(\zeta) = -\frac{1}{\pi} \iint \frac{f(z)}{z - \zeta} \frac{\partial}{\partial \bar{z}} (\sum g_k(z))dxdy = 0.$$

So

$$f = \sum G_k.$$

This decomposition has the effect of splitting the singularities of f, and dividing them among the G_k. The basic approximation procedure involves approximating appropriately the functions $G_k = G_{k\delta}$, and showing that the estimates improve sufficiently as $\delta \to 0$ to place f in $R(K)$. The following lemma illustrates the approach. The approximation technique can be adapted to a variety of situations (cf.7.4).

7.2 Lemma. Let K be compact, and let $f \in C(S^2)$. For each $\delta > 0$, let $\{g_k\}$ be a partition of unity satisfying (i)–(iv) above, and let G_k be defined as above. Suppose there is a constant $r \geq 1$ independent of δ, constants $a(\delta)$, and functions $H_k \in C(S^2)$ such that

(i) $\lim_{\delta \to 0} a(\delta) = 0$.
(ii) $\| H_k \|_{S^2} \leq a(\delta)$.
(iii) H_k is analytic off a compact subset of $\Delta(z_k; r\delta)\backslash K$.
(iv) $H_k - G_k$ has a triple zero at ∞.

Then $f \in R(K)$.

Proof. We can modify f so that f is analytic in a neighborhood of ∞—the only G_k which are altered are those which are already analytic in a neighborhood of K. Then $f = \sum G_k$. Since $H_k \in R(K)$, it will suffice to show that

$$\| \sum (G_k - H_k) \|_K \leq b(\delta),$$

where $b(\delta) \to 0$ as $\delta \to 0$.

Let $a_0(\delta) = a(\delta) + 8\omega(f; 2\delta)$. Then $\| G_k - H_k \| \leq \| G_k \| + \| H_k \| \leq a_0(\delta)$. Now $(z - z_k)^3(G_k - H_k)/r^3\delta^3$ is analytic off the disc $\Delta(z_k; r\delta)$, and it is bounded by $a_0(\delta)$ on the boundary of the disc. By the maximum modulus principle, $|z - z_k|^3 | G_k - H_k|/r^3\delta^3 \leq a_0(\delta)$ when $|z - z_k| \geq r\delta$. This inequality holds trivially when $|z - z_k| \leq r\delta$. Consequently

$$|G_k(z) - H_k(z)| \leq \frac{r^3\delta^3 a_0(\delta)}{|z - z_k|^3}.$$

Now fix the complex number $z^* \in K$. Each disc $\Delta_k = \Delta(z_k; \delta)$ passes through at least one and at most two of the circles $\{|z - z^*| = n\delta\}$, $n = 1,2,3, \ldots$. Let $N(n)$ be the number of discs which meet the nth circle. Since each point is contained in at most M discs, we can bound the sum of the areas of the discs meeting the nth circle by M times the area of the area of the annulus $\{(n - 1)\delta \leq |z - z^*| \leq (n + 1)\delta\}$. This yields the estimate

$$N(n) \leq 4Mn.$$

If $n \geq 2$, and Δ_k touches the nth circle, then $|z^* - z_k| \geq (n - 1)\delta$, so $|G_k(z^*) - H_k(z^*)| \leq r^3 a_0(\delta)/(n - 1)^3$. Using the estimate

$$|G_k(z^*) - H_k(z^*)| \leq a_0(\delta)$$

when Δ_k touches the first circle, we obtain

$$\sum_k |G_k(z^*) - H_k(z^*)| \leq N(1)a_0(\delta) + \sum_{n=2}^{\infty} \frac{r^3 a_0(\delta) N(n)}{(n - 1)^3}.$$

Hence $\| \sum (G_k - H_k) \|_K \leq b(\delta)$, where

$$b(\delta) = 4Ma_0(\delta)\left(1 + r^3 \sum_{n=2}^{\infty} \frac{n}{(n - 1)^3}\right).$$

Since $b(\delta) \to 0$ as $\delta \to 0$, the proof is complete.

7.3 Corollary. Suppose K, f, δ, g_k, and G_k are as above. Suppose there are constants $c(\delta) > 0$, and a constant $r \geq 1$, independent of δ, such that

(i) $\lim_{\delta \to 0} c(\delta) = 0$.

(ii) $|G'_k(\infty)| \leq c(\delta)\gamma(\Delta(z_k; r\delta)\backslash K)$.

(iii) $|\beta(G_k, w_k)| \leq c(\delta)\gamma(\Delta(z_k; r\delta)\backslash K)\beta(\Delta(z_k; r\delta)\backslash K)$, where w_k is the analytic center of $\Delta(z_k; r\delta)\backslash K$.

Then $f \in R(K)$.

Proof. By 6.3 there is a function H_k analytic off a compact subset of $\Delta(z_k; (r+1)\delta)$ such that $H_k(\infty) = 0$, $H'_k(\infty) = G'_k(\infty)$, $\beta(H_k, w_k) = \beta(G_k, w_k)$, and $\| H_k \| \leq 20c(\delta)$. We can arrange also that $H_k \in C(S^2)$. Now $H_k - G_k$ has a triple zero at ∞. So 7.2 applies, with r replaced by $r + 1$.

As an application of this method of approximation, we give a constructive proof of an extended version of Mergelyan's theorem II.9.1. The essential points of the proof boil down to the original proof of Mergelyan.

7.4 Theorem. Suppose K is compact, and U is a component of K^c. Every function $f \in A(K)$ can be approximated uniformly on K by functions in $A(K)$ which extend to be analytic in a neighborhood of bU.

Proof. This does not follow directly from 7.3—we must modify slightly the proof of 7.3 and then obtain the required estimates.

Extend f continuously to S^2, so that f vanishes near ∞, and construct g_k and G_k as before. The indices can be arranged so that Δ_k meets bU if and only if $1 \leq k \leq m$. Then G_k is analytic in a neighborhood of bU, for $k > m$. So it suffices to find functions $H_1, \ldots, H_m$ analytic in a neighborhood of K, such that $\left\| \sum_{k=1}^{m} (G_k - H_k) \right\|_K \leq b(\delta)$, where $b(\delta) \to 0$ as $\delta \to 0$. The desired approximators are then the functions $f_\delta = f + \sum_{k=1}^{m} (H_k - G_k)$.

Suppose that $1 \leq k \leq m$. Let $E = \Delta(z_k; 3\delta)\backslash K$. Then E contains an arc in U whose diameter exceeds δ. By 2.1 and 6.1,

$$\beta(E) \geq \gamma(E) \geq \frac{\delta}{4}.$$

Estimating α by δ in 7.1, we obtain

$$\begin{aligned} |G'_k(\infty)| &\leq 8\delta\omega(f; 2\delta) \\ &\leq 32\omega(f; 2\delta)\gamma(E) \end{aligned}$$

and

$$\begin{aligned} |\beta(G_k, z_k)| &\leq 16\delta^2\omega(f; 2\delta) \\ &\leq 256\omega(f; 2\delta)\gamma(E)\beta(E). \end{aligned}$$

If w_k is the analytic center of $\Delta(z_k; 3\delta)\backslash K$, then by 6.2,

$$|w_k - z_k||G_k'(\infty)| \le 72\delta|G_k'(\infty)|$$
$$\le 10^4\omega(f;2\delta)\gamma(E)\beta(E).$$

From $\beta(G_k, w_k) = \beta(G_k, z_k) + (w_k - z_k)G_k'(\infty)$, we obtain

$$|\beta(G_k,w_k)| \le 10^5\omega(f;2\delta)\gamma(E)\beta(E).$$

Now we have verified the estimates of 7.3 for the functions $G_1, \ldots, G_m$, with $c(\delta) = 10^5\omega(f;2\delta)$ and $r = 3$. By the proofs of 7.2 and 7.3, there are functions $H_1, \ldots, H_m$ analytic in a neighborhood of K, such that $\left\|\sum_{k=1}^{m}(G_k - H_k)\right\|_K \le b(\delta)$, where $b(\delta) \to 0$ as $\delta \to 0$. That does it.

8. Criteria for $R(K) = A(K)$

The proof of the following theorem involves essentially showing how estimates of the form 7.3(iii) for $\beta(G,w)$ can be obtained from estimates of the form 7.3(ii) for $G'(\infty)$.

8.1 Theorem. Let K be compact. The following assertions are equivalent, for $f \in C(S^2)$:

(i) $f \in R(K)$.

(ii) If g is any complex-valued continuously differentiable function which is supported on a disc $\Delta(z_0;\delta)$, then

$$\left|\iint f(z)\frac{\partial g}{\partial \bar{z}}dxdy\right| \le 2\pi\delta\omega(f;2\delta)\left\|\frac{\partial g}{\partial \bar{z}}\right\|_\infty \gamma(\Delta(z_0;\delta)\backslash K).$$

(iii) There exist $r \ge 1$ and constants $a(\delta)$ which tend to zero as $\delta \to 0$, such that

$$\left|\iint f(z)\frac{\partial g}{\partial \bar{z}}dxdy\right| \le \delta a(\delta)\left(\left\|\frac{\partial g}{\partial x}\right\|_\infty + \left\|\frac{\partial g}{\partial y}\right\|_\infty\right)\gamma(\Delta(z_0;r\delta)\backslash K)$$

whenever g is a continuously differentiable complex-valued function supported on a disc $\Delta(z_0;\delta)$.

Vitushkin [1] gives other criteria necessary and sufficient that a function $f \in C(S)^2$ belong to $R(K)$, which we state without proof. Let $S(z_0;\delta)$ denote the square with center z_0 and side of length δ. If $f \in R(K)$, then

$$\left|\int_{bS(z_0;\delta)} f(z)dz\right| \le c\omega(f;2\delta)\gamma(S(z_0;\delta)\backslash K),$$

for all squares $S(z_0;\delta)$, where c is a universal constant. Conversely, if there exist $r \ge 1$ and constants $c(\delta)$ tending to zero with δ, such that

$$\left|\int_{bS(z_0;\delta)} f(z)dz\right| \le c(\delta)\gamma(S(z_0;r\delta)\backslash K)$$

for all squares $S(z_0; \delta)$ and all δ sufficiently small, then $f \in R(K)$. It is not known whether one can replace squares by circles, in this characterization.

Proof of 8.1. First we show that (i) implies (ii). For this, we can assume that g is supported on the interior of $\Delta(z_0; \delta)$.

If f is analytic in a neighborhood of K, then

$$G(\zeta) = \frac{1}{\pi} \iint \frac{f(z) - f(\zeta)}{(z - \zeta)} \frac{\partial g}{\partial \bar{z}} dxdy$$

is analytic off a compact subset of $\Delta(z_0; \delta) \backslash K$, and $|G|$ is bounded by $2\delta\omega(f; 2\delta) \| \partial g / \partial \bar{z} \|_\infty$. Since

$$G'(\infty) = -\frac{1}{\pi} \iint f(z) \frac{\partial g}{\partial \bar{z}} dxdy,$$

we obtain the required estimate in (ii).

Now suppose $f \in R(K)$ is arbitrary, and extend f continuously to S^2. Given $\epsilon > 0$, choose $h \in C(S^2)$ such that h is analytic in a neighborhood of K, and $\| f - h \|_{S^2} < \epsilon$. Then

$$\begin{aligned} \left| \iint f(z) \frac{\partial g}{\partial \bar{z}} dxdy \right| &\leq \left| \iint (f(z) - h(z)) \frac{\partial g}{\partial \bar{z}} dxdy \right| + \left| \iint h(z) \frac{\partial g}{\partial \bar{z}} dxdy \right| \\ &\leq \epsilon \iint \left| \frac{\partial g}{\partial \bar{z}} \right| dxdy + 2\pi\delta\omega(h; 2\delta) \left\| \frac{\partial g}{\partial \bar{z}} \right\|_\infty \gamma(\Delta(z_0; \delta) \backslash K). \end{aligned}$$

Letting $\epsilon \to 0$, and noting that $\omega(h; 2\delta) \leq \omega(f; 2\delta) + 2\epsilon$, we obtain (ii).

That (ii) implies (iii) is clear.

Now suppose that f satisfies (iii). We will verify the estimates of 7.3, for appropriate constants $c(\delta)$. The notation will be that of section 7.

For $\delta > 0$ and k fixed, set

$$E = \Delta(z_k; (r + 2)\delta) \backslash K.$$

By 7.1 and 7.3, it suffices to show that

$$(*) \qquad \left| \iint f(z) \frac{\partial g_k}{\partial \bar{z}} dxdy \right| \leq c(\delta)\gamma(E)$$

$$(**) \qquad \left| \iint f(z)(z - w_k) \frac{\partial g_k}{\partial \bar{z}} dxdy \right| \leq c(\delta)\gamma(E)\beta(E),$$

where w_k is the analytic center of E, and $c(\delta) \to 0$ as $\delta \to 0$.

Now $(*)$ follows from (iii), in view of the estimate

$$\left\| \frac{\partial g_k}{\partial x} \right\|_\infty + \left\| \frac{\partial g_k}{\partial y} \right\|_\infty \leq 4 \left\| \frac{\partial g_k}{\partial \bar{z}} \right\|_\infty \leq \frac{16}{\delta}$$

and the monotonicity of γ, providing only $c(\delta) \geq 16a(\delta)$. So it suffices to obtain the estimate $(**)$.

Set $\beta = \min(\delta, \beta(E))$. Then $\beta \leq \delta$. Also,

$$\gamma(E) \leq (r + 2)\beta.$$

In fact, this follows from 6.1 if $\beta < \delta$, and it follows from the inclusion $E \subseteq \Delta(z_k; (r+2)\delta)$ if $\beta = \delta$.

Let $h_j = g_{j\beta}$ be a partition of unity as in section 7, associated with discs $\Delta(t_j; \beta)$. By $\sum'$ we will denote summation over those indices j such that $\Delta(t_j; \beta)$ meets $\Delta(z_k; \delta)$. Note that any such disc is contained in $\Delta(z_k; (r+2)\delta)$.

The quantity we wish to estimate is

$$\begin{aligned}\iint f(z)(z-w_k)\frac{\partial g_k}{\partial \bar{z}}dxdy &= \textstyle\sum' \displaystyle\iint f(z)(z-w_k)\frac{\partial}{\partial \bar{z}}(g_k h_j)dxdy \\ &= \textstyle\sum' \displaystyle\iint f(z)(z-t_j)\frac{\partial}{\partial \bar{z}}(g_k h_j)dxdy \\ &\quad + \textstyle\sum' (t_j - w_k) \displaystyle\iint f(z)\frac{\partial}{\partial \bar{z}}(g_k h_j)dxdy.\end{aligned}$$

From the properties of the g_k and h_j, we obtain

$$\left\|\frac{\partial}{\partial x}(g_k h_j)\right\|_\infty \le \left\|\frac{\partial g_k}{\partial x}\right\|_\infty + \left\|\frac{\partial h_j}{\partial x}\right\|_\infty \le \frac{8}{\delta} + \frac{8}{\beta} \le \frac{16}{\beta},$$

and

$$\left\|\frac{\partial}{\partial x}((z-t_j)g_k h_j)\right\|_\infty \le \|g_k h_j\|_\infty + \beta\left\|\frac{\partial}{\partial x}(g_k h_j)\right\|_\infty \le 17,$$

with similar estimates for the derivatives with respect to y. From (iii) we obtain

$$\left|\iint f(z)\frac{\partial}{\partial \bar{z}}(g_k h_j)dxdy\right| \le 32a(\beta)\gamma(\Delta(t_j; r\beta)\backslash K)$$

and

$$\begin{aligned}\left|\iint f(z)(z-t_j)\frac{\partial}{\partial \bar{z}}(g_k h_j)dxdy\right| &= \left|\iint f(z)\frac{\partial}{\partial \bar{z}}((z-t_j)g_k h_j)dxdy\right| \\ &\le 34\beta a(\beta)\gamma(\Delta(t_j; r\beta)\backslash K).\end{aligned}$$

Consequently

$$\begin{aligned}\left|\iint f(z)(z-w_k)\frac{\partial g_k}{\partial \bar{z}}dxdy\right| &\le 34\beta a(\beta) \textstyle\sum' \gamma(\Delta(t_j; r\beta)\backslash K) \\ &\quad + 32a(\beta) \textstyle\sum' |t_j - w_k|\, \gamma(\Delta(t_j; r\beta)\backslash K) \\ &= \sharp + \flat.\end{aligned}$$

We will estimate $\sharp$ and $\flat$ using 2.7.

By construction, each z is contained in at most M of the discs $\Delta(t_j; \beta)$. It follows that every disc of radius $(r+2)\beta$ meets at most $M(r)$ of the discs $\Delta(t_j; r\beta)$, where $M(r)$ depends on r but not on β or δ. Since $\gamma(E) \le (r+2)\beta$, every disc of radius $\gamma(E)$ meets at most $M(r)$ of the sets $\Delta(t_j; r\beta)\backslash K$. By 2.7,

$$\begin{aligned}\sharp &\le 3400M(r)\beta a(\beta)\gamma(E) \\ &\le 3400M(r)a(\beta)\gamma(E)\beta(E).\end{aligned}$$

That takes care of $\sharp$. We must work harder for $\flat$.

Choose an admissible function f_j for $\Delta(t_j; r\beta)\backslash K$ such that

$$2(t_j - w_k)f_j'(\infty) = |t_j - w_k|\gamma(\Delta(t_j; r\beta)\backslash K).$$

By 2.7, $F = \sum' f_j/100M(r)$ is admissible for E. Consequently, for R large,

$$\begin{aligned} 100M(r)\gamma(E)\beta(E) &= 100M(r)\gamma(E)\beta(E,w_k) \\ &\geq 100M(r)\beta(F,w_k) \\ &= \frac{100M(r)}{2\pi}\left|\int_{|z|=R} F(z)(z - w_k)dz\right| \\ &= \frac{1}{2\pi}\left|\sum{}'(t_j - w_k)\int_{|z|=R} f_j(z)dz + \sum{}'\int_{|z|=R}(z - t_j)f_j(z)dz\right| \\ &\geq \frac{1}{2}\sum{}'|t_j - w_k|\gamma(\Delta(t_j; r\beta)\backslash K) \;-\sum{}'\left|\frac{1}{2\pi}\int_{|z|=R}(z - t_j)f_j(z)dz\right|. \end{aligned}$$

From the estimate for a_2 in 2.5, we obtain

$$\begin{aligned} \sum{}'\left|\frac{1}{2\pi}\int(z - t_j)f_j(z)dz\right| &\leq 2er\beta\sum{}'\gamma(\Delta(t_j; r\beta)\backslash K) \\ &\leq 200erM(r)\gamma(E)\beta(E). \end{aligned}$$

Consequently

$$\frac{1}{2}\sum{}'|t_j - w_k|\gamma(\Delta(t_j; r\beta)\backslash K) \leq 100M(r)\gamma(E)\beta(E) + 200erM(r)\gamma(E)\beta(E).$$

This leads to the estimate

$$\flat \leq 10^5 rM(r)a(\beta)\gamma(E)\beta(E).$$

Combining the estimates for $\sharp$ and $\flat$, we obtain the estimate (**), with

$$c(\delta) = 10^6 rM(r)\sup_{\alpha\leq\delta} a(\alpha).$$

That completes the proof.

8.2 Theorem (Vitushkin's Theorem). The following are equivalent, for a compact set K:

(i) $R(K) = A(K)$.

(ii) For every bounded open set D, $\alpha(D\backslash K) = \alpha(D\backslash \text{int } K)$.

(iii) For each complex z, each $r > 1$, and each $\delta > 0$,

$$\alpha(\Delta(z; \delta)\backslash \text{int } K) \leq \alpha(\Delta(z; r\delta)\backslash K).$$

(iv) There exist $r \geq 1$ and $c > 0$ such that for all complex z and all $\delta > 0$,

$$\alpha(\Delta(z; \delta)\backslash \text{int } K) \leq c\alpha(\Delta(z; r\delta)\backslash K).$$

(v) For each $z \in bK$, there exists $r \geq 1$ such that

$$\limsup_{\delta\to 0}\frac{\alpha(\Delta(z; \delta)\backslash \text{int } K)}{\alpha(\Delta(z; \delta)\backslash K)} < \infty.$$

Recall that γ and α agree on open sets. Consequently if K has no interior, condition (i) states that $R(K) = C(K)$, condition (ii) states that $\gamma(D\backslash K) = \gamma(D)$ for all bounded open sets D, and condition (v) states that

$$\liminf_{\delta\to 0} \gamma(\Delta(z;\delta)\backslash K)/\delta > 0$$

for all $z \in bK$. So 9.2 is slightly weaker than 5.1, in this special case.

Proof of 8.2. First we show that (iv) implies (i). In fact, suppose (iv) is true, and suppose $f \in C(S^2)$ belongs to $A(K)$. If g is a continuously differentiable function supported on a disc $\Delta(z_0;\delta)$, then

$$G(\zeta) = \frac{1}{\pi}\iint \frac{f(\zeta)-f(z)}{\zeta - z}\frac{\partial g}{\partial \bar z}\,dxdy$$

is analytic off $K^c \cap \Delta(z_0;\delta)$, and G is bounded by $2\delta\omega(f;2\delta)\,\|\partial g/\partial \bar z\|_\infty$. Hence

$$\begin{aligned}\left|\iint f(z)\frac{\partial g}{\partial \bar z}\,dxdy\right| &= \pi\,|G'(\infty)|\\ &\le 2\pi\delta\omega(f;2\delta)\left\|\frac{\partial g}{\partial \bar z}\right\|_\infty \alpha(\Delta(z_0;\delta)\backslash \operatorname{int} K)\\ &\le 2\pi c\,\delta\omega(f;2\delta)\left\|\frac{\partial g}{\partial \bar z}\right\|_\infty \gamma(\Delta(z_0;r\delta)\backslash K).\end{aligned}$$

Thus 8.1(iii) holds, with $c(\delta) = 2\pi c\omega(f;2\delta)$. So $f \in R(K)$, and $R(K) = A(K)$.

That (i) implies (ii) follows from the localization theorem II.10.3, just as the corresponding implication of 5.1 did. The details are left to the reader.

That (ii) implies (iii) follows upon taking $D = \operatorname{int}\Delta(z_0;r\delta)$ and using the monotonicity of α. That (iii) implies (iv) is trivial. Consequently, (i)–(iv) are equivalent.

That (iv) implies (v) is trivial. To complete the proof, we will show that if the equivalent conditions (i)–(iv) fail, then (v) also fails.

Suppose $R(K) \ne A(K)$. There is $\delta_1 > 0$ and a complex number z_1 such that [taking $r = c = 10$ in (iv)]

$$\alpha(\Delta(z_1;\delta_1)\backslash \operatorname{int} K) > 10\alpha(\Delta(z_1;10\delta_1)\backslash K).$$

Let $K_1 = \Delta(z_1;2\delta_1) \cap K$. Since α is monotone,

$$\alpha(\Delta(z_1;\delta_1)\backslash \operatorname{int} K_1) > 10\alpha(\Delta(z_1;10\delta_1)\backslash K_1).$$

Since (iii) fails for K_1, $R(K_1) \ne A(K_1)$. Consequently there is $\delta_2 > 0$ and a complex number z_2 such that [taking $r = c = 10^2$ in (iv)]

$$\alpha(\Delta(z_2;\delta_2\backslash \operatorname{int} K_1) > 10^2\alpha(\Delta(z_2;10^2\delta_2)\backslash K_1).$$

In particular, $10^2\alpha(\Delta(z_2;10^2\delta_2)\backslash K_1) < \delta_2$, so $\Delta(z_2;10^2\delta_2)\backslash K_1$ cannot contain a disc of radius δ_2. Hence $\Delta(z_2;10^2\delta_2) \subseteq \Delta(z_1;2\delta_1+\delta_2)$, and $\Delta(z_2;\delta_2) \subseteq \operatorname{int}\Delta(z_1;2\delta_1)$. So $\Delta(z_2;\delta_2)\backslash \operatorname{int} K_1 = \Delta(z_2;\delta_2)\backslash \operatorname{int} K$, and

$$\alpha(\Delta(z_2;\delta_2)\backslash \operatorname{int} K) = \alpha(\Delta(z_2;\delta_2)\backslash \operatorname{int} K_1)$$

$$> 10^2\alpha(\Delta(z_2; 10^2\delta_2)\backslash K_1)$$
$$\geq 10^2\alpha(\Delta(z_2; 10^2\delta_2)\backslash K).$$

Also, $10^2\delta_2 \leq 2\delta_1 + \delta_2$ shows that $\delta_2 < \delta_1/2$.

By repeating this construction, we obtain sequences $\delta_n > 0$ and z_n such that

$$(*) \qquad \delta_{n+1} < \frac{\delta_n}{2};$$

$$(**) \qquad \Delta(z_{n+1}; \delta_{n+1}) \subseteq \text{int}\, \Delta(z_n; 2\delta_n);$$

$$(***) \qquad \alpha(\Delta(z_n; \delta_n)\backslash\text{int}\, K) > 10^n\alpha(\Delta(z_n; 10^n\delta_n)\backslash K).$$

If $m > n$, then $|z_m - z_n| \leq \sum_{k=n}^{m-1} |z_{k+1} - z_k| \leq 2\sum_{k=n}^{m-1} \delta_k \leq 4\delta_n$. So z_n converges to some point z satisfying $|z - z_n| \leq 4\delta_n$. Hence $\Delta(z_n; \delta_n) \subseteq \Delta(z; 5\delta_n)$, and $\Delta(z; (10^n - 4)\delta_n) \subseteq \Delta(z_n; 10^n\delta_n)$. By monotonicity of α, and (***), we obtain

$$\alpha(\Delta(z; 5\delta_n)\backslash\text{int}\, K) > 10^n\alpha(\Delta(z; (10^n - 4)\delta_n)\backslash K).$$

So (v) is violated at z. Evidently $z \in bK$. This completes the proof.

Using the estimate $\alpha(\Delta(z;\delta)\backslash\text{int}\, K) \leq \delta$ in 8.2(v), with $r = 1$, we obtain the following.

8.3 Corollary. If K is compact, and

$$\liminf_{\delta\to 0} \frac{\gamma(\Delta(z; \delta)\backslash K)}{\delta} > 0$$

for all $z \in bK$, then $R(K) = A(K)$.

The following corollary of 8.3 includes Mergelyan's theorem II.9.1.

8.4 Corollary. If every point on the boundary of the compact set K belongs to the boundary of some component of K^c, then $R(K) = A(K)$.

Proof. Let $z \in bK$. For sufficiently small $\delta > 0$, int $\Delta(z; \delta)\backslash K$ contains an arc whose diameter exceeds $\delta/2$. By 2.1, $\gamma(\Delta(z; \delta)\backslash K) \geq \delta/8$. Now apply 8.3.

8.5 Corollary. If K is compact, and K^c has positive lower Lebesgue density at every point of bK, then $A(K) = R(K)$.

Proof. The lower Lebesgue density of K^c at the point $z \in bK$ is given by

$$\liminf_{\delta\to 0} \frac{\text{Area}\, (\Delta(z; \delta)\backslash K)}{\pi\delta^2}.$$

Combining 8.3 and the estimate 2.3, we obtain the conclusion.

The Hartogs–Rosenthal theorem II.8.4, stating that $R(K) = C(K)$ if K has zero area, is a consequence of 8.5.

9. Failure of Approximation

Here are some examples, variations of the same idea, for which rational approximation fails.

Let J be a compact subset of the closed unit disc Δ such that J has no interior, while $\alpha(J) > 0$. Let $\{\Delta_n\}_{n=1}^{\infty}$ be a sequence of disjoint open discs in Δ which accumulate precisely on the set J, and which satisfy $\sum_{n=1}^{\infty} r_n < \infty$, where r_n is the radius of Δ_n. Let

$$K = \Delta \backslash \left(\bigcup_{n=1}^{\infty} \Delta_n \right).$$

Then

$$bK = b\Delta \cup J \cup \left(\bigcup_{n=1}^{\infty} b\Delta_n \right).$$

Let μ be the finite measure on bK which is dz on $b\Delta$ and $-dz$ on $b\Delta_n$, $1 \leq n < \infty$. If f is rational with poles off K, then $\int f d\mu = 0$. So $\mu \perp R(K)$. However, if $g \in C(S^2)$ is analytic off J and satisfies $g'(\infty) \neq 0$, then $\int_{b\Delta_n} g(z)dz = 0$, $1 \leq n < \infty$. So

$$\int g d\mu = \int_{b\Delta} g(z)dz = 2\pi i g'(\infty) \neq 0.$$

Consequently $g \in A(K)$, $g \notin R(K)$, and $R(K) \neq A(K)$.

For the first example, take J so that the area of J is zero and J is totally disconnected. Such J exist—in fact, the product of the usual Cantor set with itself has positive continuous analytic capacity (cf. the exercises). If the Δ_n are chosen with care, we obtain the following example.

9.1 Example. There is a compact set K such that int K is connected and dense in K, bK has zero area, and $R(K) \neq A(K)$.

This example shows that almost every point on bK can be a peak point for $R(K)$, although $R(K) \neq A(K)$. So the direct analogue of Bishop's peak point criterion for rational approximation II.11.5 does not extend to sets with interior.

For some time it was hoped that perhaps $A(K)$ would coincide with $R(K)$ as soon as $A(K)$ had the same peak points as $R(K)$. However, A. M. Davie has recently shown that the set J and the discs $\{\Delta_n\}_{n=1}^{\infty}$ above can be chosen so that every point of J is a peak point for $R(K)$. In other words, there is a compact set K with the properties of 9.1, and the additional property that every point of bK is a peak point for $R(K)$. Meanwhile, the problem of finding an appropriate extension of Bishop's theorem remains open.

For the second example, take J to be an arc which lies in int Δ, except for one endpoint on $b\Delta$. Any such arc of positive area will satisfy $\alpha(J) > 0$. By selecting the Δ_n carefully so that each Δ_n touches J at precisely one point, we obtain the following example.

9.2 Example. There exists a compact set K such that int K is connected and simply connected, int K is dense in K, and $R(K) \neq A(K)$.

This shows that no topological assumption on int K will assure us that $R(K) = A(K)$. In passing, we mention that the set K of example 9.2 can be chosen so that $A(K)$ is a dirichlet algebra.

9.3 Example. There is a compact set K whose interior consists of two connected components U_1 and U_2, each of which is simply connected, such that $K = \bar{U}_1 \cup \bar{U}_2$, and $R(\bar{U}_j) = A(\bar{U}_j)$, $j = 1,2$, while $R(K) \neq A(K)$. In fact, every function in $A(\bar{U}_j)$ can be approximated uniformly on $\bar{U}_j$ by functions in $R(K)$.

This example is obtained by modifying the second example so that both endpoints of J lie on $b\Delta$ (cf. fig. 5). That $R(\bar{U}_j) = A(\bar{U}_j)$ follows from 8.4, since every point of $b\bar{U}_j$ lies on the boundary of a single component of $\bar{U}_j^c$.

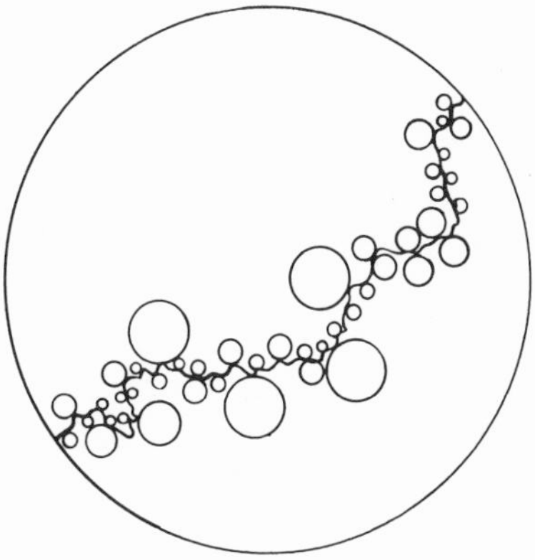

Figure 5. The stitched disc.

10. Pointwise Bounded Approximation

In this section, we consider the following problem: For which open sets U is it true that every bounded analytic function f on U can be approximated pointwise boundedly on U by rational functions?

Recall that $H^\infty(U)$ is the algebra of bounded analytic functions on U. So we ask for which U it is true that for every $f \in H^\infty(U)$, there is a sequence $f_n \in R(\bar{U})$ such that $\sup \|f_n\|_U < \infty$, and $f_n(z) \to f(z)$ for all $z \in U$.

Elementary considerations show that we may as well assume that $U =$ int K for some compact set K. Our problem can be restated in the following slightly more general form: For which compact sets K is it true that every $f \in H^\infty(\text{int } K)$ can be approximated pointwise on int K by a bounded sequence in $R(K)$?

In VI.5.3 we showed that the answer to this question is affirmative when

K^c has a finite number of components. In this section we will obtain more general results using the scheme for approximation developed in section 7.

First, observe that the desired approximation fails for some compact sets K, even though int K is connected and dense in K, and $A(K) = R(K)$. Indeed, let K be the compact set obtained from the closed unit disc by excising a sequence of open discs $\{\Delta_n\}_{n=1}^{\infty}$ such that the closures of the Δ_n are disjoint and do not meet the real axis, the Δ_n accumulate on a closed set E of positive length on the real axis, and the sum of the radii of the Δ_n is finite. Let ν be a finite measure on int K which is dz on a curve homotopic to $b\Delta$, and $-dz$ on curves homotopic to $b\Delta_n$, $1 \leq n < \infty$. Then $\int f d\nu = 0$ for all $f \in R(K)$. However, if g is the Ahlfors function of E, then $\int g d\nu = 2\pi i g'(\infty) \neq 0$. So g cannot be approximated pointwise boundedly on int K by functions in $R(K)$. [That $R(K) = A(K)$ will follow from 12.4.]

To attack our problem, we will parallel the developments of sections 7 and 8, giving indications of how the proofs there can be adjusted to cover the present case. The analogue of 7.1 is the following.

10.1 Lemma. Let f be a bounded Borel function, let g be a continuously differentiable function with compact support, and set

$$G(\zeta) = \frac{1}{\pi}\iint \frac{f(z) - f(\zeta)}{z - \zeta}\frac{\partial g}{\partial \bar{z}}\,dxdy$$
$$= f(\zeta)g(\zeta) + \frac{1}{\pi}\iint \frac{f(z)}{z - \zeta}\frac{\partial g}{\partial \bar{z}}\,dxdy.$$

Then G is bounded, G is continuous at every point of continuity of f, G is analytic off the support of g, and G is analytic wherever f is analytic. Moreover, if $g = 1$ in an open set V, then $f - G$ is analytic on V.

Proof. The integral appearing in the second formula for G is continuous on S^2, it is analytic off the support of g, and it is analytic on V. The lemma follows easily, from this remarks, except for the assertion regarding the analyticity of G wherever f is analytic. This latter assertion can be proved as in II.1. A proof can also be based on the fact that $(f(z) - f(\zeta))/(z - \zeta)$ is an analytic function of two complex variables in $U \times U$, whenever f is analytic on U.

Let K be compact. For each $\delta > 0$, choose points $z_{k\delta} = z_k$ and continuously differentiable functions $g_{k\delta} = g_k$ (again the index δ will be suppressed for convenience) such that

(i) $z_k \in bK$,
(ii) g_k is supported on $\Delta_k = \Delta(z_k; \delta)$,
(iii) $\sum g_k(z) = 1$ for z belonging to a neighborhood of bK,
(iv) $\left\| \dfrac{\partial g_k}{\partial \bar{z}} \right\|_\infty \leq \dfrac{4}{\delta}$.

(v) No complex number z is contained in more than M of the discs Δ_k, where M is a universal constant.

If $f \in H^\infty(\text{int } K)$, we set $f = 0$ off int K, and define

$$G_k(\zeta) = \frac{1}{\pi} \iint \frac{f(z) - f(\zeta)}{z - \zeta} \frac{\partial g_k}{\partial \bar{z}} dxdy.$$

Again we obtain the decomposition

$$f = \sum G_k,$$

which distributes the singularities of f among the G_k.

The analogue of 7.2 is the following.

10.2 Lemma. Let K be compact, and let $f \in H^\infty(\text{int } K)$. For each $\delta > 0$, choose z_k and g_k to satisfy (i) through (v) above, and define G_k as above. Suppose there are constants $r \geq 1$ and $a > 0$, independent of δ, and functions $H_k \in C(S^2)$ such that

(i) $\| H_k \|_{S^2} \leq a$,
(ii) H_k is analytic off a compact subset of $\Delta(z_k; r\delta)\backslash K$,
(iii) $H_k - G_k$ has a triple zero at ∞.

Then there is a bounded sequence of functions $f_n \in R(K)$ such that $f_n(z) \to f(z)$ for all $z \in \text{int } K$.

Proof. Set $f_\delta = \sum H_k$. Since

$$|G_k(\zeta)| \leq \frac{8\|f\|_\infty}{\pi\delta} \iint_{|z - z_k| \leq \delta} \frac{1}{|z - \zeta|} dxdy$$
$$\leq 16\|f\|_\infty,$$

we have

$$|G_k(\zeta) - H_k(\zeta)| \leq a + 16\|f\|_\infty$$

for all ζ. As in the proof of 7.2, we obtain

$$|G_k(\zeta) - H_k(\zeta)| \leq \frac{r^3\delta^3(a + 16\|f\|_\infty)}{|\zeta - z_k|^3}.$$

Consequently

$$|f(\zeta) - f_\delta(\zeta)| \leq \sum |G_k(\zeta) - H_k(\zeta)|$$
$$\leq (a + 16\|f\|_\infty)\left(N(1) + r^3 \sum_{n=2}^{\infty} \frac{N(n)}{(n-1)^3}\right),$$

where $N(n)$ is the number of the discs Δ_k which meet the circle $b\Delta(\zeta; n\delta)$.

As in 7.2, we obtain the estimate $N(n) \leq 4Mn$. It follows that

$$|f_\delta(\zeta)| \leq \|f\|_\infty + (a + 16\|f\|_\infty)4M\left(1 + r^3 \sum_{n=2}^{\infty} \frac{n}{(n-1)^3}\right)$$

for all ζ. So the f_δ are a bounded family in $R(K)$.

If $\zeta \in \operatorname{int} K$, and d is the distance from ζ to bK, then $N(n) = 0$ when $(n+1)\delta < d$. Hence

$$|f(\zeta) - f_\delta(\zeta)| \leq (a + 16\|f\|_\infty)4Mr^3 \sum_{n=[d/\delta]}^{\infty} \frac{n}{(n+1)^3},$$

where $[d/\delta]$ is the largest integer not exceeding d/δ. Since $[d/\delta] \to \infty$ as $\delta \to 0$, also $f_\delta(\zeta) \to f(\zeta)$ as $\delta \to 0$. That proves the lemma.

From 6.3 and some elementary estimates, we obtain the analogue of 7.3.

10.3 Corollary. Suppose K, f, δ, g_k and G_k are as above. If there are constants $r \geq 1$ and $c > 0$, independent of δ, such that

(i) $|G_k'(\infty)| \leq c\gamma(\Delta(z_k; r\delta)\backslash K)$,
(ii) $|\beta(G_k, w_k)| \leq c\gamma(\Delta(z_k; r\delta)\backslash K)\beta(\Delta(z_k; r\delta)\backslash K)$,

where w_k is the analytic center of $\Delta(z_k; r\delta)\backslash K)$, then there is a bounded sequence $f_n \in R(K)$ such that $f_n(z) \to f(z)$ for all $z \in \operatorname{int} K$.

Now the technique used in the proof of 8.1, which established the analogous implications of 8.1 and 8.2, carries over to yield the following result.

10.4 Theorem. Suppose K is compact. Suppose there exist constants $r \geq 1$ and $c > 0$ such that for all sufficiently small $\delta > 0$,

$$\gamma(\Delta(z; \delta)\backslash \operatorname{int} K) \leq c\gamma(\Delta(z; r\delta)\backslash K)$$

for all $z \in bK$. If $f \in H^\infty(\operatorname{int} K)$, there is a bounded sequence $f_n \in R(K)$ such that $f_n(z) \to f(z)$ for all $z \in \operatorname{int} K$.

10.5 Corollary. Let K be compact. Suppose that for all $\delta > 0$ sufficiently small,

$$\frac{\gamma(\Delta(z;\delta)\backslash K)}{\delta} \geq a > 0$$

for all $z \in bK$. If $f \in H^\infty(\operatorname{int} K)$, there is a bounded sequence $f_n \in R(K)$ such that $f_n(z) \to f(z)$ for all $z \in \operatorname{int} K$.

The following corollary applies, for instance, when K^c has a finite number of components.

10.6 Corollary. Let K be compact. Suppose there is $d > 0$ such that the diameter of every component of K^c exceeds d. If $f \in H^\infty(\operatorname{int} K)$, there is a bounded sequence $f_n \in R(K)$ such that $f_n(z) \to f(z)$ for all $z \in \operatorname{int} K$.

The analogue of 7.4 takes the following form.

10.7 Theorem. Suppose K is compact, and U is a component of K^c.

If $f \in H^\infty(\text{int } K)$, there is a bounded sequence $f_n \in H^\infty(\text{int } K)$ such that each f_n extends to be analytic in a neighborhood of bU, and f_n converges uniformly to f on every subset of int K at a positive distance from bU.

Theorem 10.7 can be used to handle compact sets K obtained from the closed unit disc Δ by deleting the interiors of mutually disjoint closed discs which accumulate only on $b\Delta$. One first approximates $f \in H^\infty(\text{int } K)$ by a function in $H^\infty(\text{int } K)$ which extends to be analytic in a neighborhood of $b\Delta$. That approximator extends analytically across all but a finite number of the Δ_n. So 10.6 allows us to approximate the approximator.

In order to handle examples of this type which have also a finite number of accumulation points inside Δ, we prove the following strengthened version of II.1.8.

10.8 Theorem. Let U be an open subset of the Riemann sphere S^2. Suppose $f \in H^\infty(U)$, and $z_0 \in bU$. There is a bounded sequence $f_n \in H^\infty(U)$ such that f_n extends to be analytic in a neighborhood of z_0, f_n is continuous at every point of bU at which f is continuous, and f_n converges uniformly to f on every subset of U at a positive distance from z_0. Moreover, if f is continuous at z_0, the f_n converge uniformly to f on U.

Proof. We can assume that $z_0 \neq \infty$.

Fix $\delta > 0$. Let g be a continuously differentiable function supported on the disc $\Delta(z_0; \delta)$ such that $g = 1$ on $\Delta(z_0; \delta/2)$, and $|\partial g/\partial \bar{z}| \leq 4/\delta$. Define G as in 10.1, and set $f_\delta = f - G$. By 10.1, f_δ has the properties of continuity and analyticity that we desire. From

$$|G(\zeta)| \leq \frac{8\|f\|_\infty}{\pi\delta} \iint_{\Delta(z_0;\delta)} \frac{dxdy}{|z-\zeta|} \leq 8\|f\|_\infty$$

we obtain $\|f_\delta\|_\infty \leq 9\|f\|_\infty$. So the f_δ are a bounded family in $H^\infty(U)$.

Now fix $d > 0$. If $\zeta \in U$ satisfies $|\zeta - z_0| \geq d$, and if $\delta < d$, then

$$|f(\zeta) - f_\delta(\zeta)| = |G(\zeta)| \leq \frac{8\|f\|_\infty \delta}{d - \delta}.$$

Consequently f_δ converges to f as $\delta \to 0$, uniformly on any subset of U whose distance from z_0 exceeds d.

Finally, suppose that f is continuous at z_0. We can assume that $f(z) \to 0$ when $z \to z_0$, $z \in U$. If $|\zeta - z_0| \leq \delta$, then an elementary estimate yields

$$|G(\zeta)| \leq 16\,\|f\|_{U \cap \Delta(z_0;\delta)}.$$

By the maximum modulus principle, this estimate holds for all ζ. Consequently

$$\|f - f_\delta\|_U \leq 16\|f\|_{U \cap \Delta(z_0;\delta)},$$

so that $f_\delta \to f$ uniformly on U as $\delta \to 0$.

11. Pointwise Approximation with Same Norm

In this section we show how the results of the preceding section, together with theorem VI.5.2 on bounded pointwise approximation, can be used to study the following problem: For which compact sets K is it true that every function $f \in H^\infty(\text{int } K)$ is the pointwise limit of functions $f_n \in R(K)$ satisfying $\|f_n\|_K \leq \|f\|_\infty$?

By μ_z we will denote the harmonic measure on bK for a point $z \in \text{int } K$. Let ν be any positive measure on bK which dominates μ_z for all $z \in \text{int } K$. If $f \in L^\infty(\nu)$, then the function

$$\tilde{f}(z) = \int f\,d\mu_z, \qquad z \in \text{int } K,$$

is harmonic on int K. Evidently $\|\tilde{f}\|_\infty \leq \|f\|_\infty$, so the linear map $f \to \tilde{f}$ is continuous with respect to the supremum norms.

Each μ_z is a representing measure for z on $R(K)$. So μ_z is also multiplicative on the weak-star closure $H^\infty(\nu)$ of $R(K)$ in $L^\infty(\nu)$. In other words, $\widetilde{(fg)}(z) = \tilde{f}(z)\tilde{g}(z)$ whenever $f, g \in H^\infty(\nu)$.

Observe that if both h and zh are harmonic on some open set, then h is analytic there. This follows from an elementary computation

$$\nabla^2(zh) = 4\frac{\partial h}{\partial \bar{z}} + z\nabla^2 h,$$

valid for all smooth functions h.

Since $\widetilde{zf} = z\tilde{f}$ if $f \in H^\infty(\nu)$, both $\tilde{f}$ and $z\tilde{f}$ are harmonic on int K, and $\tilde{f}$ is analytic on int K. Hence $f \to \tilde{f}$ is a continuous homomorphism of the algebra $H^\infty(\nu)$ into the algebra $H^\infty(\text{int } K)$.

11.1 Theorem. Let K be a compact set such that:

(i) Every function $f \in H^\infty$ (int K) can be approximated pointwise on int K by a bounded sequence in $R(K)$.

(ii) Each component of int K has a point z_j with a dominant representing measure ν_j (cf. VI.5.2).

(iii) $R(bK) = C(bK)$.

Let $\nu = \sum \nu_j/2^j$. Then:

(a) The homomorphism $f \to \tilde{f}$ is an isometric isomorphism of $H^\infty(\nu)$ and $H^\infty(\text{int } K)$.

(b) If $f \in H^\infty(\text{int } K)$, there is a sequence $f_n \in R(K)$ such that $\|f_n\|_K \leq \|f\|_\infty$ and $f_n(z) \to f(z)$ for all $z \in \text{int } K$.

Proof. Note first that the harmonic measure for the points in the jth component of int K are absolutely continuous with respect to ν_j, and hence to ν. So the remarks preceding the theorem apply.

Suppose that $h \in H^\infty(\text{int } K)$. Let f_n be a bounded sequence in $R(K)$

such that $f_n(z) \to h(z)$ for all $z \in \text{int } K$. Then $h = \tilde{f}$, for any weak-star adherent point f of the sequence $\{f_n\}$ in $H^\infty(\nu)$. So the homomorphism $f \to \tilde{f}$ maps $H^\infty(\nu)$ onto H^∞ (int K).

Now suppose that $f \in H^\infty(\nu)$ is such that $\tilde{f} = 0$. Let f_α be a net in $R(K)$ converging weak-star to f in $L^\infty(\nu)$. Then $f_\alpha(z_0) = \tilde{f}_\alpha(z_0)$ converges to $\tilde{f}(z_0) = 0$ for all $z_0 \in \text{int } K$. The net $(f_\alpha - f_\alpha(z_0))/(z - z_0) \in R(K)$ converges weak-star to $f/(z - z_0)$. So $f/(z - z_0)$ belongs to $H^\infty(\nu)$ for all $z_0 \in \text{int } K$, in fact, for all $z_0 \notin bK$. Consequently $R(bK)f = C(bK)f \subseteq H^\infty(\nu)$. This is absurd, unless $f = 0$.

This shows that the map $f \to \tilde{f}$ is a continuous isomorphism of $H^\infty(\nu)$ and $H^\infty(\text{int } K)$. To complete the proof of (a), it suffices to note that any continuous isomorphism of two uniform algebras is an isometry. In fact, such an isomorphism ρ is bicontinuous, by the closed graph theorem. If $||f|| = 1$, then $||f^n|| = 1$, so $||\rho(f^n)|| = ||\rho(f)||^n$ must be bounded and bounded away from zero, by the bicontinuity of ρ. It follows that $||\rho(f)|| = 1$, and ρ is an isometry.

Part (b) follows from VI.5.2. That completes the proof.

Now we make some remarks on the hypotheses of 11.1.

Sufficient criteria for pointwise bounded approximation, hypothesis (i), were developed in section 10.

Hypothesis (iii), that $R(bK) = C(bK)$, holds whenever int K is connected and dense in K. Every point of bK is then a peak point of $R(bK)$, since it is in the closure of the single component int K of $(bK)^c$. And so Bishop's peak point criterion for rational approximation II.11.5 applies.

Also, $R(bK) = C(bK)$ whenever $R(K) = A(K)$. This follows again from Bishop's theorem, and the fact that the points on bK which are not peak points for $A(K)$ have zero area. And this latter fact follows easily from 4.2, for instance, since bK has Lebesgue density one at almost all points of bK.

Both hypotheses (ii) and (iii) hold, for instance, when $R(K)$ is a dirichlet algebra, or a hypodirichlet algebra. Applying this remark and 10.6 to the case when K^c has a finite number of components, we obtain a proof of VI.5.3 which does not depend on a study of parts and the reduction to circle domains.

An infinitely connected set to which the theorem applies is provided by the champagne bubble example (cf. fig. 6).

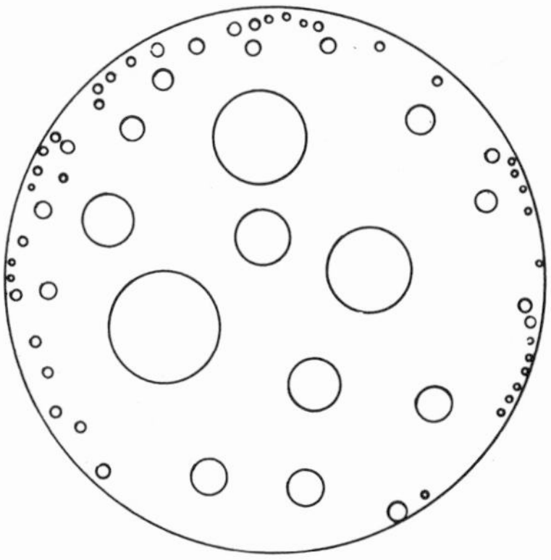

Figure 6. Champagne bubbles.

11.2 Corollary. Let K be a compact subset of the closed unit disc Δ obtained by deleting the interiors of a sequence of disjoint closed subdiscs $\{\Delta_n\}_{n=1}^{\infty}$, such that all but a finite number of the accumulation points of the Δ_n lie on $b\Delta$. If $f \in H^\infty(\text{int } K)$, there is a sequence $f_n \in R(K)$ such that $||f_n||_K \leq ||f||_\infty$ and $f_n(z) \to f(z)$ for all $z \in \text{int } K$.

Proof. As observed earlier, 10.7 and 10.8 show that the condition 11.1 (i) is met. Since int K is connected and dense in K, condition 11.1(iii) is met.

Let λ be a finite positive measure which is equivalent to the sum of arc length on the boundary circles of K and point masses at the accumulation points of the Δ_n which lie in int Δ. Recall (II.12.6) that every closed subset of zero arc length on the boundary of a disc is a peak interpolation set for the usual disc algebra. It follows that every closed subset E of bK satisfying $\lambda(E) = 0$ is a peak interpolation set for $R(K)$. By II.12.7, every measure on bK which is orthogonal to $R(K)$ must vanish on sets of λ-measure zero, and so must be absolutely continuous with respect to λ.

Fix $z_0 \in \text{int } K$. Evidently every representing measure on bK for z_0 must be absolutely continuous with respect to λ. It follows that z_0 has a dominant representing measure. In fact, any representing measure ν which minimizes $||\lambda_\nu||$, where λ_ν is the singular part of λ with respect to ν, will be dominant. So condition 11.1 (ii) is also met. That proves the corollary.

The dominant measure ν just produced need not be equivalent to harmonic measure. In fact, it may occur that ν is equivalent to λ, while μ_z has no mass at all on $b\Delta$.

There are compact sets K and functions in $H^\infty(\text{int } K)$ which can be approximated pointwise boundedly by functions in $A(K)$, but not by functions in $R(K)$. In fact, pointwise bounded approximation by $R(K)$ fails for example 9.2. However, if the arc of example 9.2 is appropriately chosen, $A(K)$ becomes a dirichlet algebra, so that pointwise bounded approximation by functions in $A(K)$ is always possible.

12. Estimates for Integrals

In this section we give an indication of some techniques involving analytic capacity which can be effectively applied to rational approximation theory. First we deal with several preliminaries.

12.1 Lemma. Let K be compact. If $f \in C(S^2)$ is analytic off K, and $|\text{Re} f(z)| \leq a$ for all z, then

$$|f'(\infty)| \leq 4a\alpha(K).$$

Proof. Since $|\text{Re}\,(f - f(\infty))| \leq 2a$, the function

$$g = (f - f(\infty))/(f - f(\infty) - 4a)$$

is bounded in modulus by 1. So $g \in \mathcal{AC}(K)$, and $|g'(\infty)| = |f'(\infty)|/4a \leq \alpha(K)$.

We will be using the anticonformal involution $z \to z^* = 1/\bar{z}$ of S^2. The image of a set E under this involution will be denoted by E^*. If $f(z) \in C(S^2)$ is analytic on U, then $\overline{f(z^*)}$ is analytic on U^*. Also, $z^* = z$ whenever $z \in b\Delta$.

12.2 Lemma. If K is a compact subset of the open annulus $\{r < |z| < 1\}$, then

$$\alpha(K^*) \leq \frac{8\alpha(K)}{r^2},$$

$$\alpha(K) \leq 8\alpha(K^*).$$

Proof. Let $f \in \mathcal{AC}(K^*)$. Set $g(z) = (\overline{f(z^*)} - \overline{f(0)})/2$. Then $g \in \mathcal{AC}(K)$, and $g'(0) = \overline{f'(\infty)}/2$. From the analogue of 2.5 for continuous analytic capacity, we obtain $|f'(\infty)| = 2|g'(0)| \leq 8\alpha(K)/r^2$. Hence $\alpha(K^*) \leq 8\alpha(K)/r^2$. That $\alpha(K) \leq 8\alpha(K^*)$ follows in the same manner from 2.5.

Let $h \in C(S^2)$ be analytic in a neighborhood of the annulus $\{r \leq |z| \leq 1\}$. Using the Laurent expansion, we can write h uniquely in the form

$$h(z) = h_0(z) + h_1(z),$$

where $h_0(z)$ is analytic for $|z| \leq 1$, $h_1(z)$ is analytic for $|z| \geq r$, and $h_1(\infty) = 0$. The function h_0 is given explicitly by

$$h_0(z) = \begin{cases} \dfrac{1}{2\pi i}\displaystyle\int_{|\zeta|=1} \frac{h(\zeta)}{\zeta - z}\,d\zeta, & |z| < 1 \\[2ex] h(z) + \dfrac{1}{2\pi i}\displaystyle\int_{|\zeta|=r} \frac{h(\zeta)}{\zeta - z}\,d\zeta, & |z| > r. \end{cases}$$

If $|z| \leq r$, then an elementary estimate yields $|h_0(z)| \leq \|h\|_\Delta/(1 - r)$. Consequently $|h_1(z)| \leq 2\|h\|_\Delta/(1 - r)$ for $|z| \leq r$. By the maximum modulus principle, we obtain

$$\|h_1\|_{S^2} \leq \frac{2\|h\|_\Delta}{1 - r}.$$

Suppose now that, in addition, h is real on $b\Delta$. The expression for h_0 can be written

$$h_0(z) = \frac{1}{2\pi}\int_0^{2\pi} \frac{1}{2}\left[\frac{e^{i\theta} + z}{e^{i\theta} - z} + 1\right] h(e^{i\theta})\,d\theta, \qquad |z| < 1.$$

Noting that $\mathrm{Re}\,[(e^{i\theta} + z)/(e^{i\theta} - z)]$ is the Poisson kernel, we see that $|\mathrm{Re}\, h_0(z)| \leq \|h\|_{b\Delta}$ whenever $|z| < 1$. Using the relation $h_1 = h - h_0$ and the maximum modulus principle, we find that

$$\|\operatorname{Re} h_1\|_{S^2} \le 2\|h\|_\Delta.$$

It follows that

$$\|\operatorname{Re} h_0\|_{S^2} \le 3\|h\|_{S^2}.$$

12.3 Lemma. If K is a compact subset of $\{r < |z| < 1\}$, then

$$\alpha(K \cup K^*) \le \frac{10^3\alpha(K)}{r^2}.$$

Proof. It suffices to show that

$$|g'(\infty)| \le \frac{500\|g\|\alpha(K)}{r^2}$$

whenever $g \in C(S^2)$ is analytic off $K \cup K^*$ and satisfies $g(z^*) = g(z)$. The general case is reduced to this case by expressing $f \in \mathscr{AC}(K)$ in the form $f = f_1 + if_2$, where $f_1(z) = (f(z^*) + f(z))/2$ and $f_2(z) = (f(z^*) - f(z))i/2$, and by applying the above estimate to f_1 and f_2.

Now suppose $g \in C(S^2)$ is analytic off $K \cup K^*$, and g satisfies $g(z^*) = g(z)$. In particular, g is real on $b\Delta$. Write $g = g_0 + g_1$, where g_0 is analytic on int Δ, g_1 is analytic outside Δ, and $g_1(\infty) = 0$. By the preceding remarks, the real parts of g_0 and g_1 are bounded in modulus by $3\|g\|_{S^2}$. Using 12.1, we obtain

$$|g'(\infty)| \le |g_0'(\infty)| + |g_1'(\infty)| \le 12\|g\|_\infty(\alpha(K) + \alpha(K^*)).$$

Now apply 12.2, to complete the proof.

12.4 Theorem (Melnikov's Theorem). There is a universal constant c_0 with the following property: If K is a compact subset of a closed disc J, and if $f \in C(J)$ is analytic on $(\operatorname{int} J)\backslash K$, then

$$\left|\int_{bJ} f(z)dz\right| \le c_0\|f\|_J\alpha(K\backslash bJ).$$

Proof. Since both sides of the inequality behave identically under affine transformations $z \to az + b$, we can assume that J is the closed unit disc Δ.

We will also assume always that f extends to be analytic in a neighborhood of $b\Delta$. In fact, by 7.4 an arbitrary f can be approximated uniformly on Δ by functions which are analytic where f is, and which extend to be analytic in a neighborhood of $b\Delta$.

Finally, we can assume that $K \subseteq \operatorname{int}\Delta$. In fact, suppose we have the estimate for the compact sets $K_r = K \cap \Delta(0; r)$, $0 < r < 1$. If $f \in C(\Delta)$ is analytic on $(\operatorname{int}\Delta)\backslash K$, and f extends to be analytic in an ϵ-neighborhood of $b\Delta$, then

$$\left|\int_{b\Delta} f(z)dz\right| \le c_0\|f\|_\Delta\alpha(K_{1-\epsilon})$$
$$\le c_0\|f\|_\Delta\alpha(K\backslash b\Delta).$$

To prove the theorem, we will first treat two special cases.

Suppose first that $K \subseteq \Delta(0; \frac{2}{3})$. Write $f = f_0 + f_1$, where f_0 is analytic on Δ, f_1 is analytic off K, and $f_1(\infty) = 0$. From the remarks preceding 12.2, with $r = \frac{2}{3}$, we obtain $\|f_1\|_{S^2} \leq 9\|f\|_\Delta$. Also,

$$\int_{b\Delta} f(z)dz = \int_{b\Delta} f_1(z)dz = 2\pi i f_1'(\infty).$$

Hence

$$\left|\int_{b\Delta} f(z)dz\right| \leq 2\pi |f_1'(\infty)| \leq 100\|f\|_\Delta \alpha(K).$$

For the second case, suppose that K is a compact subset of the annulus $\{\frac{1}{3} < |z| < 1\}$. Let

$$\begin{aligned} g(z) &= z^2 f(z) - \frac{1}{\pi i}\int_{b\Delta} \frac{\operatorname{Re}(\zeta^2 f(\zeta))}{\zeta - z} d\zeta \\ &= z^2 f(z) - \frac{1}{2\pi}\int_0^{2\pi} \left[\frac{e^{i\theta} + z}{e^{i\theta} - z} + 1\right] \operatorname{Re}(\zeta^2 f(\zeta)) d\theta, \end{aligned}$$

where $\zeta = e^{i\theta}$, and $|z| < 1$. Taking real parts and noting that $\operatorname{Re}[(e^{i\theta} + z)/(e^{i\theta} - z)]$ is the Poisson kernel, we see that $\operatorname{Re}(g)$ assumes the constant value $a = -(1/2\pi)\int_0^{2\pi} \operatorname{Re}(\zeta^2 f(\zeta))d\theta$ on $b\Delta$. The formula

$$g(z) = -\overline{g(z^*)} + 2a, \qquad |z| \geq 1,$$

extends g continuously to S^2, so that g is analytic off $K \cup K^*$.

Now $|\operatorname{Re}(g)| \leq 3$ on Δ. So $|a| \leq 3$, and $|\operatorname{Re}(g)| \leq 9$ on S^2. Also,

$$\overline{g'(\infty)} = -g'(0) = \frac{1}{\pi i}\int_{b\Delta} \frac{\operatorname{Re}(\zeta^2 f(\zeta))}{\zeta^2} d\zeta.$$

By 12.2 and 12.3,

$$\begin{aligned} \left|\int_{b\Delta} \frac{\operatorname{Re}(\zeta^2 f(\zeta))}{\zeta^2} d\zeta\right| &= \pi |g'(\infty)| \leq 36\pi\alpha(K \cup K^*) \\ &\leq b\alpha(K), \end{aligned}$$

where b is a universal constant. Replacing f by if, we obtain the estimate

$$\left|\int_{b\Delta} \frac{\operatorname{Im}(\zeta^2 f(\zeta))}{\zeta^2} d\zeta\right| \leq b\alpha(K).$$

Adding the estimates, we obtain

$$\left|\int_{b\Delta} f(\zeta)d\zeta\right| \leq 2b\alpha(K).$$

To complete the proof, we suppose now that K is an arbitrary compact subset of int Δ. Let K_1 be the set of $z \in K$ satisfying $|z| \leq \frac{2}{3}$, and K_2 the set $z \in K$ satisfying $|z| \geq \frac{1}{3}$. Let h be a continuously differentiable function such that $h(z) = 0$ when $|z| \geq \frac{2}{3}$, $h(z) = 1$ when $|z| \leq \frac{1}{3}$, and $|\partial h/\partial \bar{z}| \leq 4$. As in 7.1, set

$$G(\zeta) = \frac{1}{\pi} \iint \frac{f(z) - f(\zeta)}{z - \zeta} \frac{\partial h}{\partial \bar{z}} dxdy.$$

Then G is analytic off K_1, and $f - G$ is analytic off K_2. Also, $\| G \|_{S^2} \leq 10 \| f \|_\Delta$. Applying the cases we have treated to G and to $f - G$, we obtain

$$\begin{aligned} \left| \int_{b\Delta} f(z)dz \right| &\leq \left| \int_{b\Delta} G(z)dz \right| + \left| \int_{b\Delta} (f(z) - G(z))dz \right| \\ &\leq 100 \| G \|_\Delta \alpha(K_1) + 2b \| f - G \|_\Delta \alpha(K_2) \\ &\leq c_0 \| f \|_\Delta \alpha(K), \end{aligned}$$

where c_0 is a universal constant. That completes the proof.

12.5 Lemma: There is a universal constant c_1 with the following property: If K is a compact subset of an annulus $\{1 < |z| < R\}$, if $f(z)$ is continuous for $|z| \geq 1$, and if f is analytic on $\Delta^c \backslash K$ and at ∞, then

$$\left| \int_{|z|=R} f(z)dz - \int_{|z|=1} f(z)dz \right| \leq c_1 \| f \|_{\Delta^c} \alpha(K).$$

Proof. As in 12.4, it suffices to treat only the special cases when $K \subseteq \{2 < |z| < R\}$, and when $K \subseteq \{1 < |z| < 3\}$. The first case is handled using the Cauchy integral, as in 12.4. And so we consider only the second case.

Suppose $K \subseteq \{1 < |z| < 3\}$. Let $g(z) = \overline{f(z^*)}$. Then g is continuous on Δ, and g is analytic on $\Delta \backslash K^*$. If

$$g(z) = g(0) + g'(0)z + z^2 h(z),$$

then

$$|g'(0)| = \left| \frac{1}{2\pi i} \int_{|z|=1/3} \frac{g(z)}{z^2} dz \right| \leq 2 \| g \|_\Delta,$$

$$\| h \|_\Delta = \sup_{1/3 \leq |z| \leq 1} |h(z)| \leq 12 \| g \|_\Delta = 12 \| f \|_{\Delta^c}.$$

Hence

$$\begin{aligned} \left| \int_{|z|=R} f(z)dz - \int_{|z|=1} f(z)dz \right| &= \left| \int_{|z|=1/R} \frac{g(z)}{z^2} dz - \int_{|z|=1} \frac{g(z)}{z^2} dz \right| \\ &= \left| \int_{|z|=1} h(z)dz \right| \\ &\leq c_0 \| h \|_\Delta \alpha(K^*) \\ &\leq 12 \cdot 8c_0 \| f \|_{\Delta^c} \alpha(K), \end{aligned}$$

where c_0 is the constant of 12.4. That does it.

12.6 Theorem. There exists a universal constant c with the following property: If J is a closed annulus of conformal radius r, if K is a compact subset of J, and if $f \in C(J)$ is analytic on $(\text{int } J)\backslash K$, then

$$\left|\int_{bJ} f(z)dz\right| \leq \frac{c}{1-r} \|f\|_J \alpha(K\backslash bJ).$$

Proof. As in 12.4, we assume that $K \subseteq \text{int } J$. We also assume that J is the annulus $\{1 \leq |z| \leq R\}$, where $R = 1/r$. And we assume that $f \in C(S^2)$ is analytic in a neighborhood of bJ, and that $\|f\|_{S^2} \leq 2\|f\|_J$.

Let h be a continuously differentiable function such that $h(z) = 1$ for $|z| \leq 1$, $h(z) = 0$ for $|z| \geq R$, and $|\partial h/\partial \bar{z}| \leq 2/(R-1)$. Set

$$G(\zeta) = \frac{1}{\pi} \iint \frac{f(z) - f(\zeta)}{z - \zeta} \frac{\partial h}{\partial \bar{z}} dxdy.$$

Then $G(z)$ is ananlytic for $|z| \geq R$ and for $z \in J\backslash K$, and $f - G$ is analytic for $|z| \leq 1$. Also

$$\|G\|_{S^2} \leq \frac{10^2 R}{R-1} \|f\|_J = \frac{10^2}{1-r} \|f\|_J,$$

$$\int_{bJ} f(z)dz = \int_{bJ} G(z)dz + \int_{|z|=R} (f(z) - G(z))dz.$$

Using 12.4 to estimate the last integral, and 12.5 to estimate the other, we obtain the desired result.

Note that 12.4 can be obtained from 12.6 by letting $r \to 0$. Also, 12.5 is a special case of 12.6.

12.7 Corollary. Let K be compact, and let $z_0 \in K$. Let $0 < a < 1$, and let $E_n = \{a^{n+1} < |z - z_0| < a^n\}$. Let $f \in C(S^2)$. If $f \in A(K)$, then

$$|f(z_0)| \leq \frac{1}{2\pi} \left| \int_{|z-z_0|=1} \frac{f(z)}{z - z_0} dz \right| + \frac{c}{1-a} \sum_{n=0}^{\infty} a^n \|f\|_{E_n} \alpha(E_n \backslash \text{int } K).$$

If $f \in R(K)$, then

$$|f(z_0)| \leq \frac{1}{2\pi} \left| \int_{|z-z_0|=1} \frac{f(z)}{z - z_0} dz \right| + \frac{c}{1-a} \sum_{n=0}^{\infty} a^n \|f\|_{E_n} \gamma(E_n \backslash K).$$

Proof. Since the telescoping series

$$\int_{|\zeta - z_0| = a^N} \frac{f(\zeta)}{\zeta - z_0} d\zeta = \int_{|\zeta - z_0|=1} \frac{f(\zeta)}{\zeta - z_0} d\zeta - \sum_{n=0}^{N-1} \int_{bE_n} \frac{f(\zeta)}{\zeta - z_0} d\zeta$$

converges to $2\pi i f(z_0)$ as $N \to \infty$, we obtain

$$2\pi |f(z_0)| \leq \left| \int_{|\zeta - z_0|=1} \frac{f(\zeta)}{\zeta - z_0} d\zeta \right| + \sum_{n=0}^{\infty} \left| \int_{bE_n} \frac{f(\zeta)}{\zeta - z_0} d\zeta \right|.$$

If $f \in A(K)$, we use the estimate of 12.6 to obtain the result. For any f analytic in a neighborhood of K, and hence for all f in $R(K)$, we can replace $\alpha(E_n \backslash \text{int } K)$ in the estimate by $\alpha(E_n \backslash K) = \gamma(E_n \backslash K)$. That proves the corollary.

Now we are in a position to give the proof of the necessity of Melnikov's peak point criterion 4.5. In fact, suppose $\sum a^n \alpha(E_n \backslash \text{int } K)$ converges. If $f \in A(K)$ is such that $|f(z)| < 1$ for $z \neq z_0$, then the above estimate shows that $|f^m(z_0)| \to 0$ as $m \to \infty$. So f cannot peak at z_0, and z_0 is not a peak point of $A(K)$. Similarly, if $\sum a^n \gamma(E_n \backslash K)$ converges, then z_0 cannot be a peak point of $R(K)$.

As another corollary to 12.6 we obtain a subadditivity property of analytic capacity.

12.8 Corollary. If K is a compact subset of the plane, and J is a closed disc, then

$$\alpha(K) \leq c(\alpha(K \cap \text{int } J) + \alpha(K \cap J^c)),$$

where c is a universal constant.

Proof. We can assume that $J = \Delta$. Let $f \in \mathscr{A}\mathscr{C}(K)$. If R is large, then

$$\begin{aligned} 2\pi |f'(\infty)| &= \left| \int_{|z|=R} f(z)dz \right| \\ &\leq \left| \int_{|z|=R} f(z)dz - \int_{|z|=1} f(z)dz \right| + \left| \int_{|z|=1} f(z)dz \right| \\ &\leq \frac{c}{1-(1/R)} \alpha(K \cap \Delta^c) + c\alpha(K \cap \text{int } \Delta). \end{aligned}$$

This yields the desired estimate.

13. Analytically Negligible Sets

A subset E of S^2 is **analytically negligible** if every function $f \in C(S^2)$ which is analytic on an open set U can be approximated uniformly on S^2 by functions in $C(S^2)$ which are analytic on $U \cup E$. Evidently, countable unions of analytically negligible sets are again analytically negligible. By II.1.8, points are analytically negligible.

If E is analytically negligible, then $\alpha(E) = 0$. It is not known whether compact sets E satisfying $\alpha(E) = 0$ are analytically negligible. However, the following characterization is available.

13.1 Theorem. The following conditions are equivalent, for a compact set E:

(i) E is analytically negligible.
(ii) $\alpha(K) = \alpha(K \backslash E)$ for all compact sets K.

(iii) There exists $c > 0$ such that for all sufficiently small $\delta > 0$,

$$\alpha(\Delta(z;\delta)\backslash E) \geq c\delta, \qquad z \in E.$$

(iv) For every $z \in E$,

$$\liminf_{\delta \to 0} \frac{\alpha(\Delta(z;\delta)\backslash E)}{\delta} > 0.$$

In particular, analytic negligibility is a local property.

Proof. First one shows that (i), (ii), and (iii) are equivalent. The nontrivial implication, that (iii) implies (i), is handled by the approximation scheme developed in section 7, just as in the proof of 7.4. The details are omitted.

That (ii) implies (iv) is clear. To complete the proof, we must show that (iv) implies the equivalent conditions (i)–(iii).

Suppose that (iv) is true. Let E_n be the set of $w \in E$ such that $\alpha(\Delta(w;\delta)\backslash E) \geq \delta/n$ whenever $\delta \leq 1/n$. Then

$$\alpha(\Delta(z;2\delta)\backslash E_n) \geq \alpha(\Delta(z;2\delta)\backslash E) \geq \frac{\delta}{n}$$

whenever $z \in \bar{E}_n$ and $\delta \leq 1/n$. Applying (iii) to $\bar{E}_n$, we find that $\bar{E}_n$ is analytically negligible. Hence $E = \bigcup_{n=1}^{\infty} \bar{E}_n$ is analytically negligible. That completes the proof.

13.2 Corollary. Circles and straight lines are analytically negligible.

Proof. From 12.8 it follows that $\alpha(K) \leq 2c\alpha(K\backslash E)$ whenever K is compact and E is a circle. Hence the estimate 13.1(iii) obtains for any circle, and circles are analytically negligible. Since the family of analytically negligible sets remains invariant under conformal automorphisms of the sphere, straight lines are also analytically negligible.

At long last, we are in a position to prove that $A(K) = R(K)$ when K is the "string of beads" example (cf. VI.3, fig. 3), obtained from the closed unit disc by punching out holes centered on the real axis. In fact, any function in $A(K)$ can be approximated uniformly by functions in $A(K)$ which extend to be analytic in a neighborhood of the real axis. And such functions can be dealt with using the version of Mergelyan's theorem for finitely connected sets (cf. II.10.4 and II.10.5).

To enlarge our collection of analytically negligible sets, we prove the following.

13.3 Theorem. Let E be a subset of the closed unit disc Δ. Then E is analytically negligible if and only if every $f \in C(\Delta)$ which is analytic on an open subset U of Δ can be approximated uniformly on Δ by functions in $C(\Delta)$ which are analytic on $U \cup E$.

Proof. If E is analytically negligible, then E certainly has the property described.

Conversely, suppose that E has the property described above. Let $f \in C(S^2)$ be analytic on the open subset V of S^2. We must approximate f uniformly by functions analytic on $V \cup E$.

Applying 13.2, we can assume that f is analytic in a neighborhood of $b\Delta$. Then we can write $f = f_1 + f_2$, where f_1 is analytic inside Δ, f_2 is analytic outside Δ, and $f_2(\infty) = 0$. Replacing f by f_2, we can assume that $f(z)$ is analytic for $|z| > 1 - \epsilon$, and $f(\infty) = 0$.

By hypothesis, we can find $g_n \in C(\Delta)$ such that $g_n(z)$ is analytic for $1 - \epsilon < |z| < 1$, g_n is analytic on $(V \cap \Delta) \cup E$, and g_n converges uniformly on Δ to f. Let $g_n = h_n + k_n$, where k_n is analytic inside Δ, $h_n(z)$ is analytic for $|z| > 1 - \epsilon$, and $h_n(\infty) = 0$. Then h_n must converge uniformly on S^2 to f. That proves the theorem.

Using 13.3, and the fact that analytic negligibility is a local property, it is easy to deduce the following.

13.4 Corollary. Let g be a conformal map of a domain D onto another domain. A compact subset E of D is analytically negligible if and only if $g(E)$ is analytically negligible.

This corollary shows that any analytic curve is analytically negligible. Vitushkin [1] has shown that any twice continuously differentiable curve is analytically negligible. This is a consequence of the following important theorem, which extends 12.4 and 12.6.

Theorem. Let D be a bounded domain whose boundary consists of a finite number of piecewise twice continuously differentiable curves. There is a constant $c(D)$, depending only on D, such that for every compact subset K of $\bar{D}$, and for every function $f \in C(\bar{D})$ which is analytic on $D \backslash K$,

$$\left| \int_{bD} f(z)dz \right| \leq c(D) \|f\|_D \alpha(K \backslash bD).$$

This theorem is valid for domains containing ∞, providing $\int_{bD}$ is replaced by $\int_{bK}$.

In order to prove the theorem, it is easy to perform a reduction to the case in which D has one boundary component. In this case, one maps D conformally onto the interior of the unit disc, where one can apply 12.4. The proof depends decisively on the behavior of the mapping function at the boundary of D. For details, see Vitushkin [1].

NOTES

The main references for this chapter are the exposition of Vitushkin [1] and the notes of Zalcman [1]. For a more penetrating treatment of methods involving analytic capacity, we refer the reader to the former source. For an extensive bibliography and bibliographical notes on rational approximation, we refer the reader to the latter source.

The uniqueness of the Ahlfors function was proved for finitely connected domains by Ahlfors, and for arbitrary domains by Havinson. The proof we have presented is in Fisher [3], where other aspects of Havinson's work are simplified. Theorem 4.1, and also the proof presented here of the sufficiency of Melnikov's criterion, are due to Curtis [1]. Theorem 4.3 is due to Gonchar [1]. Theorem 5.1 is due to Vitushkin, and the definitive version is in Vitushkin [1].

The scheme for approximation goes back to Mergelyan [1]. In the case that he treated, the square of the Ahlfors function f automatically maximized β. So one could match the first two coefficients of the Laurent expansion of the approximatee, with the appropriate bounds, by a linear combination of f and f^2. This trick could have been used, for instance, to prove 7.4 (due to Mergelyan), without referring to analytic center and diameter. Vitushkin's program, which eventually led to sparkling success, involved a detailed study of the second coefficient of the Laurent series of the approximatee.

The examples of section 9 are variations of an example of Dolzhenko [1]. The application of Vitushkin's techniques to pointwise bounded approximation is from Gamelin and Garnett [1]. Section 12 is based on Melnikov [1].

EXERCISES

1. If f is analytic on $\Omega(K)$, $\|f\|_{\Omega(K)} \leq 1$, and $f(\infty) \neq 0$, then $|f'(\infty)| < \gamma(K)$.

2. If K is compact and connected, then $\gamma(K) = \beta(K)$.

3. If E is a relatively closed subset of an open set D, and $\gamma(E) = 0$, then $H^\infty(D) = H^\infty(D \backslash E)$.

4. Suppose $\{E_n\}_{n=1}^\infty$ are compact. If $\gamma(E_n) = 0$, $1 \leq n < \infty$, then $\gamma\left(\bigcup_{n=1}^\infty E_n\right) = 0$. If $\alpha(E_n) = 0$, $1 \leq n < \infty$, then $\alpha\left(\bigcup_{n=1}^\infty E_n\right) = 0$. *Hint:* Use the Baire category theorem.

5. A function $f \in A$ satisfying $\|f\|_X \leq 1$ is an extreme point of the unit ball of A if and only if whenever $h \in A$ satisfies $|f| + |h| \leq 1$, then $h = 0$. *Hint:* See the proof of 1.6.

6. If K is compact and z_0 is complex, there is a unique function $f \in \mathscr{A}(K)$ such that $\beta(f, z_0) = \beta(K, z_0)\gamma(K)$.

7. Let μ have compact support, and let $h(r) = \sup_{z \in C} |\mu|(\Delta(z;r))$. If $\int_0^1 h(r) r^{-2} dr < \infty$, then $\hat{\mu}$ is continuous on S^2.

8. If J is The Cantor Set, then $\alpha(J \times J) > 0$. *Hint:* Apply exercise 7 to $\nu \times \nu$, where ν is the usual singular measure on J.

9. If $\mu = g\,dxdy$, where $g \in L^q(dxdy)$ has compact support, and $q > 2$, then $\hat{\mu}$ is continuous. In fact, if $2 < q < \infty$, then $|\hat{\mu}(z) - \hat{\mu}(w)| \le c|z - w|^{1-2/q}$ for some constant $c > 0$.

10. If $p < 2$ and K has no interior, then $R(K)$ is dense in $L^p(K, dxdy)$. *Hint:* Apply exercise 9.

11. If P is the set of peak points on bK for $A(K)$, then $\alpha(bK \setminus P) = 0$.

12. If for some $\epsilon > 0$, $\limsup_{r \to 0} \gamma(\Delta(z_0; r) \setminus K)/r^{1+\epsilon} < \infty$, then z_0 is not a peak point of $R(K)$. An analogous result holds for $A(K)$.

13. Let K be the compact set obtained from Δ by deleting disjoint open discs whose centers lie on the positive real axis, and whose centers and radii tend to 0 (cf. II.10, fig. 2). Then 0 is a peak point for $R(K)$ if and only if the Cauchy measure dz/z on bK is finite.This result ceases to be valid if the centers are not required to lie on the real axis.

14. Suppose $a(\delta) > 0$ satisfies $\lim_{\delta \to 0} a(\delta) = +\infty$. Then there is a Swiss cheese K for which $R(K) \ne C(K)$, and such that

$$\limsup_{\delta \to 0} \sup_{z \in K} \frac{\gamma(\Delta(z;\delta) \setminus K)}{\delta^2} a(\delta) = +\infty.$$

15. If K has no interior, and $F(\zeta) = \pi^{-1} \iint_K (z - \zeta)^{-1}\, dxdy$, then $\bar{z} - F \in R(K)$. In particular, $R(K) = C(K)$ if and only if $F \in R(K)$.

16. Every $f \in A(K)$ can be approximated uniformly on K by functions in $A(K)$ which extend to be analytic in a neighborhood of the outer boundary of K.

17. If the inner boundary of K lies on an analytically negligible set, then $R(K) = A(K)$.

18. Any compact connected set E can be the inner boundary of a set K for which $R(K) = A(K)$. *Hint:* Punch out disjoint connected open sets which converge to E, and apply II.10.4.

19. Let B be a Banach space, let $C(K, B)$ be the space of continuous B-valued functions on K, and let $R(K, B)$ be the subspace of functions in $C(K, B)$ which can be approximated uniformly on K by B-valued functions which are analytic in a neighborhood of K. The following are equivalent, for $f \in C(K, B)$.

(i) $f \in R(K, B)$.

(ii) For each $L \in B^*$, $L \circ f \in R(K)$.

(iii) For every continuously differentiable complex-valued function g supported on any disc $\Delta(z_0; \delta)$,

$$\left\| \int f(z) \frac{\partial g}{\partial \bar{z}} dxdy \right\| \le \delta \omega(f, \delta) \left\| \frac{\partial g}{\partial \bar{z}} \right\|_\infty \alpha(\Delta(z_0;\delta) \setminus K).$$

20. Let Y be compact, let B be a closed subspace of $C(Y)$, and let $R(K) \otimes B$ be the closed subspace of $C(K \times Y)$ generated by functions of the form $f(x)g(y)$,

$f \in R(K)$, $g \in B$. Let S be the "slice" space of all $h \in C(K \times Y)$ such that $x \to h(x, y)$ belongs to $R(K)$ for each fixed $y \in Y$, and $y \to h(x, y)$ belongs to B for each fixed $x \in K$.

(a) There are natural identifications of $R(K) \otimes B$ and S with closed subspaces of $C(K, B)$.
(b) $R(K) \otimes B = R(K, B)$.
(c) If $h \in C(K, B)$, then $h \in S$ if and only if $L \circ h \in R(K)$ for all $L \in B^*$.
(d) $R(K) \otimes B = S$.
(e) If K_1 and K_2 are compact planar sets such that $R(K_1) = A(K_1)$ and $R(K_2) = A(K_2)$, then $R(K_1 \times K_2) = A(K_1 \times K_2)$.

BIBLIOGRAPHY

This bibliography represents a scattering of the literature on uniform algebras, together with a few textbooks and items of tangential interest. No attempt has been made to be complete.

Due to the availability of the excellent bibliography and bibliographical notes of Zalcman [1], I have limited the references concerning analytic capacity and analytic function theory to the barest essentials. The notes of Zalcman can also be consulted for references concerning uniform approximation on Riemann surfaces.

None of the individual papers in *Function Algebras* (The Proceedings of the International Conference on Function Algebras at Tulane, edited by Birtel [1]), have been referenced. In this book one can find a trove of assorted serendipities and bibliographical data, including on pp. 346–355 a list of (at that time) unsolved problems.

There is a collection of problems in the B.A.M.S. 5(1961), pp. 457–460. Even earlier, Gelfand [2] proposed a series of problems, some of which remain unsolved. (In connection with this, see Hoffman and Singer [2].)

A survey of uniform algebras, the most complete up to its time, was given by Hoffman [5]. Various notes and addresses of an expository nature have been given by Wermer [13, 14, 15, 17]. Other works of an expository nature are those of Royden [3] and Bishop [10]. For a survey of results and problems related to algebras of analytic functions of several complex variables, see "Analytic function algebras" by Rossi and Stolzenberg in Birtel [1], pp. 138–148.

AHERN, P.

1. On the generalized F. and M. Riesz theorem, P.J.M. 15 (1965), 373–376.

AHERN, P., and SARASON, D.

1. The H^p spaces of a class of function algebras, Acta Math. 117 (1967), 123–163.
2. On some hypodirichlet algebras of analytic functions, Amer. J. Math. 89 (1967), 932–941.

AHLFORS, L., and Beurling, A.

1. Conformal invariants and function theoretic null sets, Acta Math. 83 (1950), 101–129.

ALLING, N.

1. A proof of the corona conjecture for finite open Riemann surfaces, B.A.M.S. 70 (1964), 110–112.

ARENS, R.

1. The boundary integral of log $|\phi|$ for generalized analytic functions, T.A.M.S. 86 (1957), 57–69.
2. Dense inverse limit rings, Mich. Math. J. 5 (1958), 169–182.
3. The maximal ideals of certain function algebras, P.J.M. 8 (1958), 641–648.
4. The closed ideals of algebras of functions holomorphic on a Riemann surface, Ren. Cir. Mat. di Palermo 7 (1958), 245–260.
5. The problem of locally-A functions in a commutative Banach algebra A, T.A.M.S. 104 (1962), 24–36.
6. The group of invertible elements of a commutative Banach algebra, Studia Math. (Seria Spec.), Z. 1 (1963), 21–23.

ARENS, R., and Calderón, A.

1. Analytic functions of several Banach algebra elements, Ann. of Math. 62 (1955), 204–216.

ARENS, R., and Singer, I.

1. Function values as boundary integrals, P.A.M.S. 5 (1954), 735–745.
2. Generalized analytic functions, T.A.M.S. 81 (1956), 379–393.

BEAR, H. S.

1. A geometric characterization of Gleason parts, P.A.M.S. 16 (1965), 407–412.

BECK, A.

1. A theorem on maximum modulus, P.A.M.S. 15 (1964), 345–349.

BEURLING, A.

1. On two problems concerning linear transformations in Hilbert space, Acta. Math. 81 (1949), 239–255.

BINGEN, F., Tits, J., and Waelbroeck, L.

1. Séminaire sur les Algebras de Banach, Université Libre de Bruxelles, Inst. de Mathématique (1962–1963).

BIRTEL, F.

1. Function Algebras (Proc. Int'l Symp. on Function Algebras, Tulane U., 1965, edited by F. Birtel), Scott-Foresman, Chicago, 1965.

BISHOP, E.

1. The structure of certain measures, Duke Math. J. 25 (1958), 283–289.
2. Subalgebras of functions on a Riemann surface, P.J.M. 8 (1958), 29–50.

3. A minimal boundary for function algebras, P.J.M. 9 (1959), 629–642.
4. Boundary measures of analytic differentials, Duke Math. J. 27 (1960), 331–340.
5. A generalization of the Stone-Weierstrass theorem, P.J.M. 11 (1961), 777–783.
6. A general Rudin-Carleson theorem, P.A.M.S. 13 (1962), 140–143.
7. Analyticity in certain Banach algebras, T.A.M.S. 102 (1962), 507–544.
8. Holomorphic completions, analytic continuations, and the interpolation of seminorms, Ann. of Math. 78 (1963), 468–500.
9. Representing measures for points in a uniform algebra, B.A.M.S. 70 (1964), 121–122.
10. Uniform algebras, Proc. Conf. Complex Analysis, Mpls. (1964), Springer, Berlin, 1965, pp. 272–280.

BISHOP, E., and de Leeuw, K.

1. The representation of linear functionals by measures on sets of extreme points, Ann. Inst. Fourier 9 (1959), 305–331.

BOCHNER, S.

1. Boundary values of analytic functions in several variables and of almost periodic functions, Ann. Math. 45 (1944), 708–722.
2. Generalized conjugate and analytic functions without expansions, Proc. Nat. Acad. Sci. 44 (1959), 855–857.

BOURBAKI, N.

1. Théories Spectrales (Eléménts de Math. Fasc. XXXII), Hermann, Paris, 1967.

BRENNAN, J.

1. Invariant subspaces and rational approximation (preprint).

BROWDER, A.

1. Cohomology of maximal ideal spaces, B.A.M.S. 67 (1961), 515–516.
2. Point derivations on function algebras, J. Functional Analysis 1 (1967), 22–27.
3. Rational Approximation and Function Algebras, Benjamin, New York, 1969.

BROWDER, A., and Wermer, J.

1. Some algebras of functions on an arc, J. Math. Mech. 12 (1963), 119–130.
2. A method for constructing dirichlet algebras, P.A.M.S. 15 (1964), 546–552.

CARLESON, L.

1. Representations of continuous functions, Math. Z. 66 (1957), 447–451.
2. Interpolation by bounded analytic functions and the corona problem, Ann. of Math. 76 (1962), 542–559.
3. Mergelyan's theorem on uniform approximation, Math. Scand. 15 (1965), 167–175.

CHIRKA, E. M.

1. Approximation of continuous functions by functions holomorphic on Jordan arcs in C^n, Dokl. Akad. Nauk SSSR 167 (1966), 38–40. Soviet Math. Dokl. 7 (1966), 336–338.

COHEN, P. J.

1. Factorization in group algebras, Duke Math. J. 26 (1959), 199–205.
2. A note on constructive methods in Banach algebras, P.A.M.S. 12 (1961), 159–163.

CURTIS, P. C.

1. Peak points for algebras of analytic functions, J. Functional Analysis 3 (1969), 35–47.

DE BRANGES, L.

1. The Stone-Weierstrass theorem, P.A.M.S. 10 (1959), 822–824.

DE LEEUW, K. (see Bishop)

1. A type of convexity in the space of *n* complex variables, T.A.M.S. 83 (1956), 193–204.
2. Functions on circular subsets of the space of *n* complex variables, Duke Math. J. 24 (1957), 415–431.

DE LEEUW, K., and Glicksberg, I.

1. Quasi-invariance and measures on compact groups, Acta Math. 109 (1963), 179–205.

DE LEEUW, K., and Rudin, W.

1. Extreme points and extremum problems in H_1, P.J.M. 8 (1958), 467–485.

DEVINATZ, A.

1. Conjugate function theorems for dirichlet algebras, Rev. U. Mat. Arg. 23 (1966), 3–30.

DIXMIER, J.

1. Sur certains espaces considérés par M. H. Stone, Suma Brasil. Math. 2 (1951), 151–182.

DOLZHENKO, E. P.

1. On approximation on closed regions and on null-sets, Dokl. Akad. Nauk SSSR 143 (1962), 771–774.

DUNFORD, N., and Schwartz, J.

1. Linear Operators, Part I: General Theory, Interscience, New York, 1958.
2. Linear Operators, Part II: Spectral Theory, Interscience, New York, 1963.

FARRELL, O. J.

1. On approximation by polynomials to a function analytic in a simply connected region, B.A.M.S. 41 (1935), 707–711.

FISHER, S.

1. Norm compact sets of representing measures, P.A.M.S. 19 (1968), 1063–1068.
2. Bounded approximation by rational functions, P.J.M. 28 (1969), 319–326.
3. On Schwarz's lemma and inner functions, T.A.M.S. (1969).

FORELLI, F.

1. Analytic measures. P.J.M. 13 (1963), 571–578.
2. The isometries of H^p, Can. J. Math. 16 (1964), 721–728.
3. Bounded holomorphic functions and projections, Ill. J. Math. 10 (1966), 376–380.
4. Extreme points in $H^1(R)$, Can. J. Math. 19 (1967), 312–320.
5. Analytic and quasi-invariant measures, Acta Math. 118 (1967), 33–59.
6. Measures orthogonal to polydisc algebras, J. Math. Mech. 17 (1968), 1073–1086.

FREEMAN, M.

1. Uniform approximation on a real analytic manifold, T.A.M.S. (1969).

GAMELIN, T.

1. Restrictions of subspaces of $C(X)$, T.A.M.S. 112 (1964), 278–286.
2. H^p spaces and extremal functions in H^1, T.A.M.S. 124 (1966), 158–167.
3. Remarks on compact groups with ordered duals, Rev. U. Mat. Arg. 23 (1967), 97–108.
4. Embedding Riemann surfaces in maximal ideal spaces, J. Functional Analysis 2 (1968), 123–146.

GAMELIN, T., and Garnett, J.

1. Constructive techniques in rational approximation, T.A.M.S. (1969).

GAMELIN, T., and Lumer, G.

1. Theory of abstract Hardy spaces and the universal Hardy class, Advances in Math., Vol. 2, Fasc. 2 (1968), 118–174.

GAMELIN, T., and Voichick, M.

1. Extreme points in spaces of analytic functions, Can. J. Math. 20 (1968), 919–928.

GAMELIN, T., and Wilken, D.

1. Closed partitions of maximal ideal spaces, Ill. J. Math. 13 (1969).

GARNETT, J.

1. Disconnected Gleason parts, B.A.M.S. 72 (1966), 490–492.
2. A topological characterization of Gleason parts, P.J.M. 20 (1967), 59–63.
3. On a theorem of Mergelyan, P.J.M. 26 (1968), 461–467.

GARNETT, J. and Glicksberg, I.

1. Algebras with the same multiplicative measures, J. Functional Analysis 1 (1967), 331–341.

GELFAND, I. M.

1. Normierte Ringe, Mat. Sb. (N.S.) 9 (51) (1941), 3–24.
2. On subrings of the ring of continuous functions, Uspehi Mat. Nauk 12 (1957), No. 1 (73), 249–251. A.M.S. Transl. (2) 16 (1960), 477–479.

GELFAND, I. M., Raikov, D. A., and Shilov, G. E.

1. Commutative normed rings, Uspehi Mat. Nauk 1 (1946), No. 2 (12), 48–146. A.M.S. Transl. (2) 5 (1957), 115–220.

GLEASON, A.

1. Function algebras, Seminar on Analytic Functions, Vol. II, Institute for Advanced Study, Princeton (1957), 213–226.
2. Finitely generated ideals in Banach algebras, J. Math. Mech. 13 (1964), 125–132.
3. A characterization of maximal ideals, J. d'Analyse Math. 19 (1967), 171–172.

GLEASON, A., and Whitney, H.

1. The extension of linear functionals defined on H^∞, P.J.M. 12 (1962), 163–182.

GLICKSBERG, I. (see de Leeuw, Garnett)

1. Measures orthogonal to algebras and sets of antisymmetry, T.A.M.S. 105 (1962), 415–435.
2. Function algebras with closed restrictions, P.A.M.S. 14 (1963), 158–161.
3. A remark on analyticity of function algebras, P.J.M. 13 (1963), 1181–1185.
4. Some uncomplemented function algebras, T.A.M.S. 111 (1964), 121–137.
5. Maximal algebras and a theorem of Radó, P.J.M. 14 (1964), 919–941.
6. A Phragmén-Lindelöf theorem for function algebras, P.J.M. 22 (1967), 401–406.
7. The abstract F. and M. Riesz theorem, J. Functional Analysis 1 (1967), 109–122.
8. Dominant representing measures and rational approximation, T.A.M.S. 130 (1968), 425–462.

GLICKSBERG, I., and Wermer, J.

1. Measures orthogonal to a Dirichlet algebra, Duke Math. J. 30 (1963), 661–666.

GOKHBERG, I. C., and Krein, M. G.

1. Fundamental aspects of defect numbers, root numbers and indexes of linear operators, Uspehi Mat. Nauk 12 (1957), No. 2 (74), 43–118. A.M.S. Transl. (2) 13 (1960), 185–264.

GONCHAR, A. A.

1. The minimal boundary of the algebra $A(E)$, Izv. Akad. Nauk SSSR Ser. Mat. 27 (1963), 949–955.
2. Examples of non-uniqueness of analytic functions, Vestnik Moskov. Univ., Ser. I, Mat. Meh., No. 1 (1964), 37–43.

GONCHAR, A. A., and Mergelyan, S. N.

1. Uniform approximation by analytic and harmonic functions, Contemporary Problems in the Theory of Analytic Functions, International Conferences on the Theory of Analytic Functions, Erevan, 1965, 94–101.

GORIN, E. A.

1. Moduli of invertible elements in a normed algebra, Vestnik Moskov. Univ. Ser. 1, no. 5 (1965), 35–39.
2. On continuation of invertible elements of a normed algebra of functions, Vestnik Moskov. Univ. Ser. 1, no. 2 (1966), 22–25.
3. Maximal subalgebras of commutative Banach algebras with involution, Mat. Zametki 1 (1967), 173–178.

GUNNING, R., and Rossi, H.

1. Analytic Functions of Several Complex Variables, Prentice-Hall, Englewood Cliffs, N.J., 1965.

HALLSTROM, A.

1. On bounded point derivations and analytic capacity, J. Functional Analysis 4 (1969).

HARTOGS, F., and Rosenthal, A.

1. Uber Folgen analytischen Funktionen, Math. Ann. 104 (1931), 606–610.

HASUMI, M.

1. Invariant subspace theorems for finite Riemann surfaces, Can. J. Math. 18 (1966), 240–255.
2. Interpolation sets for logmodular Banach algebras, Osaka J. Math. 3 (1966), 303–319.

HELSON, H.

1. Lectures on Invariant Subspaces, Academic Press, New York, 1964.
2. Compact groups with ordered duals, Proc. London Math. Soc. (3) 14A (1965), 144–156.

HELSON, H. and Lowdenslager, D.

1. Prediction theory and Fourier series in several variables, Acta Math. 99 (1958), 165–202.
2. Prediction theory and Fourier series in several variables II, Acta Math. 106 (1961), 175–213.
3. Invariant subspaces, Proc. Internat. Symp. Linear Spaces, Jerusalem, 1960, Macmillan (Pergamon), New York, 1961, 251–262.

HELSON, H., and Quigley, F.

1. Maximal algebras of continuous functions, P.A.M.S. 8 (1957), 111–114.
2. Existence of maximal ideals in algebras of continuous functions, P.A.M.S. 8 (1957), 115–119.

HOFFMAN, K.

1. Boundary behavior of generalized analytic functions, T.A.M.S. 87 (1958), 447–466.

2. Minimal boundaries for analytic polyhedra, Rend. Circ. Mat. di Palermo, Ser. 2., 9 (1960), 147–160.
3. Analytic functions and logmodular Banach algebras, Acta Math. 108 (1962), 271–317.
4. Banach Spaces of Analytic Functions, Prentice-Hall, Englewood Cliffs, N.J., 1962.
5. Lectures on sup norm algebras, Summer School on Topological Algebra Theory, University of Brussels Press, Brussels, Belgium, 1966.
6. Bounded analytic functions and Gleason parts, Ann. Math. 86 (1967), 74–111.

HOFFMAN, K., and Ramsay, A.

1. Algebras of bounded sequences, P.J.M. 15 (1965), 1239–1248.

HOFFMAN, K., and Rossi, H.

1. The minimal boundary for an analytic polyhedron, P.J.M. 12 (1962), 1342–1353.
2. Function theory from a multiplicative linear functional, T.A.M.S. 116 (1965), 536–543.
3. Extensions of positive weak*-continuous functionals, Duke Math. J. 34 (1967), 453–466.

HOFFMAN, K., and Singer, I. M.

1. Maximal subalgebras of $C(\Gamma)$, Amer. J. Math. 79 (1957), 295–305.
2. On some problems of Gelfand, Uspehi Mat. Nauk 14 (1959), No. 3 (87), 99–114.
3. Maximal algebras of continuous functions, Acta Math. 103 (1960), 217–241.

HOFFMAN, K., and Wermer, J.

1. A characterization of $C(X)$, P.J.M. 12 (1962), 941–944.

HÖRMANDER, L.

1. An Introduction to Complex Analysis in Several Variables, Van Nostrand, Princeton, N.J., 1966.

HÖRMANDER, L., and Wermer, J.

1. Uniform approximation on compact sets in C^n, Math. Scand. (1969).

KAHANE, J.-P. and Zelazko, W.

1. A characterization of maximal ideals in commutative Banach algebras, Studia Math. 29 (1968), 339–343.

KALLIN, Eva

1. A non-local function algebra, Proc. Nat. Acad. Sci. 49 (1963), 821–824.
2. Polynomial convexity: the three spheres problem, Proc. Conf. Complex Analysis Mpls. (1964), Springer, Berlin, 1965, pp. 301–304.

KAUFMAN, R.

1. A maximality theorem in Fourier analysis, P.A.M.S. 18 (1967), 806–807.

KERR–LAWSON, A.

1. A filter description of the homomorphisms of H^∞, Can. J. Math. 17 (1965), 734–757.

KODAMA, L.

1. Boundary measures of analytic differentials and uniform approximation on a Riemann surface, P.J.M. 15 (1965) 1261–1267.

KÖNIG, H.

1. Zur abstrakten Theorie der analytischen Funktionen, Math. Zeit. 88 (1965), 136–165.
2. Zur abstrakten Theorie der analytischen Funktionen II, Math. Ann. 163 (1966), 9–17.
3. Zur abstrakten Theorie der analytischen Funktionen III , Arch. Math. 18 (1967), 273–284.

KÖNIG, H., and Seever, G.

1. The abstract F. and M. Riesz theorem, Duke Math. J. 36 (1969).

LEBESGUE, H.

1. Sur le problème de Dirichlet, Rend. Circ. Mat. di Palermo 29 (1907), 371–402.

LOOMIS, L.

1. An Introduction to Abstract Harmonic Analysis, Van Nostrand, Princeton, N.J., 1953.

LUMER, G. (see Gamelin)

1. Analytic functions and Dirichlet problem, B.A.M.S. 70 (1965), 98–104.
2. Herglotz transformation and H^p theory, B.A.M.S. 71 (1965), 725–730.
3. Integrabilité uniforme dans les algèbres de fonctions, classes H^Φ, et classe de Hardy universelle, C. R. Acad. Sci. Paris, Ser. A 262 (1966), 1046–1049.
4. Algèbres de fonctions et espaces de Hardy, Lecture Notes in Math. #75, Springer Verlag, Berlin, 1969.

MANDREKAR V., and Nadkarni, M.

1. Quasi-invariance of analytic measures on compact groups, B.A.M.S. 73 (1967), 915–920.

MCKISSICK, R.

1. A non-trivial normal sup norm algebra, B.A.M.S. 69 (1963), 391–395.

MCCULLOUGH, T.

1. Rational approximation on certain plane sets, P.J.M. 29 (1969).

MELNIKOV, M. S.

1. A bound for the Cauchy integral along an analytic curve, Mat. Sb. 71 (113) (1966), 503–515.

2. Structure of the Gleason part of $R(E)$, Funct. Anal. Appl. 1 (1967), 84–86.

MERGELYAN, S. N. (see Gonchar)

1. Uniform approximation to functions of a complex variable, Uspehi Mat. Nauk 7 (1952), No. 2 (48), 31–122. A.M.S. Transl., Ser. 1, Vol. 3, 281–286.

NAIMARK, M. A.

1. Normed Rings, Hafner, New York, 1964.

NEGREPONTIS, S.

1. On a theorem by Hoffman and Ramsay, P.J.M. 20 (1967), 281–282.

NIRENBERG, R., and Wells, R. O.

1. Approximation theorems on differentiable submanifolds of a complex manifold, T.A.M.S. (1969)

O'NEILL, B. V.

1. Parts and one-dimensional analytic spaces, Amer. J. Math. 90 (1968), 84–97.

O'NEILL, B. V., and Wermer, J.

1. Parts as finite-sheeted coverings of the disc, Amer. J. Math. 90 (1968), 98–107.

PELCZYNSKI, A.

1. On simultaneous extension of continuous functions, Studia Math. 24 (1964), 285–304.
2. Uncomplemented function algebras with separable annihilators, Duke Math. J. 33 (1966), 605–612.

PHELPS, R.

1. Extreme points in function algebras, Duke Math. J. 32 (1965), 267–277.
2. Lectures on Choquet's Theorem, Van Nostrand, Princeton, N. J., 1966.

RAINWATER, J.

1. A remark on regular Banach algebras, P.A.M.S. 18 (1967), 255–256.
2. A note on the preceding paper, Duke Math. J. 36 (1969).

RICKART, C.

1. General Theory of Banach Algebras, Van Nostrand, Princeton, 1960.
2. Analytic phenomena in general function algebras, P.J.M. 18 (1966), 361–377.
3. The maximal ideal space of functions locally approximable in a function algebra, P.A.M.S. 17 (1966), 1320–1326.
4. Holomorphic convexity for general function algebras, Can. J. Math. 20 (1968), 272–290.

RIESZ, F., and Riesz, M.

1. Uber die Randwert einer analytischen Funktion, 4e Congrès des Math. Scand., 1916, 27–44.

ROSSI, H. (see Hoffman)

1. The local maximum modulus principle, Ann. Math. 72 (1960), 1–11.

2. Holomorphically convex sets in several complex variables, Ann. Math. 74 (1961), 470–493.
3. Global Theory of Several Complex Variables, Course notes, Princeton U., 1961.
4. Algebras of holomorphic functions on one-dimensional varieties, T.A.M.S. 100 (1961), 439–458.

ROTH, Alice

1. Approximationseigenschaften und Strahlengrenzwerte unendlich vieler linearer Gleichungen, Comm. Math. Helv. 11 (1938), 77–125.

ROYDEN, H.

1. The boundary values of analytic and harmonic functions, Math. Zeitschrift 78 (1962), 1–24.
2. One-dimensional cohomology in domains of holomorphy, Ann. Math. 78 (1963), 197–200.
3. Function algebras, B.A.M.S. 69 (1963), 281–298.
4. Algebras of bounded analytic functions on Riemann surfaces, Acta Math. 114 (1965), 113–142.

RUBEL, L., and Shields, A.

1. Bounded approximation by polynomials, Acta Math. 112 (1964), 145–162.

RUDIN, W. (see de Leeuw)

1. Boundary values of continuous analytic functions, P.A.M.S. 7 (1956), 808–811.
2. Subalgebras of spaces of continuous functions, P.A.M.S. 7 (1956), 825–830.
3. On the structure of maximum modulus algebras, P.A.M.S. 9 (1958), 708–712.
4. Function Theory in Polydiscs, Benjamin, New York, 1969.

RUDIN, W., and Stout, E. L.

1. Boundary properties of functions of several complex variables, J. Math. Mech. 14 (1965), 991–1006.

SARASON, D. (see Ahern)

1. The H^p spaces of an annulus, Memoirs of the A.M.S. #56, 1965.
2. Generalized interpolation in H^∞, T.A.M.S. 127 (1967), 179–203.
3. A remark on the weak-star topology of l^∞, Studia Math. 30 (1968), 355–359.

SEGAL, I., and Kunze, R.

1. Integrals and Operators, McGraw-Hill, New York, 1968.

SHILOV, G. E.

1. On decomposition of a commutative normed ring in a direct sum of ideals, Mat. Sbornik 32 (1954), 353–364; A.M.S. Transl. (2) 1 (1955), 37–48.

SIDNEY, S.

1. The sequence of closed powers of a maximal ideal, T.A.M.S. 131 (1968), 128–148.

SIDNEY, S., and Stout, E. L.

1. A note on interpolation, P.A.M.S. 19 (1968), 380–382.

SRINIVASAN, T. P.

1. Simply invariant subspaces, B.A.M.S. 69 (1963), 706–709.
2. Simply invariant subspaces and generalized analytic functions, P.A.M.S. 16 (1965), 813–818.

SRINIVASAN, T. P., and Wang, Ju-kwei

1. On the maximality theorem of Wermer, P.A.M.S. 14 (1963), 997–998.
2. On closed ideals of analytic functions, P.A.M.S. 16 (1965), 49–52.

STEEN, L.

1. On uniform approximation by rational functions, P.A.M.S. 17 (1966), 1007–1011.

STOLZENBERG, G.

1. Polynomially convex sets, B.A.M.S. 68 (1962), 382–387.
2. A hull with no analytic structure, J. Math. Mech. 12 (1963), 103–112.
3. The maximal ideal space of functions locally in a function algebra, P.A.M.S. 14 (1963), 342–345.
4. Polynomially and rationally convex sets, Acta Math. 109 (1963), 259–289.
5. The analytic part of the Runge hull, Math. Ann. 164 (1966), 286–290.
6. Uniform approximation on smooth curves, Acta Math. 115 (1966), 195–198.

VALSKII, R. E.

1. Gleason parts for algebras of analytic functions and measures orthogonal to these algebras, Soviet Math. Dokl. 8 (1967), 300–303.

VITUSHKIN, A. G.

1. Analytic capacity of sets and problems in approximation theory, Uspehi Mat. Nauk 22 (1967), 141–199. Russian Math. Surveys 22 (1967), 139–200.

VOICHICK, M. (see Gamelin)

1. Ideals and invariant subspaces of analytic functions, T.A.M.S. 111 (1964), 493–512.
2. Invariant subspaces on Riemann surfaces, Can. J. Math. 18 (1966), 399–403.

WAELBROECK, L. (see Bingen)

1. Le calcul symbolique dans les algèbres commutatives, J. de Math. Pur et App. 33 (1954), 147–186.

WALSH, J.

1. Uber die Entwicklung einer harmonischen Funktion nach harmonischen Polynomen, J. Reine Angew. Math. 159 (1928), 197–209.

WEINSTOCK, B.

1. Continuous boundary values of analytic functions of several complex variables, P.A.M.S. 20 (1969).

WERMER, J. (see Browder, de Leeuw, Hoffman, Hörmander, Glicksberg, O'Neill)

1. On algebras of continuous functions, P.A.M.S. 4 (1953), 866–869.

2. Algebras with two generators, Amer. J. Math. 76 (1954), 853–859.
3. Polynomial approximation on an arc in C^3, Ann. Math. 62 (1955), 269–270.
4. Subalgebras of the algebra of all complex-valued continuous functions on the circle, Amer. J. Math. 78 (1956), 225–242.
5. Function rings on the circle, Proc. Nat. Acad. Sci. 43 (1957), 173–175.
6. Rings of analytic functions, Seminar on Analytic Functions, Vol. II, Institute for Advanced Study, Princeton, N.J., 1957, 298–311.
7. Function rings and Riemann surfaces, Ann. of Math. 67 (1958), 45–71.
8. Rings of analytic functions, Ann. Math. 67 (1958), 497–516.
9. The hull of a curve in C^n, Ann. Math. 68 (1958), 550–561.
10. The maximum modulus principle for bounded functions, Ann. Math. 69 (1959), 598–604.
11. An example concerning polynomial convexity, Math. Ann. 139 (1959), 147–150.
12. Dirichlet algebras, Duke Math. J. 27 (1960), 373–382.
13. Banach Algebras and Analytic Functions, Advances in Math., Academic Press, New York, 1961.
14. Uniform approximation and maximal ideal spaces, B.A.M.S. 68 (1962), 298–305.
15. Maximal ideal spaces, Proc. Int'l Congr. Math., Stockholm, 1962.
16. The space of real parts of a function algebra, P.J.M. 13 (1963), 1423–1426.
17. Seminar über Funktionen Algebren, Lecture Notes in Math. #1, Springer Verlag, Berlin, 1964.
18. Approximation on a disc, Math. Ann. 155 (1964), 331–333.
19. Analytic discs in maximal ideal spaces, Amer. Jour. Math. 86 (1964), 161–170.
20. Polynomially convex discs, Math. Ann. 158 (1965), 6–10.
21. Bounded point derivations on certain Banach algebras, J. Functional Analysis 1 (1967), 28–36.

WILKEN, D. (see Gamelin)

1. Maximal ideal spaces and A-convexity, P.A.M.S. 17 (1966), 1357–1362.
2. Lebesgue measure for parts of $R(X)$, P.A.M.S. 18 (1967), 508–512.
3. Representing measures for harmonic functions, Duke Math. J. 35 (1968), 383–389.
4. The support of representing measures for $R(X)$, P.J.M. 26 (1968), 621–626.
5. Parts in analytic polyhedra, Mich. Math. J. 15 (1968), 345–351.

YALE, K.

1. Invariant subspaces and projective representations (preprint).

ZALCMAN, L.

1. Analytic Capacity and Rational Approximation, Lecture Notes in Math. #50, Springer Verlag, Berlin, 1968.
2. Bounded analytic functions on domains of infinite connectivety, T.A.M.S. (1969).

LIST OF SPECIAL SYMBOLS

INDEX